Security in Roman Times

Using literary, epigraphic, numismatic and iconographic sources this book investigates the safety devices that were in place for the protection of the emperor and the city of Rome in the imperial age. In the aftermath of the civil wars Augustus continued to provide for his physical safety in the same way as in the old Republic while, at the same time, overturning the taboo of armed men in the city. During the Augustan age, the division of the city into regions and vici was designed to establish control over the urban space. Augustus' successors consolidated his policy but the specific roles of the various military or paramilitary forces remain a matter for debate. Drawing on the testimony of ancient authors such as Tacitus and Suetonius and on material evidence, the volume examines both the circumstances in which these forces intervened and the strategies adopted on these occasions. It also examines the pre-Augustan, Augustan and post-Augustan meaning of 'securitas', both as a philosophical and a political concept. The final section expands the focus from the city of Rome to the Italian peninsula where the security of the emperor as he travelled to his country residences required advance planning and implementation.

Cecilia Ricci is Professor at the University of Molise. Her main research concerns urban troops in the first two centuries of the empire and the relationship between the military and civilians, the 'memory of Rome' and the funeral rites of the Roman world and the presence of foreigners in the city in imperial times. She is author of a number of books, including *Orbis in urbe: Fenomeni migratori nella Roma imperiale* (2005), *Qui non riposa. Cenotafi antichi e moderni fra memoria e rappresentazione* (2006), *Soldati, ex soldati e vita cittadina: l'Italia romana* (2010) and *Venafro città di Augusto* (2015).

Security in Roman Times

Rome, Italy and the Emperors

Cecilia Ricci

Routledge
Taylor & Francis Group

LONDON AND NEW YORK

First published 2016 by Routledge

2 Park Square, Milton Park, Abingdon, Oxfordshire OX14 4RN
52 Vanderbilt Avenue, New York, NY 10017

Routledge is an imprint of the Taylor & Francis Group, an informa business

First issued in paperback 2020

British Library Cataloguing in Publication Data
A catalogue record for this book is available from the British Library

Library of Congress Cataloging in Publication Data
A catalog record for this title has been requested

ISBN: 978-1-4724-6015-8 (hbk)
ISBN: 978-0-367-59374-2 (pbk)

Typeset in Bembo Std
by Swales & Willis Ltd, Exeter, Devon, UK

Contents

Figures

Every attempt has been made to trace copyright holders for the illustrations used in this book. Any omissions are unintentional.

Preamble

Security is currently a highly topical issue. Even if the notion of 'public security', which is the product of nation states, explicitly refers to the protection, provided by the state, of persons and is a typically modern cornerstone, today the same word – whether or not linked to the adjective 'public' – refers to a wide range of issues that concern, first of all, the reliability of the places where people work and live; and then the assurance of people's safety at work and in their private lives.

In the ancient world, and particularly that of the Romans, there was not a sovereign body in charge of the security of people and places, at least not in the sense as we intend it today. However, with the conclusion of the civil wars at the end of the Republican era, a new stage begins with the *Princeps* who, in wanting to give a new order to the *Res Publica*, assumed the role (and essentially implemented the project) of guarantor of peace and safety of persons and of public places.

This book aims to investigate why and how such a plan was conceived and implemented; when; if and why it was modified or adjusted over two centuries; and whether or not, and if so, to what extent such a plan was extended to include all of Italy.

The period of time that is taken into consideration is the first two centuries of the empire, with an extension to the first three decades of the following century (the Severan age); the frame of reference is represented by the city of Rome and Italy. The reason for this choice will be made clear first, in the Introduction, and then it will be illustrated in the first two chapters.

Thus, the book is structured in the following way: in Part I, I review and comment upon studies of Roman history that, between the end of the 1960s and today, have dealt (exclusively or secondarily) with the issue of public order in Rome and in Italy.[1] This review is preceded by an overview of those studies that can be considered, for those who have dealt with security, as reference texts, namely monographs that in the last century have investigated separately the various units of the urban troops and the paramilitary forces present in Rome in the period between Augustus and Constantine. All these works, some in a more consistent way, represent points of views that often converge on the idea of security in the Roman world. I acknowledge my debt of gratitude to many of these contributions, which represent, in this area of research, a fundamental point of reference.

Nothing in my treatment is intended as definitive. It is likely that whoever reads these pages may be disappointed not to find cited those works that are considered well grounded and useful for the purposes of a study like this. Despite the undeniable methodological principle of completeness and continuity of bibliographic update, it is also true that the selection of books and articles that are considered is the result of a choice; and – with the necessary distinctions – I completely agree with what Italo Calvino writes when discussing books that each person considers critical to their education and knowledge:

> The only thing that can be done is for each person to come up with its own ideal library of classics; and I would say that half the books should be the ones we have read and that have counted for us, and half are the ones we intend to read and that we assume can count; leaving a section of empty space for those books that become surprises, occasional discoveries.[2]

In this book the word 'classic' naturally has a different meaning, as we have just seen, because of the subject which is being treated: it refers to a 'reference text', which sometimes represents a positive model, other times a negative one and more often, a middle ground between the two. Some books have not been considered and are not part of this review, mainly for lack of my knowledge; or even as a result of a motivated choice (which is often, as you will see, pointed out). Other books have been included because they have counted for me; they have been a guide and an inspiration. Moreover, other books could still count and I am more than willing, after having been informed, to fill in those empty spaces with those that will prove to be unexpected surprises or 'occasional discoveries'. If I do not pursue certain themes, it is not for lack of appreciation of their interest and importance.

After having reviewed other opinions and methods, I will try to illustrate my own point of view: my intention is to look away from both the abstract speculations on the idea of public order of the Romans and the individual events (which of course will also find space in this work) that may affect social balance; and broaden the focus on detecting the signs of an extended plan on prevention put into action starting from the last decades of the first century BC. Such an operation, as will be seen, embraced a wide range of services and contemplated a very articulated number of protagonists.

Then, I present and discuss, as far as it has been possible to reconstruct, what 'security' meant in the eyes of the Romans; how this idea had increasingly developed to slowly modify during the first century of the empire.[3] Particular emphasis is placed on the last fifty years of the history of the Roman Republic: at this time, the term *Securitas* and its corresponding adjective, along with *Tutela*, became central to Cicero's thought and found space in the work of Lucretius. Here security became the focus, together with peace, of a primary need and therefore a pressing demand that the inhabitants of Italy addressed to the new *Princeps*. Augustus, well aware of this, did not miss the opportunity to include this issue in his cultural policy, thus developing a sophisticated level of transmission of his message. Nevertheless, another eighty years at least would be spent in order for the theme, the words and the images of *Securitas* to acquire a complete formulation.

Originally, security was a philosophical theme, and not specifically Roman. It was gradually built and flushed of new tones as we can detect from inscriptions, statues and most of all coin iconography. Just when the idea of security that Augustus had created began to fade away – from the second half of the second century – the word and the images of *Securitas* started to become popular and to be used by large sections of the population, both publically and privately. The iconography of the security of the city and of the early Roman Empire would not diminish through the centuries of Late Antiquity and the Middle Ages when it would find an accomplished form of expression and exceptional effectiveness in the allegory of the Security (see Figure P.1).

Part II has chronologically the same starting point as the first chapter: the era of the last civil wars. Next to the history narrated by literature and monuments, it is the Augustan legislation, especially that concerning criminal issues, which helps us understand the security plan of the *Princeps* and to integrate aspects that, without the substance and the language typical of legal matters, would inevitably be obscure and incomplete.

Figure P.1 A medieval allegory of Securitas

Source: Ambrogio Lorenzetti, *Gli effetti del Buon Governo in campagna*, detail (Sala dei Nove, Palazzo pubblico di Siena, 1338). Open access, copyright free.

Once again, the most stimulating challenge will be associating the accounts provided by historians with epigraphic, archaeological and numismatic sources. The images and the words of propaganda[4] are the setting upon which the 'security plan' for the people was conceived. This plan – which underwent a slow development – involved the recruitment and the organisation of military and paramilitary forces intended to complement each other. The internal organisation, the hierarchical structure, the daily life of the praetorians, *milites urbani*, *vigiles*, the emperor's bodyguards, on foot or on horseback, however, will not be explored in detail. I am more interested, to make a clear example of what I mean, on how, the day after the end of the civil wars, Octavianus/Augustus continued to provide for his own safety following a certain line of continuity with the Republican era; while overturning at the same time the paradigm, which prohibited the presence of armed men in the city and also stated that each person should provide for their own security independently, without charges or interventions by the *res publica*. From this perspective, despite the 'first praetorians', which will be repeatedly mentioned, the Praetorian Guards are to be considered a novelty in terms of the number of men employed, the variety of their duties and also, already noticeable with the successor of the *Princeps*, for the grandeur and the capacity of their camp. Less clear, as we shall see, were instead the different duties that other types of soldiers, such as *speculatores* and *statores* had.

Moving on from the safety of persons to the security of places, Augustus once again intervened to define and to standardize the territorial frameworks of the city with interventions that were not less powerful, for the resulting implications, than the birth of the Praetorian Guards. In order to ensure security in the city against the most recurring risks,[5] for the first time in the history of the Republic, he created the conditions for close collaboration between old and new social and professional figures.

Without therefore wanting to submit a history of military and paramilitary troops that, all together, would make up, over time, the so-called 'Urban Garrison',[6] I will try to establish which dispositive (dispositif),[7] considering all aspects of the city Augustus implemented to ensure his own security and that of his citizens in public places; and, above all, if in doing so, whether he stuck to the typical roman criterion of binding competences, as in the case of the higher senatorial charges; or whether he chose to keep the responsibilities and the fields of action separate.

If we follow a chronological order, which is essential in understanding the implementation of the Augustan plan presented in the previous chapter, in Part III I try to shed light on what was kept and what was modified by his successors. I intend to provide a general framework of the actions taken by each emperor (from Tiberius to Severus Alexander) towards security, legislation, organisation and reorganisation of all the men employed, infrastructures of the 'security system', barracks and *stationes* of Rome and Italy; and then devote special attention to the protection of the most popular public places. Two paragraphs follow: the first focuses on the potentially dangerous spots in the city (the barracks and the *praefecturae*, places of entertainment and leisure, the posts at the gates or in the area of Rome). In a second paragraph, in order to understand how urban troops adapted to the time and circumstances – and the first century was a phase of experimentation for potential and substantial changes – I will concentrate on the activities of the urban troops between the first and second centuries. The idea is to feature how they are also, in all respects, part of the overall dispositive and can be viewed as a litmus test to see how each emperor viewed the security of the city and of Italy.

In Part IV what changes is not the chronology of events, but the geography: moving from Rome to Italy. After a brief outline of what could have been the reasons that required the presence or intervention of soldiers of Rome, I will focus on another specific theme: the safety of the *Princeps* and of his family during his travels throughout Italy, at official events or during private encounters. What will be examined are the security measures taken during the travels of the emperor's family; and, also those adopted once they arrived at their destination. The focus will be on some of the emperor's residences in Latium and Campania. Even in this case, the reconstruction of how the logistics was provided is only possible thanks to a fruitful investigation of different kinds of sources, although the lion's share is given by the epigraphs of Roman soldiers or veterans in the cities and in the territory of Italy. In fact, notwithstanding the return to their birthplaces of some soldiers, after the retreat or by reason of a premature death, some of the sources indicate instead a short or regular stay of soldiers or veterans engaged in operations of maintenance of public order in the cities of Italy, with functions of protection and escort of the *Princeps*; for supervision of road junctions, ports or other specific economic and political strategic points.

Acknowledgements

My profound gratitude to Sergio Oriente, who followed the development of the book and was a big aid for Part I, Chapter 3. Without his ideas and stimuli, that section would not be the same.

My sincere thanks to Maria Letizia Caldelli, Marta Garcia Morcillo, Victor Revilla and Francoise Villedieu. Their reading of single chapters has greatly helped with this work.

I am particularly indebted to my friends and colleagues Salvatore Monda and Everett Wheeler. I never lacked trust, advice, experience, attentive and generous help from the former. The precious suggestions and deep knowledge of the Roman army of the latter has enriched over the last ten years my outlook and my knowledge, allowing me to often look at problems with a different perspective; to Daniela Fabrizi and Amy Muschamp who have helped me with the translation.

I am particularly grateful to those who undertook the labour of reading the entire text, and in doing so saved me from typos or omissions; and to my precious and patient editors, the two Michaels (Greenwood and Bourne), Kelly Derrick and Julie Willis. Their keen work is priceless.

Lastly, I would like to thank the original editors and the curators of the journals and collective volumes from which some of the sections of this book – reviewed, updated and enriched – are taken. Their availability and generosity helped me so much.

The responsibility of what is said and any errors or omissions are of course only mine.

Notes

1 Interesting is Yannakoupolos's contribution of 2003 that investigates on the public order in the provinces. The issue, of which the deep differences according to the times and the territories are highlighted, has been identified in fact with the *Pax Romana* to be preserved in the imperial East (the idea is that of a Roman State that holds the exclusive right of the public force). Due to the historical and geographical perspective different from the one that I have adopted here, that volume is not part of my review.

2 "Non resta che inventarci ognuno una biblioteca ideale dei nostri classici; e direi che essa dovrebbe comprendere per metà libri che abbiamo letto e che hanno contato per noi, e per metà libri che ci proponiamo di leggere e presupponiamo possano contare. Lasciando una sezione di posti vuoti per le sorprese, le scoperte occasionali" (Italiani, vi esorto ai classici, in *'L'Espresso'*, 28 June 1981, pp. 56–58; now in *Perché leggere i classici*, Milano: Mondadori 1995, p. 13)

3 With reference to the progressive changes in paradigm initiated by Augustus (in the organisation of time, in the religion and for the *mores* in general), which is completed in a century, see among the others Wallace-Hadrill 2007, part. p. 58.

4 For the heuristic value of the use of the term and the alternative ones of 'self-representation' or 'representation', see Weber, Zimmermann 2003, in particular the introduction of the editor of the Acts (*Propaganda, Selbsdarstellung und Repräsentation. Die Leitbegriffe des Kolloquiums in der Forschung zur frühen Kaiserzeit*, pp. 11–40) and *ibidem* the work of A. Eich (*Die Idealtypen 'Propaganda' und 'Repräsentation' als heuristisches mittel ber der Bestimmung gesellschaftlicher Konvergenzen und Divergenzen von Moderne und römischer Kaiserzeit*), pp. 41–84.

5 It goes without saying that the 'security plan' is designed in close connection with the transformation of the image of the city that is being realized along a line of continuity with Caesar's plans. To get an idea of this, Favro 2006 with other bibliography may be seen; even before suggestions in Robinson 1992 (*infra*, p. 8). For the transformations in later times, other references will be given gradually in Part II.

6 This work has already been carried out, even if in monographs that have to deal with single corps, with different methods and outcomes (see *infra*, Part I, Chapter 1).

7 To understand what this term refers to, read the actual words of Foucault: "What I am trying to pick out with this term [the dispositive; le dispositif] is . . . a thoroughly heterogeneous ensemble, consisting of discourses, institutions, architectural planning, regulatory decisions, laws, administrative measures, scientific statements, philosophical, moral and philanthropic propositions – in short, the said as much as the unsaid. Such are the elements of the dispositive. The dispositive itself is the network that can be established between these elements" (1994, p. 299).
In order to understand if and how we can make fruitful use of this analytic concept in this investigation, refer to Part I, Chapter 3.

Notes on translation and bibliographical references

All the Greek and Latin passages, if not differently specified, are cited in the English version of the Classics Collection (Loeb).

The bibliography contains the entire set of works repeatedly referred to in the text, except the works cited only once, indicated extensively in the notes.

Part I

From public order to security

1 Studies on military forces and public order in Rome and Italy from Republic to Principate

Points of view

A question that scholars have tried to answer in the last fifty years is the following: was there a 'police' in the Roman world? The different studies, although sometimes with differing aims, have shared a common guideline that has always been to investigate which instruments and policies, according to the period of interest, the Roman state adopted to contain centrifugal thrusts and public order breakdown. The founding father of all these studies was Otto Hirschfeld who, in an article of 1891 entitled *Die Sicherheitspolizei im römischen Kaiserreich*, opened a season of studies on the issue of security and 'police' tasks carried out during the Roman Empire.[1] The term 'Sicherheitspolizei', transmitted by the Prussian administration, with minor adaptations, to the young German nation during the period in which Hirschfeld was writing, represents one of the key terms of the 'police' apparatus that had developed in European countries at the turn of the eighteenth and nineteenth centuries.[2] This was the 'security police' in charge of the governance of men (and not of things and places, which the municipal police were instead in charge of), with governing functions on issues relating to social and individual risks; and has led, as already mentioned previously and that will be reaffirmed, to modernity.[3]

Hirschfeld tried to trace the antecedents of certain figures of the Roman Empire to the Greek and Hellenistic cities. His educational background led him to consider the Digest and juridical epigraphy as the best sources, without, however, neglecting literary references and other types of inscriptions. If the sources are reported following a thematic and not a chronological order – ranging with great ease from late Republic to late antiquity – however, his synthesis is a good starting point in order to gain insight, especially at a provincial level, of the means that were adopted to quell the riots. When considering Rome, special attention is given to the *vigiles* and their officers, and to the *speculatores* and the detention areas; and to the *beneficiarii*, with outdated considerations, given the large number of inscriptions and papyri that had been unearthed and studied in the course of the following century.[4]

The most comprehensive accounts on public order, however, are given by two authors of the second half of the twentieth century: Andrew William Lintott who, since the late 1960s[5] has closely analysed the episodes in which, in

Republican Rome, violence exploded and determined the authority's reaction; and Wilfried Nippel who, in the late 1980s and mid 1990s, worked on a similar theme, although differently, drawing on a more systematic way of proceeding and coming to delineate the 'features of the new imperial order'.[6]

But before reviewing these and other works that have provided, for the present investigation, food for thought and useful interpretive tools; and before describing what I intend to do and how I will proceed, since this book focuses on the city of Rome and imperial Italy, it seems essential to consider another line of investigation, grounded in nineteenth-century historiography: namely the systematic study of the urban garrison, mostly carried out between the 1930s and the 1960s.[7] The functions of the cohorts of Rome and their changes over time will receive adequate attention in Parts II and III of this volume. The rich crop of information regarding this, taken from those monographs that are now considered classic, represent the natural premise of this work and act as the backbone to the reflections of many authors who have analysed public order and policy in the imperial age.

What must be pointed out first is that scholars have not given the same attention to all the military forces of Rome. This is due in part to the size and the political role of some of the military forces, such as the Praetorian Guards or the *vigiles*.[8] It is also in part, a result of the partial representativeness of the documents related, as in the case of *equites singulares Augusti*, to the fortuitousness of their recovery.[9]

There is no doubt that not only were the praetorians appointed as imperial guards, with their persona closely connected to the one of the *Princeps*; but they also participated in military campaigns, already in the Julio–Claudian era.[10] At the end of the 1930s two volumes, published a year apart, were devoted to the more numerous and substantial cohorts of Rome.[11] Sources and considerations contained in the works of Marcel Durry and Alfredo Passerini, despite the fact that eighty years have gone by, are still invaluable and essentially well documented when considering issues such as the number of cohorts[12] and of active soldiers[13]; the onomastics; soldiers' provenance[14] and social background; the duration of military service; the enlistment age; and the structure of careers. New valuable information – apart from the natural enrichment of the documentation[15] – regards instead: the fate and the status of veterans (praetorians, and in general of the troops of Rome); and soldiers' family and social relationships even after their military service.[16] In the latter case, on the whole, what seems to arise from an initial social isolation, is an integration, in the second century, into the social fabric mainly through family and freedmen.[17] Finally, new considerations have regarded the urban areas used for burial and the types of sepulchral monuments employed, as evidence of a will and capacity to represent themselves in relation to civilians.[18]

The purpose for which seven *vigiles* cohorts were created in 6 BC was to prevent and extinguish fires, and to patrol the streets at night. After Paul Kenneth Baillie Reynolds devoted a pioneering monograph in 1926 to this issue, *vigiles* cohorts were the subject of a major monograph at the end of the 1990s, when

Robert Sablayrolles reconstructed the size, the functions, the hierarchy and the action of the firefighters of Rome and of the ostiense *statio*.[19]

In the late 1950s, Edward Echols devoted a long article to the policing of the city of Rome. The aim of his study was to look back through time to find the beginnings of the idea of police. He argues that ancient historians showed ostentatious indifference to "the duties of Rome's imperial urban cohorts performed in the comfort and relative security of the capital of the world", which were clearly viewed as less important than the hard work of legionaries in distant and hostile lands. Echols recounts in this way the stages of a long process: during the Republic, starting from the second century BC, a reserve police force (at first composed of citizens who acted as *vigilantes*, then regular army soldiers) was introduced to support the *triumviri cis Tiberim* and, above all, the *custodes* ("members of a regular, if rudimentary, city police"). *Custodes*, reserve police and, in some rare cases, "a contingent of soldiers for emergency duties",[20] would become the protagonists of the repression of sedition and the quelling of unrest up until Augustus when the police force, according to Echols, essentially identified with urban cohorts, whose functions and features were analyzed, in 1967, in a dedicated volume by Helmut Freis.[21]

In the first century, the emperor's personal guard force was made up of soldiers of various corps, with partly differentiated roles, of which it is difficult to reconstruct the details. If in fact the first role of the praetorians was to guard the Palatium and to accompany the emperor and his family around the city and on their travels in and outside Italy, it is certain that the specific function of bodyguards, during the period from Augustus and Nero (with a brief period after the events in Teutoburg) was served mainly by a chosen body of men of German origin (the *Germani corporis custodes*).[22] The Germans' height and physical performance in addition to their reputation as skilled fighters meant that they were assigned an additional function, that of guard of honour, as Tacitus reveals when mentioning the privilege of a *statio Germanorum* being granted to Agrippina, because she was mother of the *Princeps*.[23] After more than thirty years from the publication of the book by Heinz Bellen,[24] the main issues regarding the Germanic *corporis custodes*, in consideration of the fact that the documentation in this regard is essentially the same, remain open and concern: their status; their exact size and structure; the existence and the possible location of their quartering and private burial ground. The difficulties of interpretation are also linked to the different attitudes towards the *corporis custodes* depending on the inclinations of the emperors. It is certain, for example, that with Gaius and Nero the Germans lived a golden age, which was destined to cease abruptly first with the 68–69 civil war and then with the advent of a new dynasty.

Two quite different books by Michael P. Speidel were published in the same year, 1994,[25] on the topic of the *equites singulares Augusti*, cavalry that came to bridge the gap created with the elimination of the *corporis custodes* after the civil war and, at the same time, to support with some of their functions, the Praetorian Guards. The main purpose of the creation of the *equites singulares*

Augusti was to protect the sovereign: for this reason they took part, starting in the Trajan era, in military campaigns and accompanied the *Princeps* on his travels throughout Italy.[26] Apart from their primary function, they were appointed other duties such as patrolling the area surrounding the imperial palace and escorting during *adventus* and *profectio*.

There seemed to be no more interest in the troops of Rome during the last century, at least in terms of systematic interventions; then a few years ago, Alexandra Wilhelmina Busch[27] took on the ambitious project of illustrating the life and the multifaceted activities of the soldiers of Rome and the impact on urban life, between 27 BC and AD 312. The book actually only partly keeps to its title: already in the introductory pages of the volume, Busch does not comply with the chosen title. To reconstruct how the soldiers entered the 'city scene', how the urban population considered them and how they represented themselves in relation to their status and provenance, the author draws on various types of sources, although proportionally unbalanced and with absolutely discrepant methodological rigor:[28] the historical, epigraphic and numismatic sources are in fact considered almost exclusively in the introduction, which does not bring to light any substantial new information that was not already known, neither in the choice of commented passages nor in their analysis. Busch hoped to have produced a systematic work of philological analysis of ancient sources that investigates why, in relation to the soldiers' activity, some episodes rather than others have been chosen, which topics recur and how the image of the soldiers has changed over time.[29] A work of this kind, with the type of questions raised by the author in the introduction and the sources that she claims to have drawn upon, frankly speaking, do not correspond to the author's intents and to the readership's expectations.[30] The secondary bibliography of which one can make use of is only partial and partly updated.[31]

The impression that one has is that this first part of the book is an unavoidable prerequisite to the following one (namely the technical description of the barracks of the soldiers of Rome), without considering instead that it might have been more useful to acquire a precise idea of what really happened when these soldiers arrived or, for example, of how the camps were built; in other words, the author did not offer readers a sound interpretative approach. Thus, *vigiles* are written about in a bland way and urban cohorts are described as "bodies organised only militarily without however having functions in the proper sense"(!);[32] and, again, the "field of action of the *cohortes urbanae* [that] substantially corresponded to that of the modern police",[33] showing that the author did not question herself on the fundamental issue of *whether or not* one can refer to a police force in the ancient world, while stating, in the quotes and in the bibliography, that she has read, for example, Nippel's works.

In relation to the topic of this book in which we are more specifically interested (namely the security of the *Princeps* and the safety of the city), Busch offers a rather rigid distinction of the urban military corps depending on the duties they carried out: on the one hand there were those who were in charge of protecting the emperor[34]; on the other hand, there were those in charge of watching over the city (urban cohorts and *vigiles*); still other soldiers were more or less temporarily present in Rome to carry out duties linked with the provinces. Never, just to give an example, does she mention the combined actions of *vigiles* and *milites urbani*, and she barely mentions the changes that affected the individual corps, without putting them in relation to a specific plan, except of course during the Constantinian era.

The monographs by Durry, Passerini, Freis, Bellen, Speidel and Sablayrolles have represented and continue to represent the starting point for all other research on 'public order', which, since the late 1960s has broadened the perspective and examined in depth the critical approach.[35]

At the end of the 1960s (after the publication of the majority of monographs on the military corps of the capital), from a perspective of much greater awareness compared to Hirschfeld, Andrew W. Lintott addresses questions that still today, after forty years, he repeats to himself and his readership under the title of a recent contribution.[36] The questions are the following: what level of violence was acceptable to the Roman state in the late Republican era? And what instruments were available in order to be able to keep it under control? Therefore, the focus shifts from a purely analytical and descriptive level (what forces does Augustus supply?) and from a logic of dominion (never questioned) to a completely different one, both from a qualitative (what kind of violence could be tolerated and what kind of resistance could be arranged?) and quantitative point of view (how much violence was sustainable?).

Although Lintott's work seems apparently distant from the theme and especially from the chronological horizon that is being examined here, it has, however, represented a particularly interesting stimulus for the present work for various reasons. First, for the interpretive insight and the philological rigor with which it has been conducted; then for the wealth of sources from which it draws; and finally for the fact that it has grown and been enriched over time. Therefore, rather than an analysis of Lintott's individual contributions, which would not be necessary and for which there would be no space, what seems interesting to me is to present his work as it has developed and now appears: the product of a progressive construction and some (well thought out) reconsiderations.

Lintott's intention, to shift the attention from the forces in action to the conflict taking place in the urban context, is quite clear. The very same sentence in which Lintott uses the term 'security' – "The Romans under the Republic may have seen security as the product of conflict rather than repression" – according to me, seems to summarize effectively what is meant.[37]

The main focus in Lintott's book is on the legislation with regard to private law.[38] The strategy through which, during the Republic, the state was able to maintain control is linked to a weak superstructure of public officials of different ranks and responsibilities (*praetores*, *tribuni*, *aediles* and *triumviri capitales* at a lower level; last chronologically, the 'mysterious' *quinqueviri cis Tiberim*); and to a traditional legislative system that, except for (few) corrective actions and phases of reform, continued to refer to the XII Tables. The most effectual and valid measure was therefore, according to Lintott, the introduction of law and the guarantee to be able to refer to it, and any of the preventive measures adopted to contain the explosion of violence according to the law.

Thus, conflict is at the core of Lintott's reflections; particularly analysing the first century BC, when the authorities intensified legislative measures to address issues of violence against persons and property. The legislation *de vi*, especially that of the last fifty years before the advent of the Principate, represents – in Lintott's opinion – a late attempt to extinguish a fire that has already become uncontrollable. The analysis becomes subtle when Lintott focuses on the role of the plebian tribune and on Caius Gracchus's actions, putting the outcome of the conflict in direct connection with what would happen in the *quaestiones perpetuae*, introduced with the *lex Sempronia de capite civium*. The use and proper functioning of these courts as a regular law court independent from appeals and obstructions, would represent the only alternative to the rule of the violent clash in Rome.

Tacitus' remark at the end of the Republic is famous: *corruptissima re publica plurimae leges*.[39] And yet (or, rather, because of this), the wealth of interventions in the late Republic, both in civil and criminal jurisdiction, is astonishing. These actions had the effect of decreasing, in the course of four centuries, the private force and, as happened in many other aspects of cultural life, the contemporaries were not the ones to benefit from it, but rather those who survived to see how Augustus *leges et iura populo Romano restituit*.[40]

Before moving on to the second author (Nippel) who, at the end of last century, systematically dealt with public order in Roman times, it is appropriate, for reasons of chronology and method, to pay attention to a monograph published at the end of the 1990s. The purpose of the book *Ancient Rome: City Planning and Administration* by O.F. Robinson was to understand what measures and what expedients allowed a million people in imperial Rome to live together, overcoming the obvious difficulties related to health issues and public order. In the same period, in France and in Germany, the same issues were being addressed, from different perspectives, to the physical spaces in time[41] and to the administration[42]; while we would have to wait until the new Millennium to find, again in collective works in France and Italy, an overall approach to structures and infrastructures, to administration mechanisms and daily life[43] that Robinson certainly, although in a less comprehensive way, had experienced individually ten years before.

In the scenario recreated by Robinson, *vigiles*,[44] soldiers[45] and a selection of the major incidents on public order[46] find their space. Even if, on the whole, the separation of the protagonists of the policy and of the protection of the *Princeps* from those incidents relating to them causes incomprehension; and even if, in particular, when considering the role of the urban cohorts, there is also a certain confusion with ideas and the presentation of information, one may nevertheless find some exciting observations. An example may be the digression on the control of the urban cohorts in the second century, which could be read as a clue to who (emperor or prefect) was in charge of public order in this phase.[47] The only dissonant note about the text is the often anachronistic criteria that were adopted when using the sources, with Plautus, Tacitus, Cassius Dio being cited shortly after one another, without proper distinction.

From an epistemological and linguistic perspective, Wilfried Nippel's theoretical approach[48] is more Finleyan and Weberian compared to that of Lintott. Nippel writes about a 'policing apparatus' that, between the eighteenth and nineteenth centuries, was gradually being structured by the State in modern societies; and this cannot be found in ancient societies, nor even in the Roman Republican era.[49]

The problem – Nippel reasonably points out – is in the use of the term 'policing': in the fundamental works of the nineteenth century, it is sometimes given the value of "description of a function, for example securing public order"; other times "designation of a specialized agency to fulfill the function".[50] Basically

> organizational structures, defined competences and actual functions, legal status with respects to citizens, judicial authorities, local and central government, political role, arms and uniforms, internal specializations, recruitment, public esteem, etc., vary considerably from country to country and, of course, from period to period, in response to changing political régimes and social demands within each country's history.[51]

Thus the police as we intend it today met the needs of public order in a society that faced class conflicts. If a comparison is to be made, when studying Roman republican society, a model to be considered is the English society of the Middle Ages and even more so, that of the modern era; instead, when studying specific themes, the French and the German models are mainly considered.[52]

The comparative method allows Nippel to trace a limited legislative action in the Roman Republic: there was no use of police force to regulate the lower ranks (only interventions in a few cases when the political and social order was threatened); instead there was a predominance of private initiative and the use of *arbitratus*. Nippel essentially observes a scenario that is not so different from that of Italy and of other pre-national realities in the Middle Ages.

Before examining in detail some cases, which take up a substantial part of the 1995 volume, Nippel therefore focuses on providing the framework of the men and of the tools that were available: as Lintott has already done (and would continue to do shortly after) he also presents and discusses the most excellent tools of the Republican state apparatus, namely the *coercitio* – a valuable instrument, in his opinion, in monitoring the ruling elites and also in disciplining the people – and the *provocatio*, the (re)active tool of the representatives of the people. Therefore, the powers of the tribunes, of the *aediles*, of the subordinate magistrates (*tresviri capitales* and *quinqueviri cis Tiberim*), and their collaborators' responsibilities are presented, with related examples, in a brief summary. The deterrent force of the laws and the authority of the magistrate, more than a policy apparatus, are dominant in Nippel's resulting analytical framework.

Even when the situation in the last century of the Republic degenerated, what always prevailed was private initiative and popular justice, a relic of the past, which co-existed with institutionalized procedures. However, at this stage the traditional structure was disarranged. First, normal legislative activities and criminal trials were impeded while riots for systemic grain shortages were multiplying. Then, the ruling classes were unable to adapt to emergency measures and, next to the magistrates and their collaborators, new figures, for example different types of bodyguards, intervened while soldiers regularly conscripted acted as a police force within the city. The use of these troops is something new, according to Nippel (apart from the exceptional parenthesis of the Sullan period). The personal connection of these soldiers with their commander entailed legal and psychological implications radically different from the emergency appeal to volunteers residing in the city; as well as the extraordinary position of Pompeius, which differed only slightly from that of a dictator. From this moment on, between 49 and 27, soldiers were regularly used to quell riots, employed by powerful men who did not care about constitutional laws: "The combination of autonomous organization and spontaneous articulation of the masses's material and cultural interests with cynical manipulation of their wishes in the power struggle between the Senate and certain politicians again constituted an often explosive mixture."[53]

Particularly interesting are Nippel's reflections on the relationship between the Senate and the *collegia*, which will be defined only in the Augustan age; and on Sulla's legislation, which will enable more streamlined practices and rapid procedures in relation to certain offenses. The *lex Cornelia de sicariis et veneficiis* – which persecuted those who carried arms with offensive intentions and has been considered the forerunner of the *lex Iulia de vi* (Caesarian or Augustan) – is seen as a moment of no return to the past. Its peculiarity is linked to the legislator's new point of view and to the prosecution of those participating in the riots.

Nippel, as stated above, is interested in public order in Republican Rome. Thus, one can justify the poor consistency of the paragraph 'The new forces of order',[54] which presents the features of the Augustan plan. Without any hesitation, the author expresses in a very generic way the view that the "praetorian and urban cohorts made up the new military apparatus for maintaining law and order in Rome and Italy". The author tends to minimize the importance and the effectiveness of the measures adopted by the emperors, by limiting the management of the problems to the sole reactions as a consequence of the perception that the emperor had, at a specific time, of an endangered crisis. Overall, in this section, there is nothing more than a standardized framework of the duties of the various military units, derived from the traditional works of Durry and Passerini (especially the former), without any original notes: the basic idea is that security increased during the imperial age; that the private protection of people and their property would continue to be each individual's own responsibility[55]; and that the presence of soldiers mainly had a deterrent effect.

All in all, more than a chapter, 'The new forces of order' takes on the contours of a hasty conclusion, altering, in my opinion, the lucidity of judgment and the careful examination of the facts and sources that had characterized the previous chapters. If the opinion that the praetorians' main duty was to control the city is a very generic one,[56] the tasks of the various urban military corps are not clarified, because – according to him – it is inappropriate to infer general laws from (a few) particular cases (with reference to the few events reported by the sources); and because the literary sources often refer to generic 'soldiers' without specifying to which corps they belonged to.[57]

Strictly speaking, more because of the brevity of the intervention and its schematic nature, Augusto Fraschetti's contribution on the administration of the city of Rome in the Augustan period should not be included in this review.[58] However, the subtitle that he pinpointed (*'Amministrare sorvegliando'*) and, above all, his opinions on the topic – many of which I share – have led me to do so.

Following the order of the progressive creation of the prefects who had to deal directly with the life of the city (praetorian prefect, urban prefect, prefect of the firefighters and prefect of the *annona*), Fraschetti actually reasons in a unique way that proves that the major prefectures – which he calls the 'new services' – served to meet the "obvious needs – especially in terms of public order in a city such as Rome where social pacification had never been completely restored, not even in the Augustan age"[59]: "the rationalization of the services (in the specific case of the security services), starting from the age of Augustus, took place under the domain", even personal, if we want add praetorians to the urban cohorts and to the *vigiles*. In this framework, the prefecture of the *annona* is also included, since grain shortage had always been one of the most frequent causes of riots.[60]

Therefore, a long implementation plan was developed, following what Fraschetti calls a 'subtraction strategy', which sees Republican magistrates deprived of great services, according to a precise logic so that they would have subordinate tasks (ultimately the only duty of the *aediles* was the monitoring of markets). The consequence of such a situation is quite evident and in fact coincides with the establishment of a regime. It is the progressive nature of the operation and Fraschetti's idea of a plan implemented in an indefinite time and way that largely corresponds to the same idea that I personally share and that, in my opinion, emerges only if one considers all the actions undertaken by Augustus and resumed, intensified or modified by his successors.[61]

The purpose stated by Jonathan Coulston in the chapter entitled 'Armed and Belted Men; the Soldiery of Imperial Rome' of the book dedicated to the archeology of the Eternal City[62] is "to explore a number of aspects of Rome as a 'military' city", which meant examining the military installations, the organization, the functions and the impact of the troops and, consequently, also the influence that their presence had on urban art. The latter aim is pursued unsystematically: only limited space is dedicated to the relations between the troops and the civilian population and the topic that receives more attention is the influence of the army (not only the urban troops) on Rome's policy and particularly on the emperor. The different aspects of a 'militarized' city are investigated after the usual review of the various corps, of which characteristics and consistency are analysed along with the description of military installations, cemeteries and soldiers' functions. One of the last paragraphs, similar to the author's research path, is dedicated to the soldiers of Rome in official and private art.

A section of Pallas's monograph of 2001 on the imperial city[63] deals with the 'garrison' of Rome from Caesar to Pertinax. What Robert Sablayrolles intends to give is a general framework that, although not intended as such, complements, only after a few years, Nippel's work on the order of the city in the Republican era.

Sablayrolles analyses the military and paramilitary forces of Rome individually, with a closer look at the firefighters, to whom, as we have seen, he had devoted an extensive monograph a few years before, and their antecedents. He dedicates an entire chapter to "La lente éclosion d'une garnison urbaine",[64] the result of subsequent experiments, whose ultimate aim, according to him, was to ensure the security of power without revealing its true nature.

The urban forces, according to the French scholar, never came to represent a homogeneous unit, even if 'collective habits' were established among them. In a first phase, their competences was the maintenance of order and security in the city; at the end of their existence, however, the 'shared appointments' that still maintained their specificity, the development of the *cognitio extra ordinem* and their officers' judicial skills made them a special military force and determined the development of "a more or less explicit awareness to operate collectively for the preservation of the capital city and to take part to the *salus Romae et Augusti*".[65]

The chapter that sums up the framework outlined above is the one concerning the benefits and the risks of the soldiers' presence in the city that occurred completely with Augustus' successors: the solution adopted by Tiberius and Macro[66] meant the independence of the prefects from one another in relation to the military commands (in contrast to what was happening instead in administration where a hierarchy was still clearly in force). The results will differ over time, because of the circumstances: if, on the one hand, Claudius is proclaimed emperor by the Praetorian Guards with the support of the firefighters and the soldiers of the fleet; on the other hand, in the context of the civil war of 69, the praetorian cohorts were opposed to the urban cohorts and to the firefighters. The praetorians' 'opportunist' behaviour will allow the Germans and later on, the *equites singulares Augusti*, to be considered the last troops that the emperors could retain, in order to guarantee protection for themselves and their families.

On the occasion of a series of conferences organized by the Centre d'études d'histoire de la Défense, Patrick Le Roux dealt with the topic of public order in the Roman Empire as a "normal operation . . . complying to established laws and regulations".[67] The identity between public order and 'insecurity' – clarifies shortly after Le Roux – is a modern notion; public order in ancient Rome is thus a fluid and resilient phenomenon, which cannot include the exceptional nature of certain situations such as frontier operations, civil wars and the upheaval in the immediate aftermath of an emperor's violent death. It is, in my view, a fundamental clarification that allows, in reverse, to put into perspective, for example, the works of Roy Williams Davies[68] and Jocelyne Nélis-Clement[69] who both offer a coherent vision of the modernity of an army whose many activities were adjusted to the needs of the state. If

> once what prevailed was the image of an ever more oppressive and repressive monarchy that subjugated civilians to the crudity and the pleasure of a corrupt soldiery acting without restraint, forgetting about the war which it should have, on the contrary, been seriously prepared for[70]

today this image is no longer acceptable.

What Le Roux has proposed is a categorization of the different episodes of the soldiers involved in securing order, in Rome and in Italy, but especially in the provinces,[71] in order to establish, if possible, the Roman attitude on the matter; without getting lost in the folds of the sources that are 'pleased' to describe the undeniable negative aspects. Only part of the reflection is therefore dedicated to Rome and Italy.[72] The sources, complains Le Roux, almost completely give no information about the functions of the various troops of Rome, leaving room for conventional images: the praetorians are the shadow of the *Princeps*; the soldiers of the urban cohorts were on guard during the day and dealt with security, while the firefighters were in charge of surveillance and on duty at night along with having to deal with fire hazards.[73]

The overall picture is as follows: Le Roux underwrites the theory, which I share, that the army of Rome had a preventive function rather than a repressive one.[74] The soldiers' duties mainly prevented the development of violent actions rather than address and resolve any disturbance of public order. It is difficult to distinguish within the sources a clear policy that aimed at reducing crime or preventing riots. The use of force was followed as soon as possible by the return to normality.

Le Roux believes that the imperial power did not ignore issues concerning public order and policing duties in the cities, nevertheless the empire was founded on law rather than on the use of force; and there are indications through municipal laws on the issue. Consequently, the emperor's intervention was never to appear as the use of an arbitrary power, but rather as his reaction, when he considered that the very same image of the *imperium* was at risk.

As clearly stated in the title of her contribution, Hélène Ménard recalls and analyses the different circumstances in which public order, in Rome and in the Roman Empire, was at risk between the second and the fourth centuries. The breakdown of balance, of social peace, causes a movement and Ménard's aim is to identify the "factors that cause the break-down of the social relations among groups, but also between the governor and those being governed".[75] The starting and the end points are determined respectively by Aelius Aristides' text (143–144) that celebrates the order restored by the new dynasty; and also that of Prudentius of 402, which testifies to Rome's peace and unity, due to its being the base of the Christian empire. Septimius Severus's intervention acts as a link between these two worlds with the strengthening of the urban garrison and the increased role of the prefect of the city. The new composition of the praetorian cohorts plays, according to Ménard, a determining role in the raising of social unrest, which, in the third century, reached a level that was never touched before. The focus that, inevitably, between the second and third centuries is on Rome, will move to Antioch and Alexandria during the next century. Ménard's work presents itself as a rich selection of events and occasions, without an explicitly articulated original thesis. The choice of that particular time frame appears to be a little forced and depends more on a greater range of available documentation rather than on a reasoned theory.

The list of the individual episodes is preceded by an interesting philological analysis of the language of public order.[76] The author submits a vocabulary of the language of the crowd and of its uncontrolled behaviour; and of the disorder that includes common crime and violence at and of performances. Ménard explains that the word 'order' has no direct Latin cognate: the word closest to the concept is *disciplina*, as a social behaviour of those aware of the existence of a code and adherence to a model.[77] A key term is *seditio* ("the division of the city, in opposition to those in power"), which was a serious concern in Rome between Sulla and Nero. The concept of *popularitas* is also particularly significant in relation to two main categories of disorder, namely political and religious unrest, although in the first chapter – the only one that interests us for the time frame that is being considered – only the first category seems attested.

Finally *libertas*, considered as a free(!) play of the institutions, and *securitas*, as the security of goods and the integrity of people, are, according to Ménard, the ideological basis of imperial power in the second century. The emperor is to be considered the first in charge of public order during the empire: it is no coincidence that the regular functioning of the society is severely set back when material security is no longer guaranteed; or when the legitimacy of the emperor and its officers is jeopardized.

In the same year in which Ménard's volume was issued, a rich article by Yann Rivière on episodes of guerrilla in imperial Rome[78] was published. Rivière tries to provide evidence of how such events[79] are not sufficient to prove that civilian disturbances were defeated by the soldiers' operational ability. On the contrary, the problems occasioned by them have led us to think that urban troops (mentioned generically) have essentially provoked conflict.[80] Rivière emphasizes two interesting points. The first is that the presence of soldiers in the city did not mean that their role was to suppress turmoil, but rather to discourage it (as already outlined by Lintott and Le Roux).[81] The second, that seems particularly stimulating to me, is the idea that the Augustan plan in relation to the so-called 'urban garrison' is part of a civil war logic, which was meant to keep soldiers effective in case of conspiracy or lifting of the adverse parties.[82] I believe that this, at least with regards to the organization of the Praetorian Guards, is fundamentally true. I will return to this point later.

Benjamin Kelly's research outline for his article 'Riot Control and Imperial Ideology in the Roman Empire'[83] is particularly appealing. Ancient authors who have dealt with public order, according to Kelly, almost unanimously agree in considering unrest as an act of violence both for the soldiers as well as for those involved in the uprising. Modern authors, who focus on the etiology of the disorders[84]; or on how official responses vary depending on the circumstances and were variously effective,[85] seem to share the opinion that emperors, governors and magistrates tend to act when their position, dignity or authority is threatened.[86] In this context, Kelly will analyse cases of rebellion in the imperial period, with greater attention to the triumvirate period and the first Principate,[87] trying to describe and understand the crowd through its behaviour, for what is possible through literary sources.

The philological analysis of the reports on riots is valuable[88] and allows us to highlight how the authorities considered these events. What emerges is, through literary topos (from Vergilius to Livius to the Acts of the Apostles), a lexicon of persuasion that the emperor (or magistrate in charge) used before resorting to violence in order to persuade – first by words – political agitators to desist criminal means and to allow them to change their position. Even when considering what measures to adopt, the decision was linked to the ideology of clemency and moderation: a good emperor was expected to be able to persuade and to spare the civilian population (and soldiers) from the risks connected to a direct conflict.

The use of arms was then the last resort: what is emphasized about 'cruel' emperors' conduct is their use of the *extrema saevitia*. The risk was the shedding of blood of both civilians and troops, who were capable of fighting in open space, but not in the tortuous alleys of the city. The empire's elites were well aware that an armed intervention could be a bloody and complex affair (as a matter of fact, in the sources we often find the word *bellum/polemos*). As for the crowd's verbal demonstrations, it would have been inappropriate to respond with violence to make them cease; the common idea was that it was the emperor's responsibility to listen to his people's petitions and requests, even during performances and sporting events.

In a way Kelly betrays the reader's expectations: despite the good intentions in fact, in taking into account reports on turmoil, he concludes that they cannot be considered reliable because those who give an account are indirect witnesses, often distant from the facts and ideologically biased; and, it goes without saying, it is always the elite's point of view that prevails. What follows is that, even when there is no distortion, many details that would help us understand what the popular opinion was, are omitted.

Although ancient authors are not very inclined to point out the troops' ability to contain or repress riots, Kelly's research helps to give body to a working hypothesis that others before him had put forward[89]: the authorities did not intervene only when there was a threat to the life or dignity of the emperor or his representative. The emperor's task was to keep the peace, and therefore he had a preventive role,[90] as evidenced by the repression of banditry and war against piracy.

The work is partly taken over by Kelly in his contribution to the *Cambridge Companion to Ancient Rome*[91] of 2013. In the fifteen folders that were made available to him, Kelly deals with the delicate issue of how the state, communities and individuals tried to cope with threats to people and property.[92] Public order, as one would expect in a *Companion*, is considered throughout the empire. Therefore, rather than present his own position or agree with one of those prevailing, Kelly offers a summary of the lines of discussion ranging from the traditional approach[93] to others that are more recent.[94] In the general framework, what unfortunately dominates is a chronological jumble, which disorients rather than helps the reader gain a personal idea. There are the usual, well-known episodes that have already been described and commented on by many historians, passing very easily from first century events to those of Late Antiquity, without disdaining references to papal Rome. The most interesting part is the one that deals with the authorities responsible for the repression of the riots[95]; instead the most systematic one recalls upon what has been already done in a much clearer and more coherent way a few years earlier.[96]

The book by Christofer J. Fuhrmann on police forces in the Roman provinces[97] precedes by one year Kelly's concise picture. Although the perspective is quite different from the one adopted here, Fuhrmann's book cannot but be considered for "the purpose of this book is to investigate how policing worked in the Roman world, focusing especially on the roles of soldiers".[98] The point of view that has been taken is the one that sees the direct exercise of the imperial control over the provinces and that "policing in the Roman Empire was often focused on preserving the interests of the state and of cooperative elites".[99] Fuhrmann is in fact in contrast both with the points of view of Nippel and of Virginia Hunter,[100] who both instead agree with what he calls the 'self-help model', which does not fit, in his view, with the imperial model, because state policing during the first three centuries extended significantly. The difference is in the fact that, in his opinion, it is difficult to derive anything but an aristocratic perspective from the literary sources of the Republic; in the Principate the majority of inscriptions and papyri, and the Judeo-Christian literature, open the horizon and allow a more balanced assessment of the phenomenon. The opinion expressed by Fuhrmann is highly questionable, first of all for the different topics that are dealt with: if on the one hand, Nippel is concerned with Rome and the Republic; on the other hand, Fuhrmann is interested in the provinces of the Principate. I would say that both merely touch on imperial Rome, which is, in the end, the centre of everything. To better illustrate the point: both write about Rome in imperial times by seeking and providing only conventional, not thoroughly and newly explored information.[101]

However, in the introduction that develops as a real chapter, the author poses the problem of the modern concept of police and opts for the use of the term policing, in reference to "the activities of armed groups and organized by the state to maintain law and order in the Empire".[102] Two of the central chapters explore the Augustan plan for imperial peace, and policing in Rome and in Italy with Augustus's successors, "the various solutions used to create the new political order in Italy".[103] First of all, it provides a list of persons engaged in these activities: those who perform guard duties and preventive surveillance in all possible situations; those involved in legal processes dealing with the arrest of the suspects, the delivery of warrants in order to locate culprits, torture them and execute the death sentence; and finally, those who are in charge of political assassinations and of applying repressive measures against Christians.

Despite the honorable intentions, to the eyes of the reader a rather confused scenario is unfolded (particularly in the second of the two sections mentioned above), which takes no account of the diachronic order, and a number of soldiers and functions are thrown in, tied to decisions that run together apparently without a rationale. Just to give a few examples: no distinction is made between public security and private security. According to Fuhrmann, when Augustus wanted to surround himself with guards for his own security, he chose the *speculatores*.[104] The guards, however, in which the *Princeps* most confided were the *corporis custodes*, but even "members of praetorian cohorts were also considered personal guards of the Emperor" (!).[105] Additionally, since there is no clear distinction of the different tasks of the urban militias, when Fuhrmann discusses the substantial difference between 'civilian police' and 'soldier police',[106] the same confusion occurs in describing the roles of urban soldiers and firefighters.

Notes

1 The following year, Die aegyptische Polizei der römischen Kaiserzeit nach Papyrus-urkunden will be published in: *Sitzungsberichte der Königlich Preussischen Akademie der Wissenschaften zu Berlin* 1892, II. pp. 815–824; the same journal that, the year after, hosted *Die agentes in rebus, ibidem* 1893, I. pp. 421–441. It is known that Hirschfeld's interest for these issues will remain alive and be revived periodically.

2 It is impossible not to refer to the dense pages of Foucault 2004, pp. 225–252 when considering the gradual transition from 'Polizeiwissenschaft' to the modern police apparatus. Useful for its bibliography is the valid synthesis of Campesi 2009, pp. 110–123, with chapter 4 dealing with the police apparatus (p. 125 ff.) and chapter 5 with the emergence of public safety (p. 158 ff.).

3 Maria Gabriella Zoz, referring to Mommsen of *Römische Strafrecht*, writes (2001, p. 515): "throughout the Roman period there never was a clear demarcation between the administration of justice (aimed at restoring violated legal order) and policing activities (aimed at preventing the violation)".

4 Hirschfeld 1891, pp. 846–859.

5 His most systematic intervention (Lintott 1968) was revised and reissued at the end of the 1990s (Lintott 1998).

6 From the 1960s to the present, a more or less sporadic range of interventions, often ideologically oriented, have focused on individual episodes or individual areas of the empire, where the army appears as an instrument of imperialist domination. Just to mention two examples: Isaac 1992, Alston 1994.

7 Two separate works have, in the last decade of the last century, filled in the bibliographic gaps for the *equites singulares Augusti* (Speidel 1994a although preceded by the 1965 booklet); and for the *vigiles*, Sablayrolles 1996 similarly preceded, despite the lower systematic approach of the treatment, by Baillie Reynolds' work of 1926.

8 Leaving aside, for obvious reasons, specific studies, from the 1990s to present, five monographs, in four different languages, deal with the praetorian cohorts: Rankov 1994, Jallet-Huant 2004, Luc 2004, Menéndez Arguin 2006. Sandra Bingham

(1997 and 2013) is the only one to offer some new interpretations, although not having neglected a presentation, according to traditional patterns, of the history and organization of the cohorts serving the praetorian prefect. Specific attention will be dedicated to her in Part II, Chapter 4 (pp. 93–95) on the forces involved in the emperor's security. The *vigiles*, although less 'attacked' over time, have also been the subject of great scientific interest, as the first experiment of paramilitary troops set up with the aim of preventing and extinguishing fires in antiquity. Apart from the already mentioned Baillie Reynolds 1926 and Sablayrolles 1996, there is a volume of popular character by Capponi, Mengozzi 1993. The thesis discussed in Durham in the mid 1970s by J.S. Rainbird (*The Vigiles of Rome*) has unfortunately remained unpublished; however, in the late 1980s a study on the *vigiles'* stations by the same author was released (Rainbird 1986).

9 Speidel 1965 and 1994a and 1994b.

10 For a more detailed discussion on these and other duties carried out by the Praetorian Guards in the city see Part II, pp. 89–95; see Part IV for a thorough account on the Praetorian Guards in Italy.

11 Durry 1938, Passerini 1939.

12 I point out, for particular interest, the discovery of the hotly debated honorary inscription of A. Virgius L.F. Marsus whose position of *tribunus militum in praetorio divi Augusti et Tiberi Caesaris Augusti cohortium XI et IIII praetorianarum* is recalled (AE 1978, 286, from Lecce dei Marsi), and on which I will return in Part II; and the epitaph carved on a travertine stele commemorating a soldier of a nineteenth praetorian cohort, perhaps belonging to the period of the civil wars of 69–70 (AE 1995, 227).

13 This last question remains substantially still open, even if today there is a slight tendency for the hypothesis of milliarian cohorts, even before Severus. See Angeli Bertinelli 1974; Keppie 1996 and Lelli 1999; *contra*, Menéndez Argüín 2006.

14 On the origin and the tribes of praetorians, see now Crimi 2010.

15 In particular: Panciera 1984, 2004 and Crimi 2008, 2009b.

16 Ricci 1994, 2009, 2010.

17 Panciera 1993, Ricci 1994, Makhlayuk 1996.

18 Discussed later on in note 28 and in Part III.

19 I will return on the *vigiles* and their fundamental connection with the *milites urbani* on one side, and with the *vicomagistri* on the other, at length in Part II, Chapter 5.

20 See, for example, the conspiracy against Octavian organized by Lepidus the Younger and foiled by Maecenas, as *urbis custodiis praepositus* (Vell. 2.88.1–3 and Appian. *Bell. Civ.* 4.50) presented and discussed by Echols 1958, pp. 377–380.

21 The summary of Freis' book is structured as follows: the sources and the history of the cohorts of Rome, Lyon and Carthage; their number and the number of active soldiers for each cohort; functions of the cohorts; soldier's pay, leave, origin and recruitment of soldiers; internal hierarchy; officer ranks. New documentation and new interpretations complement and partly change today the contribution of the German scholar (see Part II, Chapter 5 and Part III, Chapter 7, this volume).

22 See Part II, Chapter 4 (pp. 97–98).

23 In addition to the (conventional) one of 'soldiers' (*milites*), as Svetonius (*Nero* 33) defines them, probably Praetorian Guards. Agrippina, in truth, was "the great-granddaughter of an Emperor (Augustus), the sister of an Emperor (Caligula), the wife of an Emperor (Claudius), the mother of an Emperor (Nero)", as appropriately emphasized by A. Giardina, *Nero o dell'Impossibile*, in Tomei, Rea (ed.), 2011, pp. 10–25 (the cited page is 17).

24 Bellen 1981; preceded, without the same systematic procedure, by Speidel 1965, and followed after thirty years by Speidel 1994b.

25 An essentially archaeological book written in German (Speidel 1994a), which is the scientific edition of the monuments of these soldiers. The catalog is organized according to the type of inscriptions (the sacred dedications therefore precede the ones to the emperors and officers and the sepulchral inscriptions), and internally according to a chronological order. An appendix is intended to complement the information, certainly reduced, restored by historical sources and by the art of the 'apparatus' (Trajan's column, Antoninus's column, arch of Septimius Severus). The volume Speidel 1994b, in English, thinner than the scientific monograph, aims to provide to a wider public the basic elements required to understand the organization and the functions of the emperor's bodyguards (*corporis custodes* and *equites singulares Augusti*).

26 See Part IV, in particular Chapters 8 and 9.

27 Busch 2011. Unfortunately, the book does not provide a general or thematic bibliography, excluding the '*Abkürzungen und Hinweise zum Online Katalog*' pp. 9–10.

28 Most of the book (pp. 29–109, 114–158) is in fact dedicated to the description and technical analysis of the Praetorian Guard, *milites urbani, equites singulares Augusti, classiarii, vigiles, corporis custodes* camps (with a summary of what is already known); and of the *legio II Parthica* barracks in Albanum (with some elements of novelty); and of places of burial and funerary monuments (in essence a summary of articles which have been devoted in the past decade to these issues by the Author (Busch 2005 and 2007).

29 Busch 2011 p. 25 n. 94.

30 The same questions that in those years Kelly 2007 (*infra*, pp. 14–16) had begun to make, without Busch noticing. If it is understandable and legitimate that the eminently archaeological background induces Busch to pay more attention to the archaeological remains; it is however verifiable, for example, in the case of the '*rilievi della Cancelleria*', that the attempt of a philological analysis remains at a superficial level and/or exclusively based on the secondary sources (Busch 2011, pp. 27–28).

31 On *speculatores* (only sources in nn. 59 and 60 on p. 22, without further bibliography, existing as well), on *statores* (Durry and some inscriptions on pp. 19 nn. 22, 28 and 63), on *milites urbani* (a piece by Tacitus and, at p. 18, nn. 20 and 21), on the emperor traveling (p. 22 n. 64). Unfortunately, the bibliography is mainly in German or English; the work of French, Spanish or Italian authors is considered only when no possible alternative exists in the first two mentioned languages.

32 Busch 2011, p. 18.

33 Busch 2011, p. 19 n. 29.

34 Next to the praetorians, the *speculatores* and the *equites singulares Augusti* there is reference to the *corporis custodes*, the *evocati* and the *statores*; whereas the status, the very different conditions of service and the assigned tasks that were not always definitive for the last three mentioned groups, should have at least been mentioned.

35 A specific work on the soldiers of the Italian fleets present in Rome is still missing: a useful framework, however, is given by Reddé's volume of 1986 and by some contributions of Parma (1994, 2002).

36 Lintott 1968, 1998 and 2008.

37 See the expression used by Foucault 2004, p. 16; Lintott 2008, p. 206.

38 The juridical principle of the '*Volksjustiz*', the introduction of *vadimonium* and the formalization of the *in ius vocatio*, the protection of *possessio*. For a detailed account, refer to the review of the book edited by Cédric Brélaz and Pierre Ducrey for the *Entretiens on l'antiquité classique of the Fondation Hardt* of 2008, edited by F. Reduzzi (http://bmcr.brynmawr.edu/2010/2010-02-75.html).

39 Tac. *Ann.* 3.27.3.

40 Part II Introduction, pp. 73–78.
41 *L'Urbs. Espace urbain et histoire (Ier siècle av. J.-C.-IIIe siècle ap. J.-C.)*. Actes du colloque Rome, 8–12 mai 1985 (CEFR 98), Roma 1987, New Edition Roma 2015.
42 Eck 1979; and the papers collected in Eck 1995b.
43 Giardina 2000 (ed.); Lo Cascio 2000b (ed.); Nicolet, Ilbert and Depaule (ed.) 2000.
44 Robinson 1992, chap. 7 'Water and Fire'.
45 Robinson 1992, chap. 12 'The Forces of Law and Order'.
46 Robinson 1992, chap. 13 'Public Order'.
47 Robinson 1992, p. 187, with reference to Durry 1938, pp. 15 and 166 s. Robinson knows Lintott's work and poses some interesting questions (for example, on p. 195, how an arrest occurs concretely).
48 On the topic of public order in the late republic, in 1984 he wrote a paper to present his work (Nippel 1984), completed a substantial contribution in 1988 (Nippel 1988) and wrote a monograph of synthesis in English seven years later (1995).
49 But apparently it can be found in the imperial one: vd. *infra*, pp. 10–11.
50 Nippel's intervention of 1984, as stated, is a preliminary work: the author intends to present the problems of a work in progress (the results are in 1988 an article and in 1995 a monograph) rather than an elaboration of proven theories; and he focuses on magistrates and their functions in the late republic, reserving ten final lines to the Principate on p. 29.
51 Nippel 1995, p. 115, with reference sources.
52 Foucault 2004 p. 224–262, 229–238. The bibliography on the theme of the birth of police corps is extremely extensive and has been intensified over the last thirty years, as a result, it seems clear, of the stimuli produced by Foucault's research on prison, detention and repression. On the characteristics of the 'police corps' of the Italian pre-national states, refer to the works mentioned in Part IV Introduction.
53 Nippel 1995, p. 82.
54 Nippel 1995, pp. 90–98. The same topic has already been addressed although not systematically in the article of 1984. Very brief is also the overview in the section 'Outside the capital' (p. 100 ff.) not very interested, as in prevalence happens, in Italy, but rather in the provinces.
55 "It remained primarily task of the individual to take care of himself" (Nippel 1995 p. 97). On the self-protection activities (Selbsthilfe, Selfhelp) an excellent summary with bibliography in Cascione 2016 p. 187. On Selbsthilfe, see also Krause 2004 pp. 14–16, with a panorama on the police in the Roman world at pp. 44–51.
56 Nippel 1995, p. 91.
57 Nippel 1995, p. 93.
58 Fraschetti 2000. It is, however, clear that much of the research for the book of 1990 (*Roma e il principe*, Laterza: Bari-Roma) comes to light in this contribution; and, above all, the basic idea of Augustus' intervention plan on the space and time of the city (which forms the backbone upon which the considerations set forth in Part II are articulated).
59 Fraschetti 2000, p. 728.
60 *Res Gestae* 5.1–2. Even in this case, the measures were progressive, from the *cura annonae* undertaken personally by Augustus, to the creation of *praefecti frumenti dandi ex senatus consulto*; up to the creation of the prefect of the *annona*.
61 In relation to Augustus' action, the prevailing idea in literature is the one of a sustained experiment (rather than a 'coherent system') that, after his death, "would continue in desultory fashion along a protracted evolutionary path" (Peachin 2015, p. 497 with extensive and updated bibliography n. 1).
62 Coulston 2000.

63 'La ville de Rome sous le Haut-Empire: Nouvelles connaissances, nouvelles réflexions', (Sablayrolles 2001).
64 Sablayrolles 2001, pp. 133–135 and 143–148.
65 Sablayrolles 2001, p. 149.
66 Cass. Dio 58.9.4 (Sablayrolles 2001, p. 146 s.).
67 Le Roux 2002, p. 18.
68 Breeze, Maxfield, Davies (eds) 1989.
69 Nélis-Clement 2000.
70 Le Roux 2002, p. 19.
71 The provinces, colonies, municipalities, peregrine cities refer to the local authorities for their security. Of great interest – and with a wealth of examples and breadth of visions – is the image of the situation in the eastern provinces, where we find different figures (*irenarchi, paraphylakes, diogmitai*, etc.) engaged in preventing and limiting the disturbances to public order, in accordance with a long tradition (on the situation in the East Late Antiquity, see Lewin 1993; Loschiavo 2003; on the *irenarchi* see Zamai 2001).
   Also for the West it is, however, possible to identify local specialized functions, for example, in the fight against banditry: one may consider figures such as the *praefectus arcendis latrociniis* of Colonia Iulia Equestris (e.g. *CIL*, XIII 5010 = *ILS* 7007 = *AE* 1994, 1288 = *AE* 2002, 1052, Germania superior. The inscription is taken into account by Riess 2007, p. 211 n. 101; updated bibliography about it in Bellomo 2009, p. 257 n. 2). Or the *praefectus vigilum et armorum* of Nemausus (Nîmes), which also fought the fires and the night crime (e.g. *CIL*, XII 3002). At Vasio, however, it is known that there was a *praefectus praesidio et privatis Vocontiorum* (*CIL*, XII 1368).
72 Le Roux 2002 pp. 21–27.
73 The large number of firefighters in Rome, according to Le Roux, can be explained only by the number of assignments entrusted to them; and calls for a comparison to Suetonius (*Cla.* 18.1) and Sablayrolles 1996, pp. 380–382. For occasional use for the tracking of runaway slaves see. Dig. 1.15.5.
74 Also see further on p. 23 ff.
75 Ménard 2004, p. 12.
76 Along the line of the Appendix 'Difficultès sémantique' by Yavetz 1983, p. 189.
77 The syntagm *ordo rerum* corresponds more to 'natural order of things'.
78 Rivière 2004.
79 In the selection of the events recur: the occupation of the free places in the circus with Gaius (Suet. *Cal. 26*); and demonstrations in favor of Octavia and against Poppaea (Tac. *Ann.* 14.61.1).
80 "The military presence, rather than being an instrument of pacification, becomes a violent intrusion into the urban space where organization and decorum are redefined by the *Princeps*" (Rivière 2004, p. 77).
81 Rivière 2004, pp. 72–76.
82 Rivière 2004 p. 67 n. 12.
83 Kelly 2007.
84 And here he cites, among others, MacMullen 1966 and Yavetz 1983. He then speaks about the urban crowds reactions, the popularity of Caesar, the constitution of the *potestas tribunicia*; and finally the relationship between *plebs* and *Princeps* in the High Roman Empire, up to Nero.
85 Nippel 1984, 1988, 1995; Erdkamp 1998; Erdkamp, de Blois, Hekster *et al.* 2003 (eds).
86 Nippel 1984, 1988, 1995; Braund 1998 (ed.).

87 Although, with a prospective look, in some cases he reaches the end of the third century.
88 The implementation, in good part, of what in the volume of Busch *supra*, pp. 6–7 was hoped for.
89 Consider Lintott's works, but most of all the more recent contributions by Le Roux and Rivière.
90 Kelly 2007, p. 172.
91 Kelly 2013, pp. 410–424.
92 A similar synthetic approach (as expected in a *Companion*) is found in the contribution by G.J. Fagan for *The Oxford Handbook of Social Relations in the Roman World*, 2011 pp. 467–498), who concentrated on the Republic and considers Lintott's volume of 1999 a cornerstone about the discourse on violence in the Republican era, while 'a comparable work on the imperial period has yet to appear' (2011, p. 491).
93 Hirschfeld 1891, Echols 1958, Freis 1967.
94 Nippel 1984, 1988, 1995 and Sablayrolles 2001.
95 On *aediles* and *tresviri capitales*, from the late Republic to early Principate, an excellent selection of ancient authors are portrayed, from Plautus to Gellius: Plaut. *Rud.* 372–373; Sen. *De vita beata* 7.3; ed. and *Epist.* 86.10; Tac. *Ann.* 13.28; Varro apud Gell. *Noct. Att.* 13.13.4; Varr. *De lingua Latina* 5.81; Mart. *Epigr.* 5.84. On *tresviri capitals*, the reference to the important monograph by Cascione 1999 cannot but be mentioned. Now also Cascione 2016, pp. 188–189.
96 In addition to quoting himself (Kelly 2007, pp. 156–160), Kelly refers, for the provinces in the Imperial age, to the excellent doctoral thesis by Cédric Brélaz (Brélaz 2005, and in particular at pp. 26–39), for he "aims at clarifying which institutions and/or persons were responsible for public security in Asia Minor under the rule of the Roman Empire and how this was achieved".
97 Fuhrmann 2012.
98 Fuhrmann 2012, p. 7.
99 Fuhrmann 2012, p. 234.
100 Hunter 1994.
101 I suggest the thorough review by C. Brélaz for Bryn Mawr (http://bmcr.bryn mawr.edu/2012/2012-09-13.html).
102 Fuhrmann 2012, p. 6.
103 Fuhrmann 2012, pp. 89–121 and 123–145, which is based, as explicitly stated, on the works reviewed here (Nippel, Rivière, Ménard, Brélaz and in addition, Keppie 1996).
104 Fuhrmann says that they will then be dismissed by Nerva in 97, without citing the source of information (2012, p. 114 n. 87).
105 Fuhrmann 2012, pp. 106, 115.
106 Fuhrmann 2012, p. 29. The 'civilian police' would carry out security functions and enforcement of the law performed by civil militias, local magistrates and their subordinates; or groups of guards organized and administered by individual communities. The terminology is largely borrowed from the works of Werner Riess, who on several occasions dealt with the language and the places of violence and tumults in the ancient world. Among the other works, I would like to recall Riess 2005, 2007, 2010.

2 Between *Pax*, *Disciplina* and *Securitas*

Moving the focus

The review I have just conducted serves to offer an overview of the contributions that have, more or less directly, addressed issues regarding public order of the city and Italy and the emperor's security. Two research paths have been clearly distinguished: the first analyses in a systematic way the individual military or paramilitary units in Rome (Echols, Durry, Passerini, Speidel, Bellen, Sablayrolles, Bingham); while the second looks at the dynamics of the conflict (Lintott, Nippel, Le Roux, Rivière, Kelly, Fuhrmann and, at least in part, also Fraser).

It is clear that even though the two issues are closely linked – the people's security and that of the emperor, as the first citizen among them, in public spaces of the city – from my point of view, none of the works described in the review appears to be complete and entirely satisfying.

The works of the first group, which are fundamental, provide the starting point upon which other arguments are developed. What prevails, in this case, is a descriptive and detailed analysis, and, at least in some cases, a repetition of what is already known. For the purpose of my own work, however, I find it difficult to grasp the overall design underlying the involvement of so many forces in the city; or, the impression that one gets is likely to be that of the rapid transformation of Rome from a 'free city' to a city manned by armed forces. What is not clear is the slow progress of an underlying plan and the progressivity of the launched process; the connection between the new military force with their intervention as regards to the infrastructure; the relationships between the different figures of officers involved in the military coordination and in other tasks; and synergies that are soon created between different corps and between military and civilian forces.[1]

The contributions on public order that deal with Rome[2] are focused on the Republican age, particularly on the last century of the Republic, because of the importance of this moment and for obvious reasons due to the density and the eloquence of the sources. But there is also another reason for this chronological choice: the transition to the Principate is considered the phase where public order in Rome is no longer entrusted to the action of the law and to the management of magistrates and their assistants. It is the beginning of a new era, one in which the central government is in control. In this context, the interpretation of the Augustan policy and that of his successors for the security of the city and the imperial interventions in Italy is based on a fundamental assumption: try to understand how much was the 'desire of dominion' behind the initiative of the *Princeps* to create the 'garrison of Rome'. To investigate which instruments were adopted to prevent or minimize riots in the imperial era, the most appropriate context seems to be given by the provincial areas (rather than Rome and Italy), especially the eastern province.[3]

This view, though understandable in its general assumption, explains why when dealing with studies on public order, imperial Rome and Roman Italy have received mainly hasty consideration, sometimes accompanied by the proposal of the usual well-known episodes – inevitably always the same and not too differently interpreted – regarding 'famous' disorders of imperial Rome and Italy.[4] Yet, the definition and the content of the public order, as a relative notion, varies depending on the era and as such should not cease to be of interest with the transition from one political system to another (from the Republic to the Principate), nor should the provinces rather than Rome or Italy attract more attention.

Now, I think that if the monographic studies of the soldiers of Rome can be considered completed – on the contrary, after the systematic phase of the last century, there have recently been some repetitions[5] – studies on public order have opened new prospects for those who wish to consider the issue of security in an analytical way. Not so much from the point of view of the professionals who were involved (personnel, barracks, military command hierarchy, etc.), but rather the one involving the complex system, which is gradually implemented during the first century AD, in which the legislation precedes logistics; civil forces act by mutual agreement with the paramilitaries; and the golden rule remains, as before, that of deterrence.

The 'desire to dominate' underlying the Augustan policy in relation to security should therefore be more carefully considered and brought back within the required boundaries. If, as Rivière has rightly pointed out, Augustus' actions follow a logic of civil war, this certainly applies to the first decades in which he governs. This logic, which is strongly and necessarily affected by the recent past and the modes of conquering hegemony, very soon becomes refined policy and cannot but consider the measures that saw the birth of the praetorian cohorts, the *speculatores*, the urban cohorts and *vigiles*[6] separate from the Augustan legislation and from the reforms affecting the territory of the city.

As already pointed out, the legislation of the last century of the Republic has been repeatedly investigated[7] in close connection with the riots that occurred during that period and their causes.[8] A similar reconsideration of the legislative work of Augustus and of his early successors (especially in terms of criminal matters) is, in my view, essential.[9] This does not exclude that the legislative measures and the creation of urban troops could also be considered in the light of other Augustan interventions concerning the administration of Rome as the capital of the empire. Among these, of particular interest is the gradual introduction, along with the weak infrastructure of public magistrates of the Roman

Republic, the 'big' prefects, with an increasing legal autonomy. On the other hand, it is at the service of the prefects that thousands of men are employed, many of which armed and well trained.

To try to understand what was meant with the security of the Emperor and of the residents of the city of Rome it is necessary to reconstruct the complex apparatus that Augustus had developed in the long term and consider the many different issues (population density, food supply, materials used in construction, sewerage system, road width) that until then, in a city like Rome, had often caused deep discomfort. In this framework, the new cohorts of Rome would have primarily a preventive function,[10] and Augustus himself would deliver this message to the population. During his entire reign, the cohorts (except perhaps for the *vigiles*)[11] after fulfilling their duties every day, would leave the city because they could not reside there.[12] The aim of the emperor was to ensure the security of places and the protection of the inhabitants of Rome and himself, the *Securitas Augusti et populi Romani*, as an internal precondition for the *Pax Romana*. It was the solution to both the recurrent problems connected with security and the launch of an experimental process (completed, in my opinion, not before the end of the first century AD).[13] A plan that nevertheless has nothing to do the systematic action and planning that, in modern states, characterizes the transition from 'Sûreté' (of the *Princeps* and his territory) to 'Sécurité' (of the population and of those who govern).[14]

Among the keywords of the Augustan policy, 'security' is the one that has received more limited attention.[15] It is not an integrative virtue of *pax*, of which it is a prerequisite and with which it goes hand in hand; nor is it limited to the stoic *tranquillitas animi* of the individual who seeks refuge in himself and it does not coincide either with the "salvation of the state and the *Princeps*".[16] The actual word 'security' (and synonyms), which is often used with the meaning of 'absence of risk', 'safeguarding of people', 'maintenance of justice', deserves, in my view, further systematic investigation, as far as it would be possible through the documents that are accessible to us (literary texts, coins, inscriptions and artwork), produced and circulating between the end of the Republic and the first two centuries of the imperial age (see Chapter 3).

In reflecting on the theme of imperial virtues, Andrew Wallace-Hadrill provides a detailed analysis of the Panegyricus of Pliny, which includes a wide range, and perhaps the most complete illustration of virtues. Yet, Wallace-Hadril argues that the virtues of the Panegyricus, in the words of its author, are those of compliance with the interests of the society and of the conservation of the status quo, whose keywords are "the protection of property, of personal security (life and death), and of social standing".[17] A century earlier, Augustus had begun to think and to work in this direction.[18] Although each individual is in charge of the protection of their own private property, the State represented by the *Princeps* guarantees – as reaffirmed and better articulated by Seneca especially in the *De Clementia*[19] – the best conditions for its achievement and defense.

It might seem that 'public order' and 'security' are two different ways of defining the same substance. It is not so: public order was not always the centrepiece of Augustus' policy – and of his successors' who more systematically adopted his same action – which was defined from a more comprehensive perspective, attentive to the daily life and problems of the city. The new architecture of the buildings, the legal reforms, especially criminal law, the propaganda had *Pax* and *Securitas* as essential conditions and prerequisites in particular in Rome, its ports and more generally in the whole of Italy and Latium.[20]

The profusion of directions in which Augustus moved and the progressiveness with which he worked actually give evidence, if not of the existence of a detailed plan from the very beginning, but at least of a broad spectrum idea that during his rule and in the following periods was defined, rectified and extended. It seems no coincidence that the sources emphasize the differences with Tiberius, whose measures represented for various aspects a less experimental path compared to the one followed by his predecessor.

To understand the dispositive that remained essentially unchanged for almost two centuries, I think it is essential to follow a diachronic order, devoting considerable space to the Augustan plan (Part II), without however disregarding the events that followed until the Severan Dynasty (Part III). It is also important to try to read between the lines the provisions that were only indirectly related to security; and trying, as far as possible, to discern from the historians' reports the reasons for the imperial interventions and their adjustments, their effectiveness or failures, in order to reconstruct a picture that is as close as possible to reality.

In this book I did not want to retrace the history of the troops stationed in Rome; at the same way it is not my intention to give an historical account of the 'Verwaltung' of Rome. To be more explicit, I will not take into consideration, but I will assume as normal conditions, the complex mechanisms of the *recensus*,[21] of the *cura aquarum*[22] or of the *annona*[23] or waste disposal service[24]; in short, all those things that fell within the 'normal' administration, the plan drawn up by Augustus onwards for the regular administration of the life of the city. Not because one should neglect the connection between *securitas/aqua* or *securitas/annona*, which the same sources refer to. For example, in *De aquaeductu urbis Romae*, Frontinus describes the *curator aquarum* as having a function that regards the security of the city, clearly referring to the material well-being of its inhabitants[25]; Tacitus uses the expression *securitas annonae*[26]; instead in his Panegyricus, Pliny correlates '*securitas*' with *libertas*, and considers them the objective towards which a good emperor should aim, along with the distribution of food or money to the needy.[27]

What matters, here, is the defense of internal stability, of the emperor's security and of the places in the city in order to prevent inconveniences and to deal with emergency situations when unpredictable events occurred: the threat of a fire that could be devastating; a sudden tumult in the amphitheater or in the circus; an inconvenience during a trip or a banquet; misconduct at the baths; or an assassination attempt on the emperor. In all these cases, the presence, the immediate assistance and the professionalism of armed men, the solidity of urban infrastructure and the coordination of the forces certainly made the difference.[28]

Even the concept of public security, like that of police, is recent. But while studies dedicated to the origins of the modern police force are not few, public security, as the other side of the same coin, has received much less attention (which is now, however, growing rapidly) and studies mainly concern the modern nations of western Europe. The topics of Michèl Foucault's lectures at the Collège de France between January and April 1978 on security, territory and population[29] still remain, from this point of view, despite their seemingly unsystematic character, the greatest theoretical study that has had important repercussions for historical studies and new important frameworks of research.

It is known that Foucault's attention only in the last years of his research activity was focused on ancient societies (especially that of ancient Greece), while he was always much more interested in the peculiarity of disciplinary normalization in modern societies. Since the 1990s, studies on the dispositives (at least by those who share the value and the effectiveness of this interpretive tool) have concentrated on contemporaneity (networks, new technologies, Internet, etc.); and much less followed is the path taken by Foucault, that to the origins of the dispositives of power, according to his genealogical research.[30]

Now, there is no doubt that when talking about 'disciplinary normalization' in imperial Roman society means connecting it to an approximate and extremely inappropriate 'comparativism'. I think, however, it is clear that other considerations and Foucauldian concepts (first among all the 'safety dispositive')[31] appear remarkably effective even for the analysis of ancient phenomena, at least in the Roman Empire. The influence of Foucault's reflection in terms of 'controlling society' is on the other hand obvious in the works of Lintott, Nippel and Rivière. The actual words of the philosopher – "the problem that arises is how to maintain a type of crime within economically and socially acceptable limits"[32] – constitute the core of Lintott's reflection in his works published after the 1960s and particularly in the intervention of 2008. If one reads between the lines of Nippel's contributions, the same inspirational principle is applied in the use of the language and of the categories employed.[33]

In the series of lectures at the Collège de France, Foucault identified and described criminal law mechanisms in Western societies between the Middle Ages and the modern age with a great range of examples: legal proceedings, disciplinary measures and security scheme. Such dispositives (juridical, disciplinary and security) have not followed one another along a progressive line and, even less, along an evolutionary one. Rather, they have proven to be most effective cyclically, while presupposing the existence and the resistance of one or more dispositives that had preceded them, "what changed . . . is rather the dominant, or more precisely, the correlation between the legal-juridical, disciplinary and security mechanisms".[34]

It seems to me that the program initiated in the Augustan age – in the sense of a coherent plan of prevention and deterrent over spaces and people – is a synthesis of two models of apparatus: the first being a traditional one (the juridical apparatus), and the second, a new one, which was being elaborated (the security apparatus); this latter being considered as "a certain way of adding and operating, besides its own mechanisms, also the ancient apparatus of law and discipline"[35] (see Part II Introduction and Part IV).

Even the dispositive of discipline – which is characteristic of the modern age – with Augustus becomes part of a new paradigm, and in particular of his most innovative discourse, the one concerning security: it is a discipline that does not operate on people as individuals, but acts on the whole population and on the city as a reference area (see Part II, Chapter 5).

Imperial security is therefore configured as a dispositive in terms of manifestation of a complex and multi-faceted programme that has a variety of effects in real society. Such a dispositive consists of several action plans, where discipline is at the service of security within a deterrent and preventive framework of action. It operates as a *prevention* on existing spaces (territorial definition, location of meeting and exchange places and their maintenance, waste disposal, treatment and control of the streets and water). It operates as a *deterrent* to individuals and associations through two instruments: a traditional and rigid one the first (the law), a new and versatile one the second (law enforcement officials, ranging from those policing the streets, to firefighters, bodyguards and district magistrates), in order to "get the most from the positive elements as to circulate better, and minimize, on the contrary, risks and drawbacks, knowing that they will never be eliminated completely".[36]

In conclusion, I feel that the dispositive of security seems a far more effective analytical concept when exploring Augustus' plan and its long-term outcomes, compared to the perspective of public order. To recall an effective principle, what I will try to do is to shift the point of view from *inside the function* – which has been systematically adopted, so far, in specific studies and in those regarding public order, except for Lintott's – to *outside the strategies* and tactics.[37] I will also focus on conduct as the art of guiding, but also as a way of behaving, which is perfectly suited to Augustus' public persona; which could have seen its model, according to Seneca's intentions, in Nero and was rather fully achieved with Nerva and Traianus (see Part I, Chapter 3 and Part III Introduction).

From the ideas to the actions, what will emerge, I hope, is the birth of an apparatus both invisible (the constraints, the laws, the appeal to tradition); and perfectly clear with a strong visual impact (the *vici* and their magistrates, the cult of the altars at the intersections, the soldiers, the barracks, the *stationes*, the city guards, the offices of the prefectures). After all, the intervention on public spaces and the construction of a standing army, in ancient as in modern times, are fundamental characteristics of each system that intends to be (considered as) new.

Notes

1 The same idea that Rivière, I feel, wants to express when he speaks of studies concerning places of confinement: "L'intérêt n'est pas nécessairement donné par la découverte et la description de l'objet lui-même ... mais par ce que sa forme et sa matière révèlent des étapes de sa production, de la place à laquelle il était destiné et la

fonction qui lui était reconnue" (Rivière 2008a, p. 205). Sablayrolles' work on 'urban troops' together, is what comes closest to the aim that I want to pursue here; but it has a predominantly analytical feature and is built in large part (with the exception of the *vigiles*) by sewing together information gathered from classical repertoires.

2 I have not included in my review the studies on provincial public order because they go beyond my interest.

3 Which boasted a past of freedom for citizens and a wealth of political and administrative experience and that continues to provide primary sources with wealth. Studies have intensified over the past twenty years. Just to mention a few: Pollard in 1996 and 2000, Yannakopoulos 2003 and Brélaz 2005.

4 Nippel, Sablayrolles, Le Roux, Ménard, Rivière, Busch, Bingham, Kelly 2007; with acknowledgements, among others, of the previous authors Hirschfeld 1891 and Durry 1938.

5 From this consideration, naturally, publications reporting new archaeological and epigraphic finds are excluded.

6 I deliberately leave aside the Germanic bodyguards and the *statores* who certainly are to be classified, at this stage, as private (see Part II, Chapter 4).

7 *Supra*, p. 46 ff.

8 Lintott, Nippel, Le Roux, Rivière.

9 See Part II Introduction.

10 Le Roux, Sablayrolles, Rivière, Kelly 2007.

11 See Part II, Chapter 5 and Part III, Chapter 6, n. 11 and 13.

12 Suet. *Aug.* 49.2: *Umquam plures quam tres cohortes in urbe esse passus est easque sine castris)*; *Tib.* 37.1: *per hospitia dispersi.*

13 See Part I, Chapter 3, p. 46 ff.

14 Moving from the first to the second, in the analysis of Foucault, represents the irreversible passage from a Middle Ages extended to include the sixteenth century up to modernity (Foucault 2004, p. 57) Ménard 2004 (supra, pp. 13–14).

15 The key value of the *securitas* has been correctly identified by Ménard 2004 (*supra*, pp. 13–14), who believes it will become one of the ideological foundations of the imperial power only in the second century.

16 Sablayrolles 2001 speaks of a *Salus Rei Publicae* (p. 149). Yet this is an ancient value that will remain; whereas *securitas*, as will be seen in detail in the next section, is something completely new and different.

17 Wallace-Hadrill 1981, p. 418.

18 Also see Instinski 1952 pp. 15–20 and in particular the synthesis on p. 16: "The security of persons is assured with an orderly state, a reasonable religious practice and a guarantee of private property."

19 See Part I, Chapter 3.

20 On the different forms and means of the Augustan 'propaganda', see Galinski 1996, part. pp. 39–41.

21 Lo Cascio 1997, 1998, 2000b.

22 On *cura aquarum* created by Augustus, see Lo Cascio 1991, pp. 119–191 (part. pp. 130–131), with prior bibliography.

23 On *praefectura annonae*, Pavis d'Escurac 1976; on famine, supplies, distribution, see Virlouvet 1985 and 1991.

24 General: Dupré Raventós, Remolà (ed.) 2000; see in particular the intervention of Panciera 2000. For a general treatment of Sanitation in Roman Italy from an archaeological point of view, see now Koloski-Ostrow 2015.

25 Frontin. *De Aqued.* 1.1: *officium ad usum, tum ad salubritatem atque etiam securitatem urbis pertinens.*

26 *Ann.* 15.18, recalled by Noreña 2011, pp. 130–131.
27 Plin. *Pan.* 27.1. Also see the following paragraph p. 53 ff.
28 For the effectiveness of certain emergency measures in Italy and for escort services during the imperial travels, see Part IV.
29 Foucault 2004.
30 See the extension of the concept of dispositive in Agamben 2006.
31 Apart from the beauty of the reflections on the Christian pastorate, on the counter-conducts, on the reason of state that clearly go beyond our narrow field of interest, to which Foucault devotes exemplar analysis (Foucault 2004, pp. 105–223).
32 Foucault 2004, p. 16.
33 *Supra*, pp. 9–11. The acknowledgment, however, only in the case of Rivière 2004 (and, in a different way, by Fredrick 2003) is explicitly recognized. See now in particular Rivière 2008a in electronic version at www.cairn.info/revue-hypotheses-2008-1-page-203.htm.
34 Foucault 2004 p. 19.
35 Foucault 2004, p. 21.
36 Foucault 2004, p. 30.
37 Foucault 2004, p. 93.

3   The security of Rome and the security of the emperor

The slow development of a discourse and its transformation into a communicative instrument

Aurea secura cum pace renascitur aetas.
(Calp. Sic. *Ecl.* 1.42)

Underlying the discourse that will be further developed in Parts II and III, is the idea that Augustus and other emperors between the first and second centuries devised and implemented a series of coordinated measures to ensure their own security and that of the inhabitants of Rome. Before illustrating what these measures were and what areas were affected, it is appropriate to understand how, among the concepts in political communication that accompany the actual measures adopted by the *Princeps*, *securitas* was created and put into place. The term refers to both the safeguarding of the security of the public places of Rome, and to the subjective and objective *Securitas Principis*, in the sense of a sovereign who not only guarantees the security of his people, but at the same time can move freely because he knows he is being protected in public and private spaces.

The reflections of modern scholars on the ancient concept of 'security' are drawn from the texts of Lucretius, Cicero and Seneca, collected and analysed on several occasions, with partly shared considerations. After the article by Ludwig Moritz Hartmann on the Real-Enziklopädie of 1921, the first to have launched a systematic review of *securitas* as a political issue was Hans Ulrich Instinski, whose work I shall refer to frequently. On the basis of literary sources – and in particular Pliny's Panegyric – Instinski concentrates his studies on the first century of the empire and does not go beyond the second century. In large part, the time span considered in this work coincides with the same time span. Instinski believes that by the end of this phase, the terms and the values of *securitas* as a concept have been fully established and become part of the common lexicon of politics. However, I think it is interesting to complete the historical context to verify how, with a change in meaning, the same term – and the circumlocutory phrases that contain it – prevails in private and fully recovers the ancient meaning it initially had, that of 'absence of disturbance'.

The etymology of *securitas* from *sine cura* and its connection at a later time with the greek word ἀσφάλεια[1] is a shared idea. As for the occurrence of this word, a point of reference is represented by the work of Andrea Schrimm-Heins, who identified approximately 500 testimonials of its use throughout the Roman period until late antiquity. The scholar addressed the issue by identifying two essential meanings of *securitas*: namely 'having no preoccupations' and, as a derivative, 'protection of the individual'. The first meaning of the word is the most ancient and its roots can be traced back to the Epicurean concept of ἀταραξία, which partially corresponds to the stoic ἀπάθεια. The theme of *securitas* originated from Hellenistic philosophies and gradually reached Roman philosophical thought. Therefore the terms *ataraxia, apatheia*, and *euthymia* form the basis for the reasoning behind the concept of *securitas* (and synonymous expressions), which are found first in Cicero and then, in their highest expression, in Seneca.

In a 2008 article, Friedrich Arends offers a study on the significance of security from Homer to Hobbes; but such a wide time span allows for only a partial overview of the theme in ancient times; and Latin sources in particular are penalized since they are analysed in little detail or in too rigid a way. However, significantly, Arends puts forward the idea that *securitas*, as a political theme in the empire, has been overvalued by scholars, who are victims of a retrospection of considerations that would only emerge with Hobbes.[2] *Securitas* therefore, according to Arends, was nothing more than one of the many slogans used by imperial propaganda through images such as those found on coins. We will soon see however, that on the contrary, Instinski's intuition that *securitas* was a political problem, central to the empire, still remains valid, both in terms of communication, and in its eminently practical issue.

More recently, John T. Hamilton has written a summary of all the semantic implications that shows how the concept of security evolved from the ancient to the modern world, in a work that can be considered philosophical in spirit and philological in the setup. It is clear – and a broad range of sources confirm it – that security was marked from the beginning by an exceptional discursive versatility as well as by a peculiar ambiguity that still exists today. Hamilton's work offers insights of great interest, even for its historical background, especially for the philosophical implications of the term. In particular, I am referring to the interpretation of the *securitas* as a theme of 'care of oneself', identified by Michel Foucault as an expression of Stoic and Epicurean doctrines.[3] Literary sources are essentially the only ones taken into account; and sometimes what is missing are insights or direct evidence from the non-literary sources.[4]

Starting once again from Instinski, Alfred Kneppe covers the issue of *metus* in society and in Roman politics in his work, which is both complex and suggestively rich, and has a more psychological and anthropological approach based on, to a certain extent, the research method of Edwin Dodds.[5] Kneppe, analyses the vocabulary and expressions related to fear, from *metus Gallicus* and *Punicus* until reaching the propaganda strategies on security of the Severan period. The fifth and sixth chapters are important to set the theme of *metus* and *securitas* after Nero, when the first term is functional above all to express the fear of the cruel tyrant, a true topos from the second half of the first century AD.

The second term, that of *securitas*, highlights a new implication, which concerns safety *from* the emperor. It is accompanied by the issue of *securitas* against *libertas*: there is a price to pay for security, which implies the fear of power and apparent limitations of freedom.

After this cursory review of the main bibliography on the concept of *securitas* in ancient times, I will try now to analyse the main discourses concerning *securitas*, by reinterpreting a selection of different types of sources (literary, numismatic, epigraphic). I will attempt to take each case in turn and identify their place and role, and to understand how *securitas* was transformed during the two centuries that are the focus of this work. The review of sources that is being proposed here is rich, but without doubt the series of discourses is incomplete: that is due to the constraints imposed by an introductory chapter and to the selection of materials that seem to be the most significant examples for this reinterpretation. Particular attention will be paid to the initial phase, which concerns the conditions and the gestation of the discourse (between Late Republic and Augustus); and the phase of full adjustment (Nero's era).

When we refer to 'discourse' on *securitas*, we would like to imply, referring to Foucault, a real 'discursive event', which is a complex phenomenon, "dispersed among institutions, laws, political victories and defeats, claims, behaviour, riots, reactions".[6] To trace the threads of this event, we will start from what emerged from the philosophical and political debate of the middle of the first century BC in Rome. It is, in fact, an obvious choice, since it is in this context of communication and in this chronological period that the words *securus* and *securitas* occur for the first time. Reconstructing what 'security' meant for the Romans means considering, from time to time, the field of existence of the term, namely the set of relations that defines the word in context, and its possible functions. This means identifying sets, which are recognizable according to regular or irregular functions. From this perspective, it is necessary to create a philological analysis that takes into account the discernment of the most significant occurrences, in the belief that *"ce n'est qu'au prix d'une recherche lexicale serrée qu'on pervient à établir la 'somme des différences'"*.[7]

In this way we will see how and when the discourse on security appeared, first in philosophy and then in politics, and under what specific conditions it was modified and supplemented by the proliferation of other discourses, which have gradually defined and clarified its meaning.

The last century of the Republic was marked by profound social unrest and devastated by an almost uninterrupted succession of *bella intestina*. Up to the last decades of the second century, public order had been a problem only in special situations, but then it became an important issue that required special attention. It also became a theoretical topic that would gradually be debated and discussed. Starting from Cicero's political reflection, *securitas* was used initially to define a condition relating to the private sphere, with philosophical connotations. Its meaning then changed and it came to be considered as a fundamental attribute of a man in the public sphere and a man of the state, which brought with it a series of political consequences.

Thus, taking into consideration the time span from the middle of the first century BC to the Principate of Augustus and its crucial period, has a particular significance: at this time, the yearn for one's own 'security' and that of loved ones and personal possessions was the main concern and becomes, along with peace, the most pressing demand that the inhabitants of Italy make to the *Princeps*. It is precisely from this point, after so many revolutions and subsequent upheavals, that security starts to become necessarily one of the key themes of the Principate.

Between the first and second centuries AD, the theme of *securitas* would move from literary, philosophical and political meditations, to other means of communication, such as inscriptions and coins. As will be seen, this shift marks a substantial change in the concept, from a purely philosophical nature to a more political one. Security is no longer, at least not only, an absence of concerns or of negative events, but a pledge of political stability and social peace with the new *Princeps* as its guarantor. One can say that the imperial power gains control of the discourse on *securitas*, causing an expansion of its meaning towards an area that the term had already partially touched. However, the appropriation of the concept of *securitas* in a distinctly political sense also occurred when the senatorial class felt that its safety and the preservation of its prerogatives was threatened by emperor's behaviour. In this conflict lies the decisive transformation of the discourse on security; it begins to be interpreted as a guarantee of protection against the emperor's abuses. From this point (and above all) *securitas* takes on the meaning of being sheltered by the *Princeps*. This is not a sudden turn, but a conscious expression of what the senatorial class experienced under the Principates of Tiberius, Gaius and Claudius. The key witness to the development of this awareness is undoubtedly Seneca, and the Principate of Nero, with the situation that followed his dramatic end, being the second moment upon which I will focus more carefully.

Cicero: prologue of a discourse

Outbreaks of violence and subversion of the institutional rules that took place at the turn of the second and first centuries of the Republic broke a balance that for centuries had remained substantially unchanged and alter a system that, until then, had guaranteed the (rights of) citizens.[8] The critical step that led to the semantic change of the concept of *securitas* is linked to such profound changes and to results in the field of law. It is in fact the insecurity linked to the civil conflicts that would lead to legislative action for crimes between private individuals that could have implications for the public sphere.[9] The consequences, in terms of criminal law and, in its broadest sense, of the rules of social life, would be significant. On the one hand, individual security, from a state of mind to which to aspire to as individuals, would lean towards being recognized as a social right to be preserved. On the other hand, the security of the state did not only mean the protection of the people from foreign enemy attacks or the protection of the institutions, but also the safety of the people in a civil context, and particularly in public places in the city.

The noun *securitas* appears in Cicero's work preceded by the use of the adjective *securus*, with the meaning 'quiet, composed'. The adjective appears repeatedly in his speeches and above all in the correspondence between 59 BC and 43 BC,[10] until in a letter to Lucius Munatius Plancus the adjective, in its usual meaning of 'free from anxiety', occurs in a context that gives it a meaning that is closer to physical safety.[11]

In the same years in which Cicero uses the term, *securus* can also be found in Lucretius' work. It accompanies *aevum*, referring to the blissful state of the gods, or the *quies* of death.[12] The chronology of Lucretius, as is well known, is so problematic as to discourage a hypothesis on the exact period in which the term was used: the temptation of believing that his reading influenced the coining of the term by Cicero is not supported by evidence.[13] On the contrary, the appearance of *securus* in Cicero already in 59 BC (see note 10) appears very early compared to the famous letter of 54 BC, addressed to his brother Quinctus, in which Cicero shows all his appreciation for Lucretius' work.[14] It is therefore not in the chronological reconstruction that we will disentangle the issue of the coining of the term *secures*. Rather, as we will see later, it is the regularity with which the word appears in certain contexts that seems to suggest a Lucretian (and epicurean) conceptual priority.

In 54 BC, the adjective *securus* is used with a completely private meaning, when the Arpinas lives on the margins of political life and finds intimate comfort studying philosophy and spending time with a few dear people. In a meditation with an Epicurean content addressed, perhaps not accidentally, to his friend Atticus, Cicero says: "Providing only that my private and domestic circumstances give me pleasure, you will find my equanimity quite remarkable."[15]

However, after ten years, it is in the *Tusculanae Disputationes* that we find a complete definition of *securitas*:

> How, moreover, can anyone, about whom come or can come a throng of evils, enjoy that object of supreme desire and aspiration – security (and security is the term I apply to the absence of distress upon which happy life depends)?[16]

The term *securitas* in Cicero, in the decade between 54 BC and 44 BC, is therefore essentially defined as the absence of disease and restlessness, in which happy life lies: by staying away from evils, one can aspire to reach happiness.[17] In this first series, *securitas* corresponds to a totally philosophical and private semantic function and is, according to Andrea Schrimm-Heins, "the product of the effort made in translating and transmitting the Greek concepts of *ataraxia, euthymia* and *apatheia*".[18] This is certainly true, since, in the eclectic dimension of Cicero's reflection, the term alludes first to one, and then to the other orientation.[19] However, one cannot but emphasize that Cicero's most explicit reference is to Democritus' εὐθυμία. Similar to *apatheia* and *ataraxia*, the term refers to the conditions of the soul that in some way have negative connotations – already provided etymologically by the privative suffix – compared to an active behaviour in society. When referring to *euthymia*, *securitas* finds a balance in a state of serenity and satisfaction that does not deny the possibility of commitment.

It is likely that, in this same time span, Cicero wanted to clarify a term that was already important in his thought, excluding quietistic implications. Indeed, if we consider the use of the word in Cicero's coeval works, we notice how it was expanding its functions and its correlations, abandoning the strictly private sphere. It would appear therefore that the term *securitas* moved away from an external condition (absence of disturbances, as an accidental condition) to a more dynamic predisposition (knowing how to maintain himself *securus*).

The term *securitas* makes its way into Cicero's rhetorical and philosophical system configuring as one of the *res expetendae*, ideal states that are achieved by internal (*virtutes*) and external (such as accidental conditions) elements.[20] A careful reading of Cicero's political and philosophical works – in particular those concerning the figure of the ideal *rector* – makes it possible to follow the process that transforms *securitas* from *res expetenda* (accidental and external) to *virtus* (interior) of the statesman,[21] thanks to an expansion of the functions of the name, without, however, losing its original semantic domain.[22] It is by following this process, which is slow as much as it is recognizable, that the original meaning of *securitas*, 'lack of concern', which occupies a completely private space, touches public life and implies a shift from the sphere of 'psychological condition' to the one of personal 'quality'.

The expansion of its meaning begins to occur when Cicero identifies the *securitas* as the most important condition of a statesman, because it includes those fundamental requirements typical of a politician, which are *constantia* and *dignitas*.[23] A passage slightly after goes in the same direction, making it clear how *securitas* provides the concrete ability to cope with anxieties and preoccupations typical of an active political life.[24] Cicero is perfectly aware of an inevitable constitutional change for Rome; such awareness is accompanied by the evidence of a common desire on the part of the Roman elite, to enjoy the *otium* by distancing oneself from the political reality of the time, even under the influence of a widespread Epicureanism.[25] These two elements appear to be, in my opinion, of fundamental importance to understand not only the elaboration of the concrete concept of security in Rome, but also to view exactly in what way the meaning of the term *securitas*, as a signifier, was destined to become more similar in meaning to terms such as *tutela*.

The profile of the *Princeps* as a man able to restore the integrity of the mixed constitution that had made Rome great, is defined, as is known, in the *De re publica* (54–51 BC). It represented, for Cicero and those close to him politically, the desire to restore public order and security that Cicero himself had hardly ever experienced having been born and grown old during the time of the civil wars. Among the many figures and images that the Arpinas uses in *De re publica* to describe the ideal *rector* (*gubernator, dispensator, moderator*), one of them in relation to *securitas*, draws special attention: it is the definition of *rector* as *tutor et procurator rei publicae*.[26] In this expression, the idea of *gubernator* is enriched with new meanings through the presence of terms that refer to legal and political institutions, such as the *tutela* and the *procura*, which are both of considerable significance conveying the meaning of delegating certain civil functions on the one hand, and individual safety and security of personal property[27] on the other.

The same concepts, so central to constitute a possible definition of security in the modern sense, are explained with even more effectiveness in another passage of the *De re publica*, in which the political link between the *tutela* and the *securitas* is made perfectly clear, although still intended as having a private dimension:

> And in a State ruled by its best men, the citizens must necessarily enjoy the greatest happiness, being feed from all cares and worries, when once they have entrusted the preservation of their tranquillity to others, whose duty is to guard it vigilantly and never to allow the people to think that their interests are being neglected by their rulers.[28]

The people who are entrusted to the custody of the *principes optimi* are blessed and free from all preoccupations (*vacuos omni cura – permisso otio*): indeed, the *tutela* of the *Princeps* allows to be *sine cura*.[29] The two terms, *procura* and *tutela*, occur again in a consideration in *De officiis*.[30] Even the *procuratio rei publicae*, like the *tutela*, is established for the benefit of those who are 'entrusted' to the state.

Tutela[31] is certainly a distinctive legal institution of the Roman system, an instrument of protection and supervision for the safety and property of persons who are not personally able to defend their rights.[32] The definition of Cicero's *rector quasi tutor* is therefore of great significance in the political discourse that was taking place in Rome over the need for constitutional reform. The *rector* that was supposed to lead Rome to restore legality is juridically characterized as being the one in charge to monitor and protect the safety and property of those who are entrusted to him[33]: he is a *tutor* because he defends the citizens and their property; but he is also a *procurator* by definition and accordingly "the citizen is *securus*, because the ruler works *cum cura*".[34]

Cicero's formulation would prove to be so perfect and effective that it was to recur, almost with the same terms and substantially unchanged, more than a century later, in the Annals by Tacitus who, in the first book, outlines the conditions in which Octavian managed to retain prerogatives and to impose himself without significant obstacles.[35]

Commencing with Sulla and his communicative strategy of the *Felicitas*, within a few decades, there was a shift from the personification of abstract ideas concerning a plurality of individuals to the identification of the idea with a single person.[36] It is with Cicero that a solution to the contrast between *Virtutes Populi Romani* and *Virtutes Imperatoris*[37] was found. Through his reflection on *rector*, Cicero makes a substantial contribution to the creation of a profile and the necessary qualities of a *Princeps*, and above all he outlines the eligibility of the *Princeps* as a constitutional figure. This is a discourse – a rhetorical solution to many contradictions – that somehow ratifies a process that has been in progress for some time. The complex political reality of his time forces Cicero to put aside his cult for the republican 'constitution' and to take the risk of placing his trust in Pompey. At the same time, he participates with his authority in groundwork needed in order to justify the existence of a *consul sine collega* and the advent of the future plenipotentiary *Princeps*: by also demonstrating that the *tutela rei publicae* is a condition that only the *rector/tutor* can and must provide the community with. According to this logic, the *securitas* is a desirable asset for the private citizen and a quality of the statesman, a feature of the soul and a material condition of the individual, and after Cicero it is ready to be the central theme of Roman policy.

Maecenas, Roman epicureanism and Lucretius: genesis of a discourse

Between the thirties and the twenties of the last century of the Republic, a few decades after Cicero's views and 'political' theory were elaborated, Maecenas had the opportunity to act in Rome as *Urbis Custos*. His experience and his idea of 'order' were, in my opinion, of crucial importance for the further elaboration of a security programme for the city.

In 36 BC, after the defeat of Pompey, Octavian, in a moment of difficulty, entrusted Maecenas with the administration of Rome. Maecenas' action, quick and effective, had to be conducted on a double level, both diplomatic and military, and in the light of the day, since with Pompeians were clearly organising themselves.[38]

Cassius Dio informs us that in 31 BC, Maecenas was again charged with public order in Rome and in the peninsula, and continued to be so for at least two years, while Octavian was engaged in the war against Marcus Antonius: when Maecenas was *urbis custodiis praepositus*, and the head therefore of the city guards, he suppressed the conspiracy of Lepidus.[39] Maecenas' subtle action, both swift and fierce at the same time, not only saved the life of Caesar's son, but suffocated what Velleius rightly considers the probable beginning of a new civil war.[40] Maecenas halted Lepidus and his conspirators, after a period of covert surveillance, and applied a perfect example of surveillance and policing aimed at ensuring the safeguarding of Octavian's safety and maintaining social order.[41]

Ten years later, in 22 BC, Maecenas forestalled another conspiracy, that of Murena, brother of his wife Terentia, while giving his brother-in-law the chance to save himself; the motive leading to the subtle cooling of relations between Augustus and his right-hand man.[42]

In both the events of 36 BC and 31 BC, Maecenas must have acted with military forces at his command; it is impossible to assert if they were constituted of ordinary soldiers from Octavian's army or personal bodyguards. Furthermore, it does not seem that Maecenas had an official role on these occasions, since there are no sources to confirm it, and Maecenas's prerogatives seem to be well beyond those of the *praeraefectus Urbi*, both during the Republican era and the imperial age.[43] Ultimately, even though it is difficult to identify Maecenas' position within the magistracy correctly (this is an unforeseen circumstance and a delicate transitional phase), there are significant sources that provide evidence of the duties he undertook.

The anonymous author of *Elegiae in Maecenatem* (post 8 BC) praises the equestrian, naming him *vigil Urbis*;[44] and, later, he exalts his central role in providing security to the emperor, the city and its streets, which were finally made safe.[45] The expression *Urbis custos* anticipates what, after a few decades, would be used by Velleius (*urbis custodiis praepositus*), but there is a significant allusion to Augustus' defence plan, of which Maecenas is *obses*.[46]

Naturally, in this review particular consideration has to be given to the famous speech of Maecenas reported by Cassius Dio[47] that, positioned chronologically after the two experiences of management of the security in Rome and in Italy which have just been recalled (29 BC), may well contain elements referring to the political thought of whom we imagine had pronounced it.[48] The most important passages, for the purposes of this book, are those relating to the functions of the *praefectus Urbi*: the requirement of having few armed men so as not to be a risk to the *Princeps*; the jurisdiction in important lawsuits, and the need to suppress assemblies because they represented the leading cause of disturbance of public order. Other important passages of the long speech dwell on the equestrian ranks, referred to as the true reserve for administrators and trusted men; or on the idea of the historical necessity of a moderate Principate, capable of ensuring the management of a vast empire and preserving peace and order.[49] The extensiveness of the reasoning and arguments used in support suggest that Maecenas' experiences most likely constituted an important precedent for the development of the future 'security policy' of the city and the emperor. Furthermore, it can be said that his action was an inspiration for the creation of the urban prefecture and the *Vigiles*.[50]

The themes of the speech reported by Cassius Dio, in addition to finding validation in Maecenas' political experience, are also the results of an intellectual experience, that of political Epicureanism, or better, a Romanized version of the Epicurean tradition, because in Rome the commitments originally prescribed by Epicurus were moderately mitigated.[51] This unorthodox variant of Epicureanism had strong links with the Campanian cultural scene, which had among its regular attendees Vergil, Horatius and perhaps even Maecenas.[52] In this environment, the philosopher Philodemus participated, with his treaty 'On the Good Monarch according to Homer',[53] in the debate on the figure of the *Rector* of which Cicero had been an advocate, and in which Timagenes of Alexandria[54] would also take part. We know, moreover, that even Epicurus wrote a treatise 'On the Monarchy', of which, unfortunately, very little has been preserved.[55]

It is certain that Epicurean philosophers, fleeing upheaval and seeking stability, considered anarchy as a risk for the fulfillment of the *ataraxia*; and that the theme of external security, the *asphaleia*,[56] was the focus of their reflection.[57] If, as has been demonstrated, Roman Epicureanism did not contain a total refusal of political activity, we may suppose that thinkers like Philodemus supported and approved the conduct of that part of the ruling class that showed interest in Epicureanism, without having to necessarily postulate an estrangement from orthodoxy.[58]

A recent interpretation allows us to establish a more direct relationship between Lucretius – as a conscious interpreter of the Roman epicurean experience – and contemporary politicians, including most likely Maecenas himself. Jeffrey Fish having read the fragment KD 7 by Epicurus,[59] considering the passages of Lucretius in which the poet criticizes political ambition and compares it to the torture of Sisyphus,[60] deduced logically that the Epicurean critical approach to politics, rather than being against the whole Roman political class, might well have been directed towards the so-called *homines novi*.[61] If this is true, Epicureanism, from a quite conservative perspective, above all, harboured strong reservations about the 'mad' careers of the emerging classes, rather than those of families that traditionally were called to govern.[62]

Furthermore, we cannot forget that Lucretius, evidently taking part in the drama of his time, wrote for Memmius, whose noble birth (*Memmi clara propago*) he recognized along with his significant political commitment in the difficult moments that Rome was experiencing (*patriai tempore iniquo*)[63]; and is fully aware that, to gain *asphaleia* and be *securi*, first of all *salus communis*[64] must be assured.

In the first century BC, the Roman followers of Epicurus, due to both the state of constant insecurity and the continuously frustrated desire for *otium*,[65] were quite aware of the fact that the establishment of a Principate would allow the wise man to fulfil his desire of a secluded life, in a guaranteed condition of security and tranquility. In this sense, the figure of *Optimus Princeps* was certainly worthy of great consideration, because the 'sacrifice' of personal tranquility guaranteed the tranquility for others.

It is not difficult to imagine therefore, that Epicureanism, thanks to its typical utilitarian conformation, provided themes and words to justify the advent of a Principate, finding a natural harmony in the aspirations of those potential emperors who were evoked in this role, but also in the desire of the civil society (and its elites) to find a delegate, a leader whose decisive action would have been seen like that of a saviour.[66] The Epicureanism, moderate or imperfect, of Maecenas and of his literary 'circle', which undoubtedly received influence from the scene in Campania, helped to provide philosophical and ideological support for the suitability, the necessity and the absolute legitimacy of the Augustan regime; and the action of restoring public order for the purposes of restoring peace.

Augustus: the words of the discourse

Augustus' concrete programme regarding the security of places is clearly outlined by Suetonius[67]: in order to halt the diffusion of armed bands of brigands,

Augustus planned the installment of *stationes* in strategic points on the Italian peninsula with the purpose of preventing robberies and kidnappings. He requested that servile *ergastula* were put under control since they often hid the kidnapped; he dismantled the *collegia*, which were often responsible for violence and riots (but left the *antiqua et legitima* ones intact). The *Princeps* was presented as the one who, in addition to peace, was concerned with security when referring to criminal bands. Indeed it is evident that both Ovid[68] and Suetonius (*He made it safe too for the future, so far as human foresight could provide for this*)[69] also refer to these measures.

These initiatives determined by Augustus were included in a series of reforms,[70] which as is documented, covered a very wide range of measures relating to the *cura urbis*: ranging from the prevention of fires and flooding of the Tiber River to urban reorganization and the restoration of religious traditions, to the control of time by way of the calendar.[71] The management and the control of space was also a focal point when ensuring security. As Wallace Hadrill writes, under the rule of Augustus "mapping from being an irrelevance becomes an obsession".[72]

The *Princeps* devoted serious attention to rule and discipline and intervened effectively through laws and the service of justice. Suetonius reveals that Augustus, without any limits, personally dealt with both areas, by examining by himself, until late into the night, even quite ordinary cases.[73] An exceptional confirmation of this can be given by an intervention of Augustus with a letter, dated 6 BC, aimed at resolving a private dispute.[74]

The situation was the following: a married couple tired of the insults and threats by two brothers, Philinus and Eubulus, had ordered a slave to drive them away; to threaten them, the slave was supposed to have emptied the contents of a chamber pot onto their heads. Unfortunately the slave lost his grip and the pot struck Eubulus's head, killing him. The city governors were responsible in giving the verdict; in spite of that, Augustus' opinion was that the spouses should not be considered guilty. The reason why Augustus was called to intervene is perhaps linked to the consequences that the two protesters' behaviour could have had on public order.[75] This is a significant case for two reasons. First, it demonstrates that the disruption of social peace is not only triggered by collective needs, which can provoke unrest, but also by the improper behaviour of a single person. Second, this episode took place in the eastern part of the empire where the community's safety is entrusted to the local authorities and the *Princeps* does not avoid defining the principles that guide his actions in terms of safety.

Augustan's actions will be addressed shortly. What is interesting here is to focus on the words that helped to expand its effectiveness and scope. If the theme for the *securitas* that, as already argued, seems to have appeared in Rome in the Epicurean sphere and lit up its political and practical aspects thanks to its philosophical genesis, it is, however, through the language of communication (figurative, literary, epigraphic, numismatic) that it gradually assumes shape and legitimacy.

Provided that the political perspective of Epicureanism has as its ultimate goal the realization of that *secura quies* (the *ataraxia*), it does not seem a coincidence that the occurrences of the noun *securitas* and the adjective deriving from it show a surge in literary works in the Augustan age. The term recurs consistently, according to various declensions and, along with other expressions by then explicitly referring to the conditions of peace and serenity guaranteed by the *Princeps*, thus contributing to the expansion of an endless Discourse on security. For reasons of space and for the purposes of this work, it will be sufficient here to offer a quick exemplification of the occurrences of the term in the works of the so-called Augustan poets.[76]

The adjective *securus* is often used to refer to *quies*[77] or *otium*,[78] and particularly interesting are the references to *secura otia* of the Golden Age, which, both in Virgil and in Ovid, show how *securitas* had already become available for and perhaps even under the domain of the power.[79] *Securi* are the deities, or the time that they spend, in full continuity with Lucretius' model[80]; *secura* is the death or metaphoric elements that represent it, such as the waters of Lethe (Lucretius) or sepulture; in Tibullus there is evidence of the use of the adjective (*securus/-a*) in the phrases used to officiate a funeral, which will become very popular in the second century until the late Roman Empire.[81]

One can be *securus* in the sense of careless, forgetful, not at all worried over someone or something[82]; *securus* can also be a table or a meal, in the sense of 'frugal' or even a serene moment that drives away evil thoughts[83]; and finally *secura* is the peace.[84]

In the period that runs from Lucretius to Ovid, from the mid-first century BC until the first quarter of the first century AD, *securus* therefore retains a distinct spiritual value and signifies the absence of concern, even with very obvious philosophical implications: it is a typical attribute of the deities, of their *quies* and *otium*, synonyms of a blessed life; it is also the liberating condition of death.

But it is in the same time frame that *securus* starts to have a broader range of functions, which implies the preservation of public order and the protection of people and property. There is no doubt that the literature of that time expresses with continuity the aspiration for peace and personal security: one need only think of the first Eclogue by Vergil, which is exactly a meditation on the conditions of security that can only be guaranteed by the will of a *Princeps*; and with time the identification of *Princeps-Securitas*, as the one who is able to remove all concerns, is clarified by using the terms *rector quasi tutor*, which were already ciceronian.

Horatius calls Augustus "thou mighty guardian of Italy and imperial Rome",[85] an expression that cannot but be placed alongside another equally meaningful one: "Neither civil strife nor death will I fear, while Caesar holds the earth"[86]; so in the anonymous *Consolatio ad Liviam*, Augustus is called guarantor, protector of mankind (*tutela hominum*).[87] There is a persistent recurrence in Ovid's verses that praise the tireless work of the *Princeps* as a universal *tutor*.[88]

In Ovid, and again in an eulogy for Augustus, the connection between *tutela* and *securitas* finally becomes explicit: "by our native land which is safe and secure under the fatherly care, of which I was one among the people was recently part".[89]

In the Augustan propaganda and in that of his successors, the theme of *securitas* would have, as we shall see, a much greater success than the one of *tutela*: it actually seems possible to affirm that, perhaps already in the Tiberian age, *securitas* supported and partly integrated *tutela*,[90] in the same way that it is placed alongside the theme of *salus*. In both cases, the major success of *securitas* as a communicative topic is also linked to the original meaning that it continues to evoke through its signifier: the assurance of not being afflicted by any kind of concern.

In a rich and complex panorama as the one offered during the Augustan age, literature and philosophy, however, are not isolated communicative spheres in which the discourses on security are expressed. In addition to literary attestations, the field that has offered scholars the most extensive documentation is certainly that in which *Securitas* (type and legend) can be seen on coinage, which definitely expands from the time of Nero.

It must be said that none of the coins of the Augustan age have *Securitas* as an inscription, other language spies tell us that, in the lexicon of communication during the years between the end of the civil wars and the first decades of the first century, there was utmost attention on the theme. On a series of coins of 16 BC (see Figures 3.1a and 3.1b), the legend on the reverse, framed in a laurel wreath, indicated the votes that the Senate and the people of Rome expressed for Augustus' salvation: "because through him the commonwealth is in a greater and more tranquil state": *quod per eu(m) r(es) p(ublica) in amp(liore) atq(ue) tran(quilliore) s(tatu) e(st)*.[91] The obverse features an altar, and the initials SC (*senatus consulto*) at its side, on the front can be read the words of the dedication: *Imp(eratori) Caes(ari) Augu(sto) comm(uni) cons(ilio)*.

This numismatic emission and its legenda were compared with the one from 32 BC, with the difference that on the latter the Princeps's action for the extension and peace of the *Res publica* is evoked in detail. The origin of the adoption of the legenda on the coin from 16 BC can rather be traced back to Augustus' coinage *civibus servatis*[92] and the motto recalls the censors' prayer to the immortal gods indicated in a passage by Valerius Maximus,[93] with some significant adjustments: compared to the prayer, on these coins the comparative *amplior* becomes the focal point, while *tranquillior* specifies the ambiguous *melior*.[94]

The close connection between the two concepts of *Pax* and *Securitas* during the Augustan age appears in the iconographic and epigraphic programme achieved in the ancient city of Lazio *Praeneste* (now Palestrina): the decurions and the people of the city dedicate a richly decorated altar to the *Pax Augusti*,[95] and another one to the *Securitas Augusta*.[96] The *Securitas*, usually a sort of ancillary motif, compared to the great theme that includes all of the *Pax Augusta*,[97]

Figures 3.1a and 3.1b A Denarius of Mescinius Rufus, 16 BC. On the reverse, in
a wreath, first letters of a prayer (*vota*) of the Roman senate
and people for Augustus (RIC I² 68 no. 358)

here is given equal dignity. Both altars are definitely dated to the Augustan age, and have been interpreted as one of the first rare examples of the spread of the imperial cult in Latium. In the mid 1970s, however, Fausto Zevi, through both a brilliant and sophisticated interpretation of the reliefs that decorate the two altars, observed that, in the prenestine inscriptions,

> the two concepts [Securitas and Pax] complement each other, one by clar-
> ifying the value of the other, almost in an interpretation of the Augustan
> policy from an Italian point of view . . . finding a parallel most in contem-
> porary literature than in urban monuments.[98]

I believe we can rightly assert that the altars of *Praeneste* reflect the final out-come of a process that saw first the coexistence and then the integration of values conveyed by *Securitas* and *Salus*. Still at the beginning of the Principate of Augustus, *Salus* is worthy of the same consideration as *Concordia* and *Pax*. Consequently, when the Senate and the Roman people contributed to the financing of statues for the emperor, Augustus commissioned two of them: one for the *Salus Romana*, and the other for the *Concordia mitis*.[99] The traditional cult of *Salus*, considered the "absolute assurance of the stability of the state, as the essence of its integrity",[100] is contaminated and enriched, over time, with the soteriological components of Hellenistic influence, which tended to identify collective salvation with individual salvation. It will assume a more and more private character and will therefore coexist with *Securitas*, which will gradually assume those public features that *Salus* was losing.[101]

The terms *salus* and *pax*, but also *securitas*, are present in the discourses that coincided with the birth of the Principate. Furthermore, the measures taken by Octavian and his collaborators, Maecenas *in primis*, created a connection between these keywords and the *Princeps*, who had guaranteed their return. Octavian then Augustus represented the point of convergence of a series of discourses on security whose premise dated back to a few decades before the emperor's rise. The lament over the *confusio Temporum* of the ruling senatorial class of the age of Cicero, as Karl Galinsky has rightly indicated, is nothing more than preparation for the imminent constitutional metamorphosis of Augustus, who found among the regrets and the dissatisfaction, the ideal conditions for his rise to power.[102]

The figure of Augustus extended and strengthened his authority through action and communication that allowed him to stand as the *restitutor* of all those lost truths of which the previous generation lamented. The term *Securitas* coexisted with *Salus* and is accompanied by other concepts such as *Tutela*, as evidenced by Ovid's verses and the altars of *Praeneste*, and over time they will come to represent a synthesis: 'the absence of preoccupation' implied by its etymology, and would assume the role of a connector between the vari-ous needs that the citizens of Rome had expressed.[103] Ultimately, *securitas* also meant delegating a whole range of rights and freedoms[104]: this was the price to be paid for security and this contrast would emerge shortly.

From Tiberius to Nero: the affirmation of a discursive strategy

The empire of Augustus, despite the wealth and the extent of its communicative language, represented a phase of beginnings for discourse on security. It is not possible, at this moment, to talk of a true 'propaganda' on the issue of *securitas*; and it is no coincidence that the legend of *Securitas* does not appear on any of the Augustan coins. In fact, during the time span analysed so far, there is a clear predominance of the use of the adjective *securus* over the noun *securitas*. The latter was limited to Cicero's invention and clearly its usage had not yet expanded.

It was at the end of Augustus' reign, during a particularly delicate and feared period of change,[105] that the noun reemerged, enriched in its function by many new nuances. At the same time, it is perfectly legitimate to agree with the opinion according to which, as will be seen, the appeal to *securitas* occurred frequently during periods of instability, and it is clear that the Augustan age did not correspond to such.

In relation to the adoption of Tiberius, Velleius longs for a *perpetua securitas*, that is, a continuity of such peace that had characterized the empire of Augustus: "and the hopes which they entertained for the perpetual security and the eternal existence of the Roman Empire".[106] In the cited passage and in the paragraph that describes the *cura Urbis* of Maecenas, the historian outlines, with deliberate and rhetorical tones, Augustus' work of restoring *securitas* to the empire: "Agriculture returned to the fields, respect to religion, to mankind freedom from anxiety, and to each citizen his property rights were now assured."[107] John T. Hamilton identifies in the two passages the first definitive evidence of the coexistence of *asphaleia-securitas*, in the sense of both the person's physical safety, and the security of the space where one lives.[108]

We realize, in fact, that with Velleius Paterculus the overlap is now fully accomplished: he gives particular attention to security, especially urban security. Therefore, it is not a coincidence that the author himself gives a detailed account of Maecenas' actions as *urbis custodiis praepositus*[109]; and that, with the same attention, when talking about the urban prefect, Lucius Calpurnius Piso, – who during the absence of Tiberius dealt with the preservation of public order in the city – Paterculus used the expression *securitatis urbanae custos*.[110]

The passage on the succession of Tiberius should be read alongside the one cited earlier on the restoration of Augustus,[111] where the adjective *perpetua* implies nothing more than an extension and a confirmation of the Augustan *status quo*; and security is guaranteed by the order of the empire, religious practice, and security of private property. In accordance with Instinski, in Velleius's words concerning Maecenas, despite the emphasis, we can perceive how "*Security, as an objective condition and a subjective feeling, belongs to the values of the Augustean order.*"[112]

In a phase that lies between the establishment of the Augustan regime and the rise of Tiberius, the term *Tutela* still coexists with *Securitas*: in imperial propaganda the latter coincides in fact with *tutela*, but it is enriched, in its psychological result, by the original meaning of its signifier, namely by the philosophical foundation of the *tranquillitas animi*. The prevalence of the term *securitas* over *tutela* – which, as we will see, however, survives in the imperial propaganda and would be particularly popular at the end of the sixties[113] – is due strictly to its effective evocative capacity as well as its greater acceptability from a psychological point when compared to *tutela*, whose strong juridicial connotations could be unpalatable to the Roman senatorial class.

During the Julio-Claudian dynasty, *securitas* and *securus* (the latter still occurring more than the noun) became very common in use with three basic meanings: the first two, already identified, recalled either safety or absence of concerns; the third, which was new, was developed from the meaning of 'care-lessness' and resulted in indicating 'superficiality, overconfidence'.[114]

One particular testimony suggests that the diffusion of the term was due to its use for propaganda: in one of the letters from Saint Paul to the Thessalonians, in which derides the expectations of peace and security of the population facing the upheaval of the next coming of Christ,[115] from the words εἰρήνη καί ἀσφάλεια can be inferred a sarcastic reference to the slogans of Roman imperialism.[116] Without rejecting such a suggestion in principle, is a passage of Calpurnius Siculus, which celebrates the *Aurea aetas* restored by the young Nero, and shows the third meaning of the word which introduces us to a new phase of life for the concept: *Aurea secura cum pace renascitur aetas*.[117] The formula *pax secura*, synthesis of two concepts – which, in the Augustan age had been a sign of change, one term next to the other (even if the first term had a decisively more prominent role) – would acquire a particularly intense value, especially if we consider that, in the age of Nero, *securitas* would have its most complete definition and final consecration as a keyword of imperial propaganda.

Securitas would become, in fact, one of the pivots of Seneca's philosophical reflection and through him it would become a central word of an entire epoch.[118] Compared to Cicero's speech, there is a certain continuity of thought. First, even for Seneca the real definition of a blessed life includes security as its first element.[119] Despite the fact that Nero's tutor uses the term mainly in reference to a happy life – inextricably grounded in the conditions of spiritual peace of Stoicism and also Epicureanism – the practical aspect of physical security is also clear. In particular, in works addressed to Nero, Seneca presents *securitas* as a condition of safety based on mutual trust between the sovereign and the people.[120]

The *De Clementia* is, in this sense, a work of extraordinary interest, since it continues the discussion on the 'good king', and was considered by Nero a manual of good conduct.[121] Seneca's main argument to convince the emperor of clemency is centered entirely on the theme of security, regarded as safety. The safety of the *Princeps* depends on his ability to ensure the same condition for his people; when this is assured, everyone is ready to sacrifice for their ruler.[122] But the sovereign must be willing to renounce some of his unlimited power, he must moderate himself, he must agree to be unpopular and not envy others, he must be willing to lose some of his liberty in favour of the freedom of his people: this is the price to be paid for *securitas*. If a man can easily walk

around the city without an escort, it is only thanks to the *Princeps*, who, however, could never give up his own:

> How many things there are which you may not do, which we, thanks to you,
> may do! It is possible for me to walk alone without fear in any part of the city
> I please, though no companion attends me, though I have no sword at my
> house, none at my side; you, amid the peace you create, must live armed.[123]

However, the ruler may certainly avert greater risks such as attempts on his life
if he can demonstrate clemency: the cruel ruler instills fear and fuels people's
hatred, thus jeopardizing his own existence.[124] Seneca therefore recommends
that Nero always keeps to the political lines of his government, which guarantees the *securitas* of all, because the ruler's power is mitigated by his respect
for the rights and freedoms of his people. As a result, constitutional guarantees
represent the only real foundation for everyone's security.[125]

> "A security deep and abounding, and justice enthroned above all injustice;
> before their eyes hovers the fairest vision of a state which lacks no element
> of a complete liberty except the licence of self-distruction."[126]

The relation between *securitas Augusti* and *securitas Publica* thus represents the
final outcome of a slow process: if, with Augustus, the first was the absolute
guarantor of the second,[127] in the words of Seneca it is possible to perceive how
it now represents the main threat. *Securitas* meant total delegation that finally
freed citizens from anxieties regarding the protection of their lives and their
property; now it is determined by the conduct of the *Princeps*, and is linked to
a delicate exchange of ideological and material guarantees with his subjects.[128]

   Seneca is clear when expressing the terms of what is a necessary agreement,
designed to ensure the security of both parties:

> "For if anyone thinks that a king can abide in safety where nothing is safe from
> the king, he is wrong; for the price of security is an interchange of security."[129]

In this passage, highly emphasized by figures of speech, it is above all the use of
pacisco, a verb that typically indicates the conclusion of a contract, that is note-
worthy. Equally explicit, not without a veiled sense of threat, is the meaning of
the entire manual, finally clarified in the concluding summary, that clemency not
only makes the emperors *honestiores*, but above all *tutiores*, which finally coincides
with their own *salus*. Clemency, therefore, not only makes one more honour-
able, but also safer, and is both an ornament and real health for the empires.[130]

   The prophetic suggestions of the wise counsellor were destined to clash
with a different reality, which soon unveiled serious disagreements between
the policy of the *Princeps* and the expectations of the Roman ruling class; disagreements that culminated in the episode of Piso's conspiracy, in AD 65. It
is precisely this conspiracy that has been traditionally placed in connection
with the mintage of the first coins bearing the image of *Securitas* and the legend Securitas Augusti (see Figures 3.2a and 3.2b),[131] to be interpreted in all

Figures 3.2a A Dupondius of Nero, AD 63, Rome mint. On the reverse, the Securitas
and 3.2b Augusti is stripped to the waist, draped on the sides, sitting on a throne
facing to the right. The left hand holds a short sceptre; the right hand
is brought to the head, the elbow leaning back off the seat; in front, a
flaming altar, against which is a lighted torch (RIC² p. 160 nos 112–114)

probability both in an active and passive sense: the emperor's quality, which ensures safety; but also the virtue that protects the emperor, who is safe.

On the coins (*dupondia*) of the years AD 64–66, referring to the mint of Lugdunum and Rome, *Securitas* is represented as a matron with naked breasts, sitting comfortably on the throne, facing the right, with one hand supporting her head and the elbow resting on the back of the throne; she carries a scepter in her left hand; in front of her there is an altar upon which a sacrifice has just been offered; on the ground there is a *bucranium*. The quiet satisfaction of the goddess *Securitas* is carefully designed to communicate profound serenity.[132]

It has recently been underlined how, in addition to the conspiracy of Piso, there are not few circumstances (between AD 63 and 66) that could have justified the emission of coinage carrying the message *Securitas* in order to reassure the public about the health of the state, as well as that of the emperor.[133] It is certain, however, that 'the invention' of the Securitas (*principis et populi Romani*) would be greatly impoverished if it were not able to connect it to Seneca's *De Clementia*, and to those themes so succinctly brought to the fore by the philosopher.

Regardless of the individual events, we can imagine that *Securitas*, as of the sixties, represented for Nero an attempt to revive, at least in propaganda, that social pact that Seneca had presented to him in such a clear and defined way, well aware of the fact that *securitas* constituted "the fundamental political problem for the Romans under the Principate".[134] At the beginning of his period in office as tutor, Seneca was well aware that Nero was expected to eliminate the negative experience that the Roman senatorial class had had with his predecessors, and invited the young emperor to follow in the footsteps of someone within the family, namely Augustus.[135]

As much in *De Clementia* as in the tragedies, the tutor provided the young *Princeps* with numerous examples of despicable tyrants destined to a tragic end.[136] It is from this point on that the 'topos of the tyrant' starts to emerge, intended as a rhetorical expression of that political fear that permeated the existence of the ruling class in Rome.[137] The political addresses that Nero made after the so-called *quinquiennium* ruined relations with the Senate and had the result of confirming and keeping alive the *topos* of the mad and cruel tyrant to the point that the *metus Temporum* evoked by Tacitus mainly coincided with the Principates of Nero first, and Domitianus after.[138]

Securitas et Tutela: the adaptations of a discursive strategy

With the death of Nero, the short period of time between AD 68–69 – albeit dense with civil wars – saw security as a focal point of political communication and *Securitas* in abundance on coins types.[139] For our purposes, this short time span is a privileged vantage point, for the fortunate coincidence of different types of sources.

Galba, in relation to the use of themes that exalt the *libertas restituta* or *publica*, the *Roma renascens* and the *Salus generis umani* – which had to immediately signal the end of Nero's tyranny – was quick to renew the legend of *Securitas*, now *Populi Romani*.[140] Galba's wish to specify the public nature of security meant that not only did he calm his people of their fears of a new civil war, but at the same time was able to communicate his desire, unlike Nero, to be a 'constitutional *Princeps*'.

Galba's great attention for the political issue of security is also reflected in the Acts of the Arvals of 69, with the sacrifice of a calf in honour of the *dea Securitas* on the occasion of the adoption of Piso.[141] Just like Velleius Paterculus had evoked *securitas* during the delicate transition between Augustus and Tiberius, now Galba wanted to invoke *Securitas* as a personification with fully divine attributes, and inserted into the mechanism of the official religion of Rome. Even in Tacitus' account the theme is not absent when he refers to Galba's speech to Piso.[142] The *securae res* are as difficult to obtain as they are to maintain: for this reason the *securitas* is identified fully with the theme of the continuity of governance in a way which is viewed favourably by the ruling class of Rome.

The coinciding of these sources shows that, in this delicate phase, security was still regarded as a key word of political propaganda and essentially became pivotal to a multi-functional communication strategy. Far from fearing the parallel with Nero's language of communication, Galba would exploit the full semantic power of *Securitas* to revive the Senate's expectations after the fall of the 'tyrant'. But above all, he tried to exploit it, in order to guarantee the continuity of a filo-senatorial Principate, through the adoption of Piso. *Securitas* is confirmed to be a polycephalous word and rich in connotations: it denotes, as always, the removal of concern (and the iconography of its personification promptly recalls it); it ensures the preservation of public order and legality; it confirms the stability of the state and evokes the good health of the *Princeps;* it guarantees respect for the *ius*, the senate, the property and the safety of the Roman people; and finally it declares continuity of good governance.

It is in this difficult climate, following the failure of Galba's plan, that we find on coin types the legend and the image of *Securitas populi romani* by Otho[143] and Vitellius[144]; the latter also offers another type with a particular legend *Securitas Imperatoris Germanici*.[145] The protagonists change, but the communicative system remains the same, enriched by an old word and some new images. Vitellius is the first to introduce a new personification: on two coins (*dupondia*) the *Tutela Augusti* is represented as a woman seated on a throne, with the particular feature of two children by her side, upon which she poses thoughtfully her hands[146]: "the children might represent citizens and Tutela would then be the emperor's ward over his subjects".[147] In this sense, *Tutela* had the same use as *Securitas*, but with a sharper meaning of 'protection', expressed in particular by the maternal figure associated with the children. Both in the case of Vitellius and in that of Vespasianus (who significantly retrieves the coin type),[148] the message is unequivocal: the security of the empire is founded on the secure position of its emperor's power, therefore also on the continuity of rule.

In this troubled phase that saw the succession of civil wars, *Tutela*, like *Securitas*, both having shared a significant part in political communication, received its own personification and gained, as it were, a propagandist space. A space destined to be maintained over time, although not in coin images: a new form of protection, strongly connected with the idea of economic stability, would return as a political issue at the end of the century, in the politics of Nerva first and that of Trajan later. To the legal value of the term (with reference to the idea of control over children and incapacitated persons) is due in large part the surrendering of the *Securitas* motive to its philosophical, religious and social implications now typical.

Among the members of the Flavian dynasty, as is well known, Titus is the man identified as a model of behaviour for the *Optimus Princeps*, bringing the old century to an end and opening a new era. In the *Panegyricus* by Pliny, Titus is Trajan's direct antecedent, because he had provided *ingenti animo* to *securitati nostrae ultionique* ("to our security and revenge").[149] In the sixth chapter of the Suetonius' *Life of Titus*, what surfaces, however, is a common Tacitanian theme: the future emperor is described as being entirely absorbed, almost like an obsession, by the issue of safety, which makes him unpopular[150] and evokes the ghost of Tiberius.[151] However, also as a result of the short duration of his Principate, Titus left a good memory impressed in the senators' minds, preserved not only in the Panegyric, but also in the image of a *Princeps* as *amor et deliciae humani generis*,[152] who did not put his hands on other people's property, sentence anyone to death, and, in essence, made everyone feel secure.

The coin type depicting *Securitas* does not undergo particular changes under the Flavian dynasty.[153] It is proposed again, with slight variations, in AD 71 on coins (*dupondia*) of Vespasian, perhaps issued to celebrate the triumph of Vespasian and his son and the closing of the temple of Janus.[154] In AD 78, it appears with the name of Titus Caesar on the aforementioned coins,[155] and it is presented by the *Princeps* at the time of his accession to the throne, with the legend of *Securit(as) August(-)*, to be interpreted as *August(i)*, thus referring to the *Princeps*; or as *August(a)* referring to the *virtus*.[156] The usual image is also accompanied now by *modius*, as a new attribute that alludes to securing the supply of wheat.[157] Domitian instead does not leave us with any coins bearing the image of *Securitas*, preferring instead, in his long and autocratic reign, other legends such as *Salus*. If we did run the risk of resorting to a facile interpretation, we could be tempted to consider this eloquent absence as a voluntary act of breaking dialogue with the senatorial order.

The first burial inscriptions where the *Securitas* of oneself and one's family is evoked as peaceful rest[158] can be traced back to the last decades of the first century, in continuity with the Epicurean philosophical legacy.[159]

Of particular interest is the proliferation, in the same years, of the term *securitas* in rhetorical exercises (in both the *suasoriae* and *declamationes* by Seneca the Elder and in the *corpus* of oratories attributed to Quintilian).[160] Almost in all occurrences, *Securitas* has a purely political value: in some cases, it is the other face of *metus*, fear mostly generated by the *Princeps* himself. In my opinion, it is an interesting literary indicator of how at this point the word had become part of the political lexicon and how the concept it conveyed remained essential; as well as how, in the Flavian era, *Securitas* had become a word symbolizing the fragile balance between the power of the *Princeps* and the freedom of his subjects.

The second century: from *metus Temporum* to *securitas Temporum*

After the slow development that, over the course of a century, defined and therefore enhanced its political value, 'security' is now ready to become a communicative instrument of other finely balanced relations, first, that between the *Princeps* and the Senate.

Tacitus' reflection on the nature of the Principate takes place, between *Agricola* (see later in this chapter) and the *Annales*, during a period in which there is an increase of awareness: if even, in the preamble of the *Historiae*, the historian recalls that, after the battle of Actium, it was necessary to establish a Principate because "in the interest of peace it was worthwhile putting power back into the hands of a single man"[161]; at the beginning of the *Annales*, after twenty years, the memory of the Augustan Empire had become a model of *securitas*.[162] With the passing of the years what had been proven was that the finely balanced relationship between the *Princeps* and the Senate could endure.

In the Panegyric Pliny praises the *securitas Temporum* as the emblem of the new era,[163] as a result of the emperor's ability to combine the different virtues and values, and promote a policy of assurance of privileges and rights.

The first assurance is stability and continuity of governance; the means to guarantee this is through adoption. In this way, Nerva follows Galba's policy: just as Galba adopted Piso after Nero's death, so did Nerva adopt Trajan after Domitian' assassination.[164] The public and shared manner in which the adoption is proclaimed, through the words of Pliny, reinforces its value and increases its importance:

> And thus the adoption took place not in his bedroom and by his marriage-bed but in the temple before the couch of Jupiter Best and Highest, the adoption which was to be the basis of no servitude for us, but of security, happiness, and freedom.[165]

In the incipit of Tacitus's *Agricola* we find Nerva's exaltation for creating harmony between opposites – the Principate and freedom; and Traianus' for producing the *Felicitas Temporum*:

> Now at last hearth is coming back for us; from the first, from the very outset of the happy age, Nerva has united things long incompatible, the Principate and liberty; Trajan is increasing daily the happiness of the times; and public confidence has not merely learned to hope and pray, but has received assurance of the fulfilment of its prayers and so has gained strength. Though it is true that from the nature of human frailty cure operates more slowly than disease.[166]

Common security, thanks to the successive action of two *Princeps*, is no longer just a hope or a prayer (*nec spem modo ac votum*), but it becomes absolute certainty, because the emperor himself acts as its guarantor. The combination of two aspects that are considered irreconcilable – the *res olim dissociabiles* – is the result of the *Princeps'* ability to advance the common safety before his own,[167] namely the implementation of the project of compatibility between *Securitas Augusti* and *Securitas Publica* that Seneca had theorized for Nero.

The same ability of mixing and rearranging differing elements is transmitted from Nerva to Trajan:

> For you have two extremes combined and blended in your person, a beginner's modesty and the assurance of one long accustomed to command[168]. . . We have celebrated with appropriate rejoicing, Sir, the Day of your accession whereby you preserved the empire; and have offered prayers to the gods to keep you in health and prosperity on behalf of the human race, whose security and happiness depends on your safety.[169]

Just like the *Libertas* can coexist with the Principate, even *Securitas*, in the practice of government, does not exclude *Pudor* and has *Tutela* as an instrument. The latter acts as a guarantee for wills,[170] for the end of lese-majesty trials, for the removal of traitors,[171] and leads to economic measures. In the Panegyric by Pliny, *securitas* and *libertas*, once again united, are celebrated as the conditions that allow initiatives to aid the most vulnerable, such as the *alimenta* and the *congiaria*. Therefore, political stability becomes closely joint to economic stability in the certainty that the emperor guarantees the well-being of citizens.[172]

It is no coincidence that Nerva, rather than returning to the *Securitas* coin types, uses the *Tutela*[173] type, with a legend that refers to the policy of the *alimenta*; while *Securitas* is used again on the coins issued during the empire of Trajan. The iconographic innovation of the cornucopia associated with the Personification reaffirms the message of prosperity: *Securitas*, who is standing, presents this feature that explicitly emphasizes the close link between peace and economic prosperity, already acquired and guaranteed. On gold coins of Hadrian (AD 134–138), the matronly figure (Securitas Aug(usti/-a), is sitting on a throne; the altar is missing.[174] In other cases, *Securitas* continues to be represented seated, but without any explanatory legend,[175] which is evidently no longer required for its identification. In general, during the second century, *Securitas* coin types are a remarkable triumph, which are gradually enriched with features[176] and adjectives[177] (see Figure 3.3).

At the dawn of the new century, Pliny became the prophet of the 'new security', and its paradigm was Trajan, the *optimus Princeps*. From this point on, there is a decisive amplification in the use of this word, both in purely numerical sense of the multiplication of attestations, both in its use in different contexts and areas. This proliferation is a sign, in my view, of a progressive reduction of the political meaning that the term had acquired at the beginning of the imperial era, and reaffirmed in the aftermath of the civil war. After the

Figure 3.3 The reverse of an Aureus of Hadrianus, AD 145–161. A matronly
figure of Securitas is sitting on a throne, with two cornucopiae
(BMCREmpire IV p. 276 nos 1713–1714 = RIC III p. 126 no 782)

triumph of the imperial virtues during the epoch of Trajan and of Hadrian,[178]
during the second half of the second century, the term *Securitas* goes from being
a key word of imperial propaganda to losing its communicative meaningfulness:
on the one hand, in order to return to its original meaning of 'no concerns';
and on the other, in the aftermath of the civil conflict following the removal
of Commodus, so as to become the slogan of a new dynasty in search of legitimacy.

In the words of Epictetus, the theme of peace resurfaces alongside that of
security: it is a tangible peace known as the absence of wars and the freedom to
walk along the streets and sail the seas; it is a relative peace, at a time expecting
turbulence. This is supported by the fact that Epictetus, in the continuation of
his argument, states that only the philosopher can give absolute peace:

> For you see that Caesar appears to furnish us with great peace, that there are no longer enemies nor battles nor great associations of robbers nor of pirates, but we can travel at every hour and sail from east to west. But can Caesar give us security from fever also, can he from shipwreck, from fire, from earthquake or from lightning? Well, I will say, can he give us security against love? He cannot. From sorrow? He cannot. From envy? He cannot. In a word then he cannot protect us from any of these things. But the doctrine of philosophers promises to give us security even against these things.[179]

Therefore, only philosophy can save man from the preoccupations of the spirit and from the unthinkable, because it is free from the fear of evil; while the only really safe haven, the *Perpetua securitas*, is death.[180] It is a return to that which has never totally vanished and has certainly not been isolated: *securitas* with its original meaning of *quies* and to its purely philosophical and private dimension.

Thus, in second century inscriptions, *Securitas* (in some cases as *Aeterna* or *Perpetua*) triumphs, in private funerary dedications, accompanied by the *Dei Manes* and by the *Somnus Aeternalis*.[181] Other times they include all the *dei Securitatis* (namely *Securitas, Quies, Tranquillitas* and *Somnus Aeternalis*).[182] Sacred dedications[183] are rare; while in the imperial inscriptions of the second century, *securitas* appears again, but rarely, as a guarantor of financial protection on behalf of the emperor. An interesting dedication to Hadrian by the Senate and people of Rome, dated between December AD 118 and January AD 119 and placed in the Traian's Forum, praised the emperor who had made his citizens safe (*hac liberalitate securos*), not only his contemporaries, but those who would come after him, thanks to his generous donation to cover a *debitum fisci*[184]; and, at the end of the century, Commodus "provides for the security of his provincials" (*securitati provincialium suorum consulens*), in the celebration of a construction initiative.[185]

At the beginning of third century, in the language of Severan propaganda, *Securitas* is at times connected to the emperor's *Indulgentia*, and other times to Pliny's theme of continuity of governance as a guarantee for the well-being of the Roman state.[186] Within the iconographic and epigraphic plan designed by *Marcus Caecilius Natalis*[187] at *Cirta*, a statue[188] is dedicated to the *Securitas Saeculi* as evidenced by one of the inscriptions placed on an arch in honour of the emperor's *Indulgentia*.[189] And *Securitas* and *Indulgentia*, at times linked to the *Virtus*[190] and other times to *Gloria*,[191] are keywords of the political lexicon for the celebration of the *decennalia* by Septimius Severus in AD 204.

The motif of *Securitas*, of which we have followed the events through its multiple occurrences (literary, epigraphic and numismatic), once the transition was made, from the late Republican era to the beginning of the Principate, from the strictly philosophical dimension to the political one, was destined to become a key word of imperial propaganda (in sculpture, coinage, epigraphic language); and heritage of the funerary culture with a prominence that would continue until Late Antiquity.

Securitas is the only unprecedented *virtus* in the Republican era, and in the outline that I have drawn so far, on the contrary to what has been suggested by Arends,[192] it cannot be identified as the rallying cry of Augustus' policy; nor did it appear (as did *Salus*, *Concordia* and, most of all, *Pax*), at this phase, in iconographic plans or in the promotion of the construction activity. It did not take on monumental forms, nor was there a surge in its occurrences,[193] even if it was during the Augustan period that the foundations were laid for it to become a fully entitled political *virtus*. Such foundations, as is well known, consist of a series of legislative and urban measures, designed to celebrate the end of the civil wars, the advent of political stability and the restoration of law.[194] In all likelihood, the political experience, as well as the wealth of literary and philosophical culture, of the *custos Urbis* Maecenas is not unfamiliar to all of this.

The slow development of the political concept of *securitas* would terminate at around the middle of the first century AD: from Cicero's first reflections one must wait for Seneca's works for the conception and the consecration of *Securitas* as a *virtus* of the emperor. It is at this stage, I believe, that it became the key word of the Principate and it is reflected on coin types with its own legends. From this moment on, despite and perhaps because of the failure of Seneca's project, the theme of *securitas* would be used at times when there is most fear for security (during the civil wars of AD 68–69 and before the Severi), and in phases of affirmation of a new dynasty. After the death of Domitian, who had preferred 'salvation' to 'security', Nerva and his son Trajan had the task of recovering the concept, extending its range of meanings, accompanying with *Tutela*, *Pudor* and mostly *Libertas*, transforming it into *Securitas Temporum*. Their message aimed to guarantee economic prosperity and recover a paternalistic figure of the *Princeps* as someone who acts on behalf and with the support of the Senate.

The two moments that have been selected for the opening and the closing of this volume not by chance are those in which security is a crucial issue in the (re)organization plan of military forces, also in the city. With Augustus, security, which is not propaganda yet, was a multifaceted and concrete plan of action. As I have previously mentioned, the more a period becomes uncertain, and the more an old motto becomes the most appropriate instrument upon which to ground the empire's stability. With Severus, *Securitas Saeculi* – which cannot but be related to the intense period of military reforms – masked the rupture of a balance of relations among the emperor, the Senate and the people that had been painstakingly, intermittently, built, over two centuries.

Notes

1 Hartmann 1921, with an update by Binder 2001.
2 Arends 2008, p. 259 n. 25.
3 Hamilton 2013, p. 54.
4 Hamilton 2013, p. 58: this is the case of a referenced Neronian altar of the *Securitas*, which could not be accounted for through research. It is probably a misunderstanding of the closing part of Pera's article 2012, to which we will make further extensive reference.

5 Kneppe 1994.
6 Foucault 2015, p. 211. This is the lecture of 17 March 1971, taken from the collection of lectures at the Collège de France from 1970 to 1971. The definition of 'discursive event' is a linguistic and conceptual novelty with respect to the content of the other fundamental text on the same topic, which is the *Order of Discourse*, inaugural lecture at the Collège de France given in 1970 (Foucault 2004).
7 Rivière (2008a p. 204) in which the distance that separates the modern economy of punishment ('prison') from the ancient system of penalty (*carcer et vincula*) is measured.
8 Recalled by Lintott and Nippel (see Part I, Chapter 1, this volume).
9 Brélaz 2007, p. 229. On the consequences of this significant shift in the Caesarian and Augustan legislation on security, see Part II Introduction, this volume.
10 For 59 BC, see the *Pro Flacco* 46.14–16: *Securus Hermippus Temnum proficiscitur, cum iste se pecuniam quam huius fide sumpserat a discipulis suis diceret Fufiis persoluturum*; for 56 BC, see the famous letter to *Lucceius*, in which Cicero asks to celebrate the events of their own consolate (*Epist. ad famil.* 5.12.4.1: *Habet enim praeteriti doloris secura recordatio delectationem*); then, again, in 40 BC, in a passage of a letter to Atticus (*Epist. ad Att.* 12.52.3.1) Cicero writes *de lingua latina securi es animi*.
11 *Epist. ad famil.* 10.24.6.11: *Quod si aut Caesar se respexerit aut Africanae legiones celeriter venerint, securos vos ab hac parte reddemus*.
12 *De Rerum Natura*, 3.211 and 939 (*secura quies* of death); 5.82 *didicere deos securum agere aevom*.
13 Arends 2008, p. 269.
14 *Epist. ad Quinct.* 2.9.
15 The whole text certainly makes perspicuous an atmosphere similar to the theme of the Epicurean ataraxy, however denied by Arends 2008, p. 268. *Epist. ad Att.* 4.18.2.5–18 (author emphasis): *Recordor enim quam bella paulisper nobis gubernantibus civitas fuerit, quae mihi gratia relata sit. Nullus dolor me angit unum omnia posse; dirumpuntur ii qui me aliquid posse doluerunt. Multa mihi dant solacia, nec tamen ego de meo statu demigro, quaeque vita maxime est ad naturam ad eam me refero, ad litteras et studia nostra. Dicendi laborem delectatione oratoria consolor. Domus me et rura nostra delectant. Non recordor unde ceciderim sed unde surrexerim. Fratrem mecum et te si habebo, per me isti pedibus trahantur; vobis ἐμφιλοσοφῆσαι possum. Locus ille animi nostri stomachus ubi habitabat olim concalluit. Privata modo et domestica nos delectent, miram securitatem videbis; cuius plurimae mehercule partes sunt in tuo reditu. Nemo enim in terris est mihi tam consentientibus sensibus.*
16 *Tusc.* 5.42.1–6: *Qui autem illam maxime optatam et expeditam securitatem – securitatem autem nunc appello vacuitatem aegritudinis in qua vita beata posita est – habere quisquam potest, cui aut adsit aut adesse possit multitudo malorum.*
17 More or less similar is the definition found in the coeval work *De finibus* (3.23, tra 45 and 44): *Democriti autem securitas, quae est animi tamquam tranquillitas, quam appellant εὐθυμίαν, eo separanda fuit ab hac disputatione, quia [ista animi tranquillitas] ea ipsa est beata vita; quaerimus autem, non quae sit, sed unde sit.* Still in *De natura deorum* we find a very clear definition: *nos autem beatam vitam in animi securitate et in omnium vacatione munerum ponimus* (*De Nat. Deor.* 1.53). In *De Amicitia* the *securitas* is used instead as a clear reference to the stoic *apatheia*.
18 Schrimm-Heins 1991, p. 134.
19 For these references, see Hamilton 2013, pp. 51–58 and Arends 2008, pp. 263–269. The reflections of both depend on the work of Schrimm-Heins 1991, pp. 123–134.
20 Cf. Cic. *Leg.* 2.11.28: *rerumque expetendarum nomina, Salutis, Honoris, Opis, Victoriae, quoniamque exspectatione rerum bonarum erigitur animus.* Grounded on such passage,

Harold Mattingly (1937, pp. 103) has identified the fundamental distinction, then clarified in a philosophical sense more extensively by Wallace-Hadrill (1981, p. 309).

21 Wallace-Hadrill suggests that the political and philosophical treatises on the ideal statesman should be regarded as an important guide to comprehend what had to be considered virtue and what not (1981, p. 309).

22 On the transformation of *Salus* and then of *Securitas* from *res expetendae* to *virtutes*, see Mattingly (BMC II.74 and I.21 on *Salus*), who considers the additional indication, on the coin legends, of the adjectives *Augusta* or *Augusti* as discriminating.

23 *De officiis* 1.69: *Vacandum autem omni est animi perturbatione, cum cupiditate et metu, tum etiam aegritudine et voluptate nimia et iracundia, ut tranquillitas animi et securitas adsit, quae affert cum constantiam tum etiam dignitatem* (on the passage, see Hamilton 2013, p. 52; Arends 2008, p. 269).

24 *De officiis* 1.72: *Capessentibus autem rem publicam nihil minus quam philosophis, haud scio an magis etiam, et magnificentia et despicientia adhibenda sit rerum humanarum, quam saepe dico, et tranquillitas animi atque securitas, si quidem nec anxii futuri sunt et cum gravitate constantiaque victuri.* See Arends 2008, p. 269.

25 Pani 2013, pp. 22–29.

26 Cic. *De re publica* 2.51.

27 On the emergence of the historical necessity of a delegation of powers in Cicero's writings, see Pani 2013, p. 25. On the concept of protection in Cicero and on the centrality of the discourse regarding *security* in his works see Wood 1988, pp. 185–193.

28 *De re publica* 1.52 (the section translated in our text is underlined): *Sic inter infirmitatem unius temeritatemque multorum medium optimates possederunt locum, quo nihil potest esse moderatius; quibus rem publicam tuentibus beatissimos esse populos necesse est vacuos omni cura et cogitatione aliis permisso otio suo, quibus id tuendum est neque committendum, ut sua commoda populus neglegi a principibus putet.*

29 For comments and explanation of the passage see Pani 2013, p. 27.

30 Cic. *De off.* 1.85: *Ut enim tutela, sic procuratio rei publicae ad eorum utilitatem, qui commissi sunt, non ad eorum, quibus commissa est, gerenda est.*

31 To which *tutus* is related etymologically, that we would not hesitate to translate with 'safe'. Arends 2008, p. 268 believes that the use of the adjective *securus*, already present in Lucretius, corresponds to *tutus*; actually this interpretation seems a bit forced, since *securus* mainly recalls to the 'lack of concern' and the meaning of safety is only secondary. As already said, on the contrary, it is in Cicero that the adjective *securus* begins to extend its meaning to *tutus*.

32 Dig. 26.1 (*De tutelis*): *Tutela est, ut Servius definit, vis ac potestas in capite libero ad tuendum eum, qui propter aetatem sua sponte se defendere nequit, jure civili data ac permissa. Tutores autem sunt qui eam uim ac potestatem habent, exque re ipsa nomen ceperunt: itaque appellantur tutores quasi tui[t]ores atque defensores, sicut aeditui dicuntur qui aedes tuentur.*

33 Zarecki 2014 in a recent work discusses extensively for the first time on the figure of the Ciceronian rector. According to Zarecki, the tutor that Cicero cites is close to the philosopher as educator, a definition that, however, does not explain all the implications of the term. For a more recent summary of the history of Cicero's political thought, see Atkins 2008, who does not give any space to the figure of the rector.

34 Hamilton 2013, p. 65.

35 Tac. *Ann.* 1.2.1: *Ceteri nobilium, quanto quis servitio promptior, opibus et honoribus extollerentur ac novis ex rebus aucti tuta et praesentia quam vetera et periculosa mallent.*

36 Fears 1981, p. 877 and *passim*.

37 Cicero is driven, by the urgency of civil wars, to support Pompeius and to present him as champion and savior of the *res publica* (Zarecki 2014, p. 72). According to Zarecki, Pompeius is probably not the true model of the *rector* in Cicero's ideal; rather, the apology of the monarchy in *De re publica* could somehow reflect the need to justify his sine collega.

38 See Watson 1994. Also see Graverini 1997, pp. 233–234 and 2006, p. 58.

39 Vell. *Hist. Rom.* 2.88.2. Cf.Tac. *Ann.* 6.11: *Ceterum Augustus bellis civilibus Maecenatem equestris ordinis cunctis apud Romam atque Italiam praeposuit.* Echols 1958, pp. 379–380. *Infra*, nn. 41, 43 and 45.

40 Vell. *Hist. Rom.* 2.88.3: *Hic speculatus est per summam quietem ac dissimulationem praecipitis consilia iuuenis et mira celeritate nullaque cum perturbatione aut rerum aut hominum oppresso Lepido, immane noui ac resurrecturi belli ciuilis restinxit initium.*

41 On the episode of 31–30, also see Liberati, Silverio 2013, pp. 93–94: according to the authors, with whom I disagree, the notion of *urbis custodis praepositus* refers to the *custodia urbis* "used to indicate a military surveillance. In it, however, is also contained the set of information aimed at the protection of public security: the use of the verb *speculor* is significant in this regard."

42 Graverini 2006, p. 59.

43 Watson 1994, pp. 102–103. For a discussion on *praefectus*, *custos* and *cura urbis*, refer to Ruciński 2009, pp. 66–71, which may also be consulted for its bibliography.

44 *Eleg. in Maec.* 1.14.

45 *Eleg. in Maec.* 1.27–28: *Vrbis erat custos et Caesaris obses, num tibi non tutas fecit in Vrbe uias?* It is worth mentioning also Horace's famous portraits of his protector occupied in the management of politics and citizen order (Hor. *Od.* 3.29; 3.8).

46 On the passages by Velleius, Hartmann 1921, coll. 1000–1003; André 1967, pp. 64–66. The expression recalls another passage by Velleius (2.98) in which, in respect of the urban prefect Lucius Calpurnius Piso, who during Tiberius' absence had to deal with preserving order in the city, we read *securitatis urbanae custos*.

47 Gabba (who writes in 1994) asserts that "Maecenas' speech has been the subject of an endless series of analysis and research" (p. 149). I shall therefore limit myself to refer to G. Zecchini, *Il pensiero politico romano. Dall'età arcaica alla tarda antichità*, La Nuova Italia Scientifica: Roma, 1997, pp. 121–123 and its bibliography. On the issue of the historical accuracy of the dialogue of Agrippa-Maecenas and on the relationship with the events of the Severan age, see F. Millar, *A Study on Cassius Dio*, Oxford University Press: Oxford 1964, pp. 102–118; U. Espinoza-Ruiz, *Debate Maecenas en Dio*, Madrid 1982; Id., *El problema de la historicidad en el debate Agrippa-Mecenas de Dion Casio*, in *Gerión* 5, 1987, pp. 289–316. Also see the bibliography in the next note.

48 Scholars today, almost unanimously, consider Maecenas as Dion's spokesman. Vd. A. Favuzzi, *Osservazioni su alcune proposte di Mecenate nel libro LII di Cassio Dione*, in A. Piani (ed.), *Epigrafia e territorio, politica e società: temi di antichità romane*, Bari 1996, pp. 273–283, with extensive bibliography at page 273 n. 19; and Graverini 1997, pp. 231–289.

49 Peace, which was requested urgently by the citizens of the empire who were battered by the succession of civil wars, was the main aspiration of the members of *equestris ordo*, who only in a situation of general security could hope for a booming development of commerce (André 1967, p. 68).

50 André 1967, p. 64 talks about the urban praefectura and the praetorian one.

51 André 1967, pp. 75 and following. On Epicureanism during Republican Rome, see Sedley 2008; during the empire, see Timpe 2000. For a summary of the critical

positions of Maecenas' Epicureanism, see Graverini 1997, pp. 243–246. For relations between the Hellenistic philosophies and politics (also Roman), see now, among the others, E. Brown, *Politics and Society*, in J. Warren (ed.), *The Cambridge Companion to Epicureanism*, Cambridge University Press: Cambridge 2009, pp. 179–198.

52 Avallone 1962, p. 115. The text by Philodemus has been interpreted considering it not only as a manual on Homeric criticism, but also as a real political treatise: lately, Fish recovering the original setting of Marcello Gigante, tends to consider the work by the philosopher of Gadara as a written piece with a political content, a *speculum principis*, addressed to the Roman noble class, well represented by Calpurnius Piso, Caesar's father-in-law, recipient of the treatise. On Philodemus and his relations with the intellectuals and the politics of his time see: Gigante 1990; D. Obbink, *Philodemus in Italy*, The University of Michigan Press: Ann Arbor 1995; D. Armstrong, J. Fish, P.A. Johnston and M.B. Skinner (eds), *Vergil, Philodemus, and the Augustans*, University of Texas Press: Austin 2004. For bibliographies commented on the text of Philodemus and its political implications, Fish 2011, pp. 72–74.

53 Dorandi 1982.

54 Timagenes seems to have written a treatise 'About the Monarchs' (Sen. *De ira* 3.23, 4–8). On the figure of the historian during the reign of Augustus with whom he later fell in disgrace, see R. Laqueur, s.v. (Timagenes 2), in *RE VI* A.1, 1936, coll. 1063–1071 and following.

55 It is difficult to reconstruct what the position of the philosopher and his disciples was in regards to the monarchy, although it is likely that it was not totally negative. For this interpretation see Fish 2011, in particular pp. 91 and 95 and following. *Contra*, V. Buchheit, *Epicurus's Triumph of the Mind*, in M.R. Gale (ed.), *Oxford Reading in Classical Studies: Lucretius*, Oxford University Press 2007, pp. 104–131.

56 See in this regard how Colotes of Lampsacus, one of Epicurus' disciples, has presented in a positive way the laws and the monarchy (Plut. *Adv. Colot.* 1124d). For the commentary, see A. Corti, *L'Adversus Colotem di Plutarco. Storia di una polemica filosofica*, Leuven University Press 2014, p. 103. Cf. Grimal 1986.

57 G. Roskam, *Live Unnoticed (Lathe biosas): On the Vicissitudes of an Epicurean Doctrine. Philosophia Antiqua, v. 111*, Brill: Leiden/Boston 2007, pp. 36–41; A. Barigazzi, Sul concetto epicureo della sicurezza esterna, in *Syzetesis. Studi sull'epicureismo greco e romano offerti a Marcello Gigante*, Biblioteca di Parola del Passato: Napoli 1983, pp. 73–92.

58 Also Pani 2013, pp. 41–42. For a negative view (called upon by Fish 2011, pp. 74–75) see D.P. Fowler, *Lucretius and Politics*, in J. Barnes, M. Griffin (eds), *Philosophia Togata: Essays on Philosophy and Roman Society*, Oxford 1989, pp. 120–150, reprinted in M.R. Gale (ed.), *Oxford Reading in Classical Studies: Lucretius*, Oxford University Press 2007, pp. 397–431. To clarify the attitude of the Roman Epicureanism towards politics, the impulse given by deciphering the Herculaneum papyri containing Philodemus' works was crucial; and in particular the reading of fragments of the Treaty on *The Good King according to Homer*, as already mentioned, the launch of this research is due to the figure of Marcello Gigante, see n. 52. See Arrighetti 2000 and Benferhat 2005.

59 Diog. Laert. KD VII. Ἔνδοξοι καὶ περίβλεπτοί τινες ἐβουλήθησαν γενέσθαι, τὴν ἐξ ἀνθρώπων ἀσφάλειαν οὕτω νομίζοντες περιποιήσεσθαι. ὥστ' εἰ μὲν ἀσφαλὴς ὁ τῶν τοιούτων βίος, ἀπέλαβον τὸ τῆς φύσεως ἀγαθόν· εἰ δὲ μὴ ἀσφαλής, οὐκ ἔχουσιν οὗ ἕνεκα ἐξ ἀρχῆς κατὰ τὸ τῆς φύσεως οἰκεῖον ὠρέχθησαν. [Some people conceived a wish to become famous and held in high honour, thinking that they would thus acquire security from men. Consequently, if the life of such men

is safe, they received a natural good; but if it is not safe, they do not possess that for the sake of which from the start they conceived a desire which was in accord with what is suitable to nature] (translated by Fish 2011).

60 Lucr. 3.995–1002; 5.1120–1134.
61 Fish 2011, pp. 81 and following.
62 On Lucretius and the Campanian Epicureanism, see Dorandi 1997.
63 Lucr. 1.40–43: *Funde petens placidam Romanis, incluta, pacem; nam neque nos agere hoc patriai tempore iniquo possumus aequo animo nec Memmi clara propago talibus in rebus communi desse saluti.*
64 Fish 2011, p. 87.
65 Pani 2013, pp. 22–23.
66 Pani 2013, pp. 30–36.
67 Suet. *Aug.* 32: *Grassaturas dispositis per opportuna loca stationibus inhibuit, ergastula recognovit, collegia praeter antiqua et legitima dissoluit.* On this theme, see Lafer 2001, pp. 125–134; and, in this volume, Part II, Introduction.
68 When he sings the praises of *patria tuta et secura*, see *supra*, n. 68.
69 Suet. *Aug.* 28: *Tutam vero, quantum provideri humana ratione potuit, etiam in posterum praestitit.*
70 Suet. *Aug.* 29–31.
71 Wallace-Hadrill 2007, pp. 58–64 (at n. 68): see further, in the text.
72 Wallace-Hadrill 2007, pp. 76–78, following the work of Nicolet 1991. The security measures implemented by Augustus demonstrate a logical idea that is not too dissimilar from the idea of security in modern age, being close to what Foucault calls 'security technologies': those tools that point to securing the social space by reducing the risk factors and limiting as much as possible the undesired events (Foucault 2004, p. 47).
73 Suet. *Aug.* 33–34. For legislative action on criminal matters, refer to Part II, Introduction. It is also known that in the *Res Gestae* the *Princeps* himself claims to have accepted the appointment of *cura annonae* in order to personally deal with the terrible famine, resolved in a few days, thus freeing the entire city *metu et periclo*, with his care and at his own expense (*Res Gestae* 1.33–35).
74 Sherk 1970, nr. 67.
75 Kelly 2013, p. 420.
76 On Campanian Epicureanism and Vergil, see Gigante 2004. On Roman Epicureanism in the Late Republic, and in general, its influences on Augustan poets, also see now Rocco 2016, dedicated to the theme of *praesens deus*, with detailed bibliography.
77 The lucretian *secura quies* reappears in Vergil, with the praise of rural life in the second book of the Georgics: *At secura quies et nescia fallere vita/ diues opum uariarum, at latis otia fundis* (*Georg.* 2.467). In Ovidius, *secura quies* belongs to the gods, but this time it is polemic against the Epicurean vision (*Ars Amatoria* 1.639–640); *secura* is again the *quies* in *Fasti* 6.734.
78 In Ovidius *securum* is also the *otium* (*Fasti* 1.163–170). Also see *Tristia* 2.9.
79 The *otia* of the *aurea aetas* in *Met.* 1. Cf. Appendix Vergiliana, *Aetna* v. 9: *aurea securi quis nescit saecula regis.*
80 In Vergil *secura* is Venus, when she accepts, by the will of Jupiter, that Octavianus raised to divine honors (*Aen.* 1.290). Horatius recalls Lucretius with the expression *securum aevom*, the untroubled life of the blessed (*Serm.* 1.5 v. 101). In Ovidius, Vesta is safe when taking a placid rest (the *placidam quietem* of *Fasti* 6, vv. 331–332.
81 A strong inspirational motive is found in the Tibullian verses *Et 'bene' discedens dicet 'placideque quiescas,/ terraque securae sit super ossa levis* (Tib. 2.4, vv. 49–50).

For beverages that give oblivion, also see Tib. 1.3, vv. 45–46 and 2.1, vv. 45–46. In Vergil the waters of Lethe (*Aen.* 6.15: *securos latices*); Ovidius also on Lethe (*Epist.* 2.4 vv. 23–24).

82 In Vergil, *securus amorum*, means careless or indifferent to passions (*Aen.* 1.350; 10.326). Horatius is *securus*, meaning he is not interested in political events and in the fate of the empire (*unice securus* in *Carm.* 1.26 v. 5–6); in Horatius *securus* also has the meaning of careless in arranging dinner (*Serm.* 2.4.50); with the same nuance, *Epist.* 2.2.17.

83 In Horatius for a legumes soup, *securum holus*: in this case the adjective is used in the sense of 'simple, frugal' (*Serm.* V 2.7. 30); in *Corpus Tibullianum* 3.6. vv. 30–32 the meal is assured, without worries, on a lovely day.

84 Liv. 42.13.5.

85 *Carm.* 4.14 vv. 43–44: *Tutela praesens / Italiae dominaeque Romae.*

86 *Carm.* 3.14 v. 14: *Ego nec tumultum / nec mori per vim metuam tenente / Caesare terras.*

87 *Cons. ad Liv.* 473

88 *Fasti* 1 vv. 529–533: *Tempus erit cum vos orbemque tuebitur idem, / et fient ipso sacra colente deo, / et penes Augustus patriae tutela manebit: / hanc fas imperii frena tenere domum; Vrbs quoque te et legum lassat tutela tuarum / et morum, similes quos cupis esse tuis.*

89 *Tristia* 2 v. 157: *Per patriam, quae te tuta et secura parente est, cuius, ut in populo, para ego nuper ego.*

90 *Infra*, pp. 51 and 54.

91 RIC I² 68 no. 358. BM Photograph. The coin is taken from Galinsky 1996, p. 55 (who has done the cited translation).

92 Köhlr 1957, p. 151; the same idea has already been expressed by Hartmann 1921, col. 1001.

93 Val. Max. 4.10: *Ut populi Romani res meliores amplioresque facerent.*

94 Of course the censors' prayer is not the only source of inspiration: the adjective *tranquillus* placed next to *amplior* (with reference to the territorial expansion of Rome) alludes to the general situation of stability and peace (thorough exemplification, even literary, in Suspène 2009).

95 *CIL*, XIV 2898 (p. 494) = *ILS* 3787 = SupplIt Imagines – Latium 1, 600, with photo (M.G. Granino Cecere): *Paci August(i)/sacrum/decuriones populusque/coloniae Praenestin(ae).* The same text appears on the back, but with a different abbreviation.

96 *CIL*, XIV 2899 = *ILS* 3788 = SupplIt Imagines – Latium 1, 601, with photo (M.G. Granino Cecere): *Securit(ati) Aug(usti)/sacrum/decurion(es) populusque/coloniae Praenestin(ae).*

97 It is known that peace is one of the recurring themes of the Augustan communication from the *cistofori* of Ephesus to the Ara Pacis (Zanker 1988).

98 "I due concetti si integrano a vicenda, l'uno chiarendo il valore dell'altro, quasi in un'interpretazione della politica augustea secondo un angolo visuale italiano. . .che trova un corrispettivo piuttosto nella letteratura contemporanea che nei monumenti ufficiali urbani" (Zevi 1976, pp. 38–41). A decade later, Paul Zanker will point out how the altars of *Securitas* and *Pax*, offered jointly by the *decuriones* (the city senate) and the *populus*, could immediately communicate their message to anyone who would observe them (Zanker 1988, p. 297).

99 Recalled with the Ara Pacis by Ovid (*Fasti* 3.881–882): *Ianus adorandus cumque hoc Concordia mitis et Romana Salus Araque Pacis erit.* The dedication most likely occurred in AD 10 (cf. Cass. Dio 54.35.2).

100 Granino Cecere, in the introduction to Cattaneo 2011.

101 According to Hamilton 2013 p. 59, during the Republic *Salus* was defined as protection against internal strife and external aggression; from the age of Augustus, the

political and civil salvation was clearly separate from the psychological *Securitas* of stoic derivation. The *Salus* as physical preservation had been restored to the people, "self-treatment had produced the securitas; state therapy had generated the salus". In reality, as seen above, the process is not so schematic and the values at stake are more numerous than the pair *Securitas – Salus*. Also see Köhler 1957.

102 Galinsky 1996, p. 70.
103 Schrimm-Heins 1991, p. 139 s. As of the first century AD, *Securitas* indicates the agreement reached in the Principate between peace and tranquility, and in particular the political stability guaranteed by Augustus.
104 Pani 2013, p. 53.
105 It is Instinski's insight 1952, resumed, among the others, by Gottschall 1997. See *infra*, n. 125.
106 *Hist. Rom.* 2.103.3: *spem conceptam perpetuae securitatis aeternitatisque Romani imperii abunde persequi poterimus.*
107 *Hist. Rom.* 2.89.1: *Rediit cultus in agros, sacris honos, securitas hominibus, certa cuique rerum suarum possession.*
108 Hamilton 2013, p. 58.
109 *Supra*, p. 38.
110 *Hist. Rom.* 2.98. On Italy and Rome in the *Historia* by Velleius, see in particular E. Gabba, *Italia e Roma nella 'Storia' di Velleius Patercolo*, in *Critica storica* 1, 1962, pp. 1–19, then included in *Esercito e società nella tarda Repubblica romana*, Firenze 1973, pp. 347–360.
111 *Hist. Rom.* 2.103.3 (*supra*, n. 106).
112 Instinski 1952, p. 16.
113 *Infra*, pp. 46–47.
114 For example Sen. *De Benef.* 5.12.2; later on common in Quintilian, for example *Inst. Orat.*, 10.6.2.; or in Tac. *Hist.* 3.83.14. Already in Cicero, *Securitas* proves to be able to have a character that strikes us in the passage of the *De amicitia*, in which Cicero blames stoic *securitas-apatheia* when it means neglecting the search for friendship in order to avoid worries (*Laelius de amicitia* 45 and 47. For the commentary see Hamilton 2013, pp. 61–62).
115 *Tess.* 1.5.3. The letter is datable to the years AD 49–50 (Harrison 2011, p. 47).
116 Harrison 2011, p. 61.
117 Calp. Sic. *Ecl.* 1.42. Vd. Lana 2001, p. 37.
118 Max Pohlenz even identifies the words *securus* and *securitas* as the philosopher Seneca's favourite. Fundamental to this section and for the understanding of the topic Lana 2001, to whom also refer to for the relative bibliography.
119 *Quid est vita beata? Securitas et perpetua tranquillitas. Hanc dabit animi magnitudo, dabit constantia bene iudicati tenax* (Sen. *Ad Luc.* 92.3).
120 Lana 2001. Also Arends 2008, p. 270.
121 For the idea of Seneca to found, through the Treaty, 'a new metaphysics of the Principate', see E. Malaspina, *La teoria politica del De Clementia: un inevitabile fallimento*, in A. De Vivo, E. Lo Cascio (eds.), *Seneca uomo politico e l'età di Claudio e di Nerone*, Edipuglia: Bari 2003, pp. 139–157 (the above cited expression is at p. 148).
122 *De Clem.* 1.4.1. For this interpretation, following on Lana, recently Beltrami 2008.
123 *De Clem.* 1.8.2: *Quam multa tibi non licent, quae nobis beneficio tuo licent! Possum in qualibet parte urbis solus incedere sine timore, quamvis nullus sequatur comes, nullus sit domi, nullus ad latus gladius; tibi in tua pace armato vivendum est. Aberrare a fortuna tua non potes; obsidet te et, quocumque descendis, magno apparatu sequitur.*
124 The same theme has already been presented more or less in the same terms by Philodemus (Fish 2011, p. 89).

125 Lana 2001, p. 45. The scholar also notes the connection with the verses of
     Calpurnius Siculus, who reaffirm how the *pax* is ensured only by respecting the
     law: *Ecl.* 1.71–73: *sed legibus omne reductis ius aderit moremque fori vultumque priorem
     reddet et afflictum melior deus auferet aevum.*
126 *De Clem.* 1.1.8: *Securitas alta, adfluens, ius supra omnem iniuriam positum; obversatur ocu-
     lis laetissima forma rei publicae, cui ad summam libertatem nihil deest nisi pereundi licentia.*
127 As well as peace, as recalled by the Praenestinian altars (*supra*, pp. 43–45). For the
     practical implications of the security program, see Part II, this volume.
128 Lana 2001, p. 37: *Alla pace politica, al benessere, manca qualcosa di essenziale: manca la
     garanzia che tali beni non siano precari, non siano legati esclusivamente all'arbitrio di chi di
     volta in volta detiene il potere: manca, in una parola, la securitas.*
129 *De Clem.* 1.19.5: *Errat enim, si quis existimat tutum esse ibi regem, ubi nihil a rege tutum
     est; securitas securitate mutua paciscenda est.*
130 *De Clem.* 1.11.4: *Clementia ergo non tantum honestiores sed tutiores praestat ornamen-
     tumque imperiorum est simul et certissima salus.* See commentary by Beltrami 2014,
     p. 12.
131 BMCRE I, p. 179, where Mattingly believes there is a reference to the Pisonians
     or the to grain supply. See also Köhler 1957 and Pera's useful synthesis (2012) with
     prior biography.
         Prior to Nero, the *securitas* coin type could only have a single precedent, attested
     to the Gaius kingdom. This is the famous and discussed *sextertius* with the repre-
     sentation of the emperor's three sisters: Julia with the cornucopia and the rudder
     would represent *Fortuna*; Drusilla with the cornucopia and the *patera* would
     represent *Concordia*; Agrippina leaning to the column would represent *Securitas*.
     The hypothesis of Agrippina/*Securitas* is not, however, in any way demonstrable:
     already questioned by Gottschall 1997, pp. 1091–1093, not even considered by
     Pera 2012, p. 348.
132 Pera 2012, pp. 348–349, 353. According actually to this Author, the type with the
     throne facing right would mean that a protective plan in favour of the emperor or
     the people was implemented; the same type would be suggested, in the coinage of
     the various emperors, functionally for this same purpose.
133 Pera 2012, pp. 353–354 in the wake of Instinski 1952, p. 20 brings forward the idea
     that any such action would have become typical of those moments for which it is
     not easy to talk about stability. The emperor who more than anyone else showed a
     preference for this coin type, after Nero, is Costantinus, whose intent (as in the case
     of Valentinianus, Valens and Gratianus, who made abundant use of it) is to sym-
     bolically allude to his own triumph and to the return to stability in the West after
     usurpers' destabilizing attempts. The last appearance of *Securitas* in the so-called
     traditional form – namely standing and leaning against the column – seems to be
     on silver multiples of the epoch of Magnentius and Decentius, at the mid-fourth
     century; although the legend will continue to accompany other types. For the
     Securitas on imperial coins, in particular of the third century, see Manders 2012.
134 "Il problema politico di fondo per i Romani sotto il principato" (Lana 2001, p. 39).
135 *De Clem.* 1.9. Augustus still represents for Seneca a model to whom refer to as the
     restitutor of the *securitas* to the people, more for an utilitarian need rather than for a
     sincere admiration for the founder of the Principate.
136 Lana 2001, p. 42.
137 Kneppe 1994, pp. 165–216.
138 Kneppe, 1994, pp. 167–169.
139 On *Securitas* during the civil wars, see Pera 2012, pp. 354–355.
140 RIC I² p. 256 and plate 29, no. 504; BMCRE I p. 361 no. 266. Pera 2012, p. 354, n. 47.

141 *CIL*,VI 2051.1.30.
142 Tac. *Hist.* 1.16: *Ne ipse quidem ad securas res accessi, et audita adoptione desinam videri senex, quod nunc mihi unum obicitur.*
143 RIC² I p. 260 no. 7 = BMCRE I, p. 366, no. 13.
144 RIC² I p. 277 no. 175 = BMCRE I, p. 383 (without a number!).
145 RIC² I p. 268 nos. 11–12 and plate 30 n. 12, referring to BMCRE I, ccxxix for the interpretation of the legend. Pera 2012, p. 355.
146 E.F. Krupp, The tutela type of Vitellius, in *Numismatic Chronicle* 100, 7/1 (1961) pp. 129–130, pl. 16. These two *dupondii* of Vitellius were erroneously omitted in Sutherland's revised RIC I on 1984.
147 BMCRE II, p. xliv.
148 See previous note. The woman on the throne would in this case represent Domitilla and the two children are Titus and Domitian, suggesting the stability guaranteed by the continuity of the dynasty. The same purpose pursued, on the other hand, the *dupondi* with the legend *Securitas* coined on behalf of Titus Caesar of 77–78 (Pera 2012, pp. 355 s.).

In support of his hypothesis, Mattingly quotes two significant citations: the first, by Suetonius (*Vesp.* 5.4: *Desertam rem publicam civili aliqua perturbatione in tutelam eius ac velut in gremium deventuram*); the second by Martial (*Epigr.* 5.1.7–8), addressed to Domitianus: *O rerum felix tutela salusque/sospite quo gratum credimus esse Iovem.* Also interesting is the reference to *CIL*, II 3349 = *ILS* 3786 and C. González Román – J. Mangas Manjarrés, *Corpus de Inscripciones Latinas de Andalucia*, Sevilla 2002, I 334, Ossigi – *Baetica*, with dedications to Augustus, to Pax Perpetua and Concordia Augusta, by a *sevir*, Q(uintus) Vibius Felicio and his spouse (or daughter), Vibia Felicula, which is *ministra Tutelae Augustae*.
149 Plin. *Pan.* 35.4.1: *Ingenti quidem animo diuus Titus securitati nostra ultionique prospexerat, ideoque numinibus aequatus est: sed quanto tu quandoque dignior caelo, qui tot res illis adiecisti, propter quas illum deum fecimus.*
150 Suet. *Titi* 6.2: *Neque ex eo destitit participem atque etiam tutorem imperii agere . . . Quibus rebus sicut in posterum securitati suis cavit, ita ad praesens plurimum contraxit invidiae, ut non temere quis tam adverso rumore magisque invitis omnibus transierit ad principatum.* It is a personal security, or rather addressed, in Suetonius' actual words, more to the future than to the present.
151 On Tiberius and Security, see Part III Introduction.
152 Eutr. *Brev. ab Urbe cond.* 7.21.
153 Pera 2012, p. 349, n. 19 where reference is made to Mattingly's comments, according to which, under the Flavian dynasty, *Securitas* as a legend on coins would indicate more a passively enjoyed state, rather than an actively exerted action (BMCRE II, p. xlviii with note 4).
154 Pera 2012, p. 355.
155 See n. 147.
156 Pera 2012, p. 356.
157 As pointed out by Instinski 1952, p. 22.
158 A big urban altar, dating to the last decades of the first century AD, carries a dedication *Securitati cognationis suae* (*CIL*,VI 1887 = *ILS* 1944 = SupplIt Imagines – Roma 3, 3495 = EDR 123188).
159 The indication of the grave as *templum Securitatis* has a long duration in chronological terms, but a limited geographical diffusion, restricted to Sardinia and Africa. For the first, the reference is to the renowned series of rock inscriptions from *Caralis*; among which the main one, *CIL*, X 7719, bears the dedication of T(itus) Vinius Beryllus for a *templum Se[cu]ritati suq[e]*. For a correct overview,

even chronologically, refer to A. Mastino, *Le Iscrizioni rupestri del templum alla Securitas di Tito Vinio Berillo a Cagliari*, in Rupes loquentes. *Atti del Convegno Internazionale di Studio sulle iscrizioni rupestri di età romana in Italia* 1989, Studi pubblicati dall'Istituto italiano per la storia antica: Roma 1992, pp. 541–578, part. p. 562, no. 1; and the collection by P. Floris, *Le iscrizioni funerarie pagane di Karales*, Cagliari 2005, pp. 144–146, nr. 36a, both with extensive bibliography.

    There are two African attestations and both belong to a later period (Severan era): *ILAfr* 357, *Carthago*, seriously incomplete; and *ILTun* 1715, from an unknown location of Proconsular Africa.

160 For example, Sen. senior, *Controversiae*, 9.4.3; Quint. *Declamationes Minores*, 249.20; 252.14–15; 255.2; 258.8; 269.12; 274.4; 322.10; 348.2–12; 351.3; 1.4; 305.18; 348.7; 353.3; 377.10; *Declamationes Maiores* 1.15; 4.10; 7.10; 8.22; 16.8. On the topic, Kneppe 1994, pp. 40–43.

161 Tac. *Hist.* 1.1 where the origin of the political conflict with the senatorial class is reported retrospectively to the beginning of the Principate.

162 Tac. *Ann.* 1.9: *Non regno tamen neque dictatura, sed principis nomine constitutam rem publicam [. . .] ius apud cives, modestiam apud socios; urbem ipsam magnificio ornatu; pauca admodum vi tractata quo ceteris quies esset.* The values at stake mentioned by Tacitus are basically always the same and concern respect, at least formal, for the constitutional structure; the exercise of rights of citizens; the moderation on the part of the imperial power in the repressions, implemented only for public safety.

163 Plin. *Pan.* 50.7: Such is our Princeps' goodness of heart, such the security of our times, that he believes us worthy of princely possessions and we have no fears about seeming so (*Tanta benignitas principis, tanta securitas temporum est, ut ille nos principalibus rebus existimet dignos, nos non timeamus, quod digni esse videmur*).

164 Instinski 1952, pp. 40–43.

165 Plin. *Pan.* 8.1.5.

166 Tac. *Agr.* 3: *Nunc demum redit animus; et quamquam primo statim beatissimi saeculi ortu Nerva Caesar res olim dissociabiles miscuerit, principatum ac libertatem, augeatque cotidie felicitatem temporum Nerva Traianus, nec spem modo ac votum securitas publica, sed ipsius voti fiduciam ac robur adsumpserit, natura tamen infirmitatis humanae tardiora sunt remedia quam mala.*

167 Plin. *Epist.* 10.58.7, by quoting a decree of Nerva: *Illum intellegi satis est, cum hoc sibi civium meorum spondere possit vel non admonita persuasio, me securitatem omnium quieti meae praetulisse, ut et nova beneficia conferrem et ante me concessa servarem.*

168 Plin. *Pan.* 24.2.1: *Iunxisti enim ac miscuisti res diversissimas, securitatem olim imperantis, et incipientis pudorem.*

169 Plin. *Epist.* 10.52.1.3: *Diem, domine, quo seruasti imperium, dum suscipis, quanta mereris laetitia celebrauimus, precati deos ut te generi humano, cuius tutela et securitas saluti tuae innisa est, incolumem florentemque praestarent.*

170 Plin. *Pan.* 43. 1: *In eodem genere ponendum est, quod testamenta nostra secura sunt, nec unus omnium nunc quia scriptus, nunc quia non scriptus heres.*

171 Plin. *Pan.* 34.8: *Abirent, fugerentque vastatas delationibus terras: ac, si quem fluctus ac procellae scopulis reservassent, hic nuda saxa et inhospitale litus incoleret: ageret duram et anxiam vitam, relictaque post tergum totius generis humani securitate, moereret.*

172 Plin. *Pan* 27.1: *Magnum quidem est educandi incitamentum, tollere liberos in spem alimentorum, in spem congiariorum; maius tamen, in spem libertatis, in spem securitatis.* Cf. *Epit. de Caes.* 12.4: *Iste quicquid antea poenae nomine tributis accesserat, indulsit; afflictas civitates relevavit; puellas puerosque natos parentibus egestosis sumptu publico per Italiae oppida ali iussit.* See Noreña 2011, pp. 130 and following, p. 192 n. 7.

173 BMCRE II p. 229 no. 92.

174 The coin type with throne facing to the right, according to Pera, is recuperated and coincides with the conclusion of the Jewish revolt and with Lucio Aelio's adoption (Pera 2012, pp. 357–358).

175 Pera 2012, pp. 356–357.

176 Pera 2012, p. 358.

177 During the Principate of Marcus Aurelius, exactly in 181, the legend of the *Securitas Publica* reoccurs, in occasion of Commodus' return to Rome (Pera 2012, pp. 358–359; Manders 2012, p. 210, n. 75; p. 245, n. 78). Between the second and third centuries there is an alternate (or interchange) use of *Securitas Imperii*, *Securitas Orbis*, *Securitas Perpetua*, *Securitas Rei Publicae* . . . preparing what, in the fourth century, will become an unremitting insistence.

178 Fears 1981, pp. 910–924.

179 Arrian. *Epict. Diss.* 3.9.2.

180 *Altercatio Hadriani Augusti et Epicteti Philosophi* 23. The work is dated between the late second and early third centuries, according to the editors (L. W. Daly and W. Suchier, *Altercatio Hadriani Augusti et Epicteti Philosophi*, III Urbana 1939), but it could also be much later.

181 *CIL*, VI 7102, from vigna Aquari; *CIL*, VI 4877, from vigna Codini (columbarium 2), external wall; and 5164 (nearby the columbarium 1). On a sarcophagus dated to AD 165 the *Securitas* is *perpetua* (*CIL*,VI 18378); while *Aeterna* is the one of *Domitia Patruini*, daughter or, perhaps but less likely, freedwoman of the *consul suffectus* of AD 136 (*AE* 1997, 152).

In Rome, a holy precinct was dedicated to *Securitas* where, on a marble puteal a *negotiator*, probably foreigner, praises her (N.M. Nicolai, *Della basilica di San Paolo*, Roma 1815, pp. 67–68, no. 25; C. Ricci, *Consepti Securitatis*, in *Epigrafia della produzione e della distribuzione*, Roma 1994, pp. 721–725 = *AE* 1994, 296I); another merchant praises the *Dei Mani* and *Securitas* (*CIL*, VI 5929 = *ILS* 7579). On the adjective *securus* and on the dedications *Perpetuae Securitati* from the Middle of the Second century onwards, see Sanders 1997.

182 At Saint Paul outside the walls are preserved two sepulchral dedications, respectively to the *Dei Securitatis* and to *Securitas* and *Quies* (*CIL*,VI 2268 and *CIL*,VI 22337). An urban funeral inscription is addressed to the *Dei Securi* of the young M(arcus) Salluvius Felicissimus Heraclitianus, who died at the age of twelve (*CIL*, VI 10217 = *ILS* 6060).

183 A sacred inscription of Rome has as its object the *Securitas Caelestis*; and its dedicant (*Glitius or Clitius Felix*), who says to be a 'Vergilian poet', offers a twin column to *Silvanus* (*CIL*,VI 639). According to P.F. Dorcey (*The Cult of Silvanus: A study in Roman Folk Religion*, Brill: Leiden, New York, Koln 1992, p. 31 n. 83) *Caelestis* is not the Punic deity, but rather idiosyncratic of the devotee who refers also to *Silvano Caelesti* in *CIL*,VI 638, cf. p. 3006 = *ILS* 2954.

184 *CIL*, VI 967 = *ILS* 309. Cf. Cass. Dio 69.8 that, according to Marietta Horster (*Bauinschriften römischer Kaiser. Untersuchungen zur Inschriftenpraxis und Bautätigkeit in Städten des westlichen Imperium Romanum in der Zeit des Prinzipats* (Historia Einzelschriften 157), Franz Steiner: Stuttgart 2001, p. 67), is questioned by the dedication. Also see H. Bellen, SAEC(ulum) AVR(eum). *Das Säkularbewußtsein des Kaisers Hadrian im Spiegel der Münzen* in H. Bellen, *Politik – Recht – Gesellschaft. Studien zur alten Geschichte*, edited by L. Schumacher, Stuttgart 1997, p. 139 and pl. 8.

185 He commissioned the construction of new towers and restored the previous ones, at the municipality of Auzia in the Mauretania Caesariensis (*CIL*,VIII 20816 = *ILS* 396). Cf. A. Saastamoinen, *The phraseology and structure of Latin building inscriptions in Roman North Africa*, Helsinki 2010, nr. 282).

186 On Severan coins, the *Indulgentia* appears more frequently after AD 201. By the mint of Rome and on Caracalla's coins, we find legends as: *Securitas imperii* (BMCREV2, p. 258 n. 516A tav. 41.3, AD 206–210; cf. RIC IV.1 p. 236 n. 168), *Securitas Perpetua* (RIC IV.1, p. 212, n. 2, between AD 196 and 198?; p. 244 n. 229, AD 210–213; p. 276 nn. 399 and 403, AD 196–197; p. 296 s. nn. 512, 515 and 520, AD 210–213; p. 301 n. 536, AD 213–214); *Securitas Orbis* (RIC IV.1, p. 214, n. 22, between AD 198 and 199?; p. 219 nn. 43 and 44, between AD 199–200). *Securitas* appears also simply as an image (e.g. RIC IV.1, p. 216, n. 26 A, AD 199 and n. 29; p. 227, n. 92 AD 207; p. 293 n. 492 A and 494 AD 212). On the images and legends of the Securitas in the third century, see, Manders 2012, pp. 205–211.

187 Fulvia Lovotti (*L'arco di* Cirta: *considerazioni sulle epigrafi onorarie*, in *Africa romana*, 13, 1998, pp. 1603–1612) studied the celebatory program of the age of Caracalla and considers the *Securitas* as a characteristic personification of the Severan era.

188 *CIL*, X 6996 cf. p. 1847 = *ILAlg* II.1, 562.

189 *CIL*, VIII 7095, 7096, 7097, 7098 recalls Caecilius's career and his donations, while a fifth inscription (probably set on the attic) recalls the dedication of the arch (*CIL*, VIII 7094).

190 Once again in Africa, at *Bisica*, the *Securitas Urbis* appears together with the *Virtus Divina*, in an incomplete inscription with an archaic language (*CIL*, VIII 12293, Bisica, *Africa proconsularis*).

191 *CIL*, XI 5631 = *ILS* 432 = *AE* 1961, 209 = SupplIt, 6, 1990, p. 64 ad no. (S.M. Marengo) and SupplIt, 22, 2004, p. 164 ad no. (S.M. Marengo), Camerinum. Cf. Panciera 1976–1977, pp. 210–212 = Panciera 2006, pp. 775–776, with bibliography).

192 "*Securitas* first connected exclusively to private life and philosophical dimension, with Augustus acquired a foundation in political reality" (Arends 2008).

193 The Praenestinian altar (*supra*, nn. 95 and 96) is from this point of view almost a testing ground.

194 Hamilton 2013 refers to policing rather than policy.

Part II

The birth of a dispositive

Introduction: Augustan criminal legislation and military reforms

At the end of the last, unrestrained phase of civil war, the great-nephew of Caesar had before him an Italy that was devastated, not only materially but even more so politically and ethically. The wars had compromised the idea of the *res publica* that had been slowly developing over the centuries. The new idea, promoted by the new *Princeps*, with the help of his collaborators and supporters, had the same name, but was completely different in substance. The secret of his success, as is well known, was the complex mix of innovation and tradition, of authoritarianism and formal respect for the organs of government, which came to characterize his political action and that could be extended from Rome to the provinces and to all fields of government.

In his redesigning of the fragments of the upturned *res publica*,[1] Augustus paid particular attention to the reorganization of the territories and the structures of Italy and Rome. Essential to his programme was the transmission of a message of social pacification, which often assumed the tone of propaganda: the conflicts in the territory of the peninsular must cease and a climate of serenity must be re-established in which the economies of the cities and of the countryside could once again return to prosper.

Although impossible, a return to the past, to an Italy of small properties and small- and medium-scale production for self-consumption and contained commercialization, was promoted. At the same time, emphasis was placed upon the promotion of the image of the *civis Romanus*. This became an almost obsessive *leit motif*, which could be found in works of art harking back to a former classical style. The calm and even-tempered peasant-soldier ready to sacrifice himself and his own existence, architect of the triumph over the populations of Italy and the Mediterranean, an aware citizen, ever more able to appreciate the artistic beauty and multiple stimuli which had come to surround him, was the recurring protagonist of literary production. His activities and his values were represented not only in the so-called 'high' art production but also in jewellery making and the finest examples of handcrafted tableware.

Therefore, art celebrated innovation and change. At the same time, a series of interventions in urban planning transformed the fabric of the *Urbs* and of the cities of Italy, through restoration, new construction, and the development of infrastructure and its protection.[2] Running parallel to this intense renovation activity and interest in restoring /renewing the image of the city, was a policy of intervention, widely investigated in its most diverse artistic manifestations, on the imaginary level promoted through the language of propaganda,[3] where *Pax* became the rallying cry for the new (yet old) myth that would advance.[4]

How were the inhabitants of Italy, traumatized by insecurity, to be reassured? Which discourse was to be adopted to ensure the necessary tranquility for a peaceful series of exchanges and meetings, without bringing to mind the fear of violent repression and a reduction of individual liberty? We do not know with certainty how these questions appeared before the eyes of the victor of Actium. What we do have are the results, if partial, of his resolution: what Augustus declared in his *Res gestae*, what has been reconstructed by the literature (from the time and after); the operative solutions; and a few material results of his intervention.

It has been well documented how Augustus operated in order to transmit his reassuring message regarding the removal of instability both for the individual and the world in which he lived.[5] What is of interest here is a series of actions, which from the point of view of the city and of Italy, came to represent a systematic response to the needs of order, both in terms of the regulatory framework and the implementation of military and paramilitary deployment. Therefore, this section hopes to highlight the various initiatives (in the field of criminal legislation and in the reorganization and control of the territory), which, through different types of language and instrumentation, would answer the same purpose and converge towards a common aim. Indeed, these differing acts would, over time, help to construct what would become a plan for the security of the city.

The Augustan programme of reforms and new laws, interpreted in strict continuity in respect to the Caesarian era, was in fact brought to fruition following the same logic. If the recovery, the rectifications and the expansion of law were a response to a desire to give new life to the *mores antiqui*, the creation of military and paramilitary units and the reorganization of the territory of the city represented an absolute change. The idea is that the underlying principle could have been a far more thought-out construction and a new rallying cry: the security of the city and the *Princeps* as its supreme guarantor. Measures to protect peace and internal security will therefore serve as a premise and framework into which the precise actions that will be discussed in the following paragraphs and chapters can be inserted.[6]

Therefore, the starting point for this analysis will be the Augustan legislation in criminal matters and the relationship between the Triumvir and his soldiers. After having established the analytical framework, attention will be moved to the organization of the armed forces, protectors of the *maiestas Principis*, and on the guards responsible for his private safety (Part II, Chapter 4). This will be followed by an examination of ad hoc measures regarding civilian and military forces, active in the urban territory, in order to safeguard its security.

Persistence and innovation in Augustan criminal legislation

Ten years after the end of the wars, the emperor (*iustissimus auctor*) began to systematically dedicate himself to legislative action, which had come to be considered fundamental.[7] Through this legislation, Augustus sought to create such conditions that would not allow a repeat of the social disorder of the Late Republic; individual and collective behaviours would be sanctioned if deemed detrimental to social harmony.[8]

His activity (*leges Iuliae*) was carried out between 18 and 16 BC, and successively – primarily promoted by the consuls – between 2 BC and AD 9.[9] While the *lex de iudiciis publicis e privatis* of 17 BC[10] ratified the definitive disappearance of trial through *legis actiones* and redefined the system of *quaestiones perpetuae*, the changes introduced during his government regarded the recovery or implementation (*lex Iulia de ambitu, maiestatis* and probably *di vi* and *peculatus*) and creation (*de adulteriis, de annona*) of *leges*. Regarding this impressive activity, it is important to underline the continuity or eventual innovation in Augustus' approach in respect to who preceded him in the previous century. Above all, it is also essential to understand the value that such legislation gave to the field of security.

The general structure of criminal legislation that the *princeps* inherited at the end of the Republic[11] was that which was enacted between the end of the second century and the first ten years of the first century BC. This period saw the creation of *questiones perpetuae*, permanent law courts that judged certain crimes. Their gradual growth and the modifications that concerned them made sure that acts of violence, until then considered private, were introduced into the category of public crimes, given that their occurrence could affect more than the private sphere and have an impact on collective security.[12]

The Sullan legislation, in particular the *lex Cornelia de sicariis et veneficiis*[13] and after this, the Caesarian legislation, followed the same direction, with the principle aim of containing the actions of armed mobs and neutralising political agitators, particularly during the Sullan era.[14] At the end of the Republican age, nine permanent courts were in place, of mixed composition, five of which were political in nature (*de maiestate, de repetundis, de ambitu, de peculatu, de vi*) and four were reserved for common offences.[15]

From Cicero we know that a *lex de maiestate* already existed under Caesar.[16] Augustus therefore gave a new form to the measure, which is clearly expressed in the Digest. The first *caput*, which can be indebted to Ulpianus, reads: "The crime of lese majesty is committed against the Roman people, or against their Safety, and he is guilty of it by whose agency measures are maliciously taken for the death of hostages, without the order of the emperor."[17]

Acts that constituted an offence for the security or the safety of the state and the integrity of its institutions were essentially confirmed in line with their definition and their content. However, as guarantor of security and active protector of order and social harmony, alongside the *res publica* and its representatives, there was now the emperor. Indeed, this change was introduced by the phrase *iniussu principis:* from this moment on, a series of types of behaviour and action, not previously contemplated by the law, were gradually incorporated into the concept *maiestas*. Among these were: "having pronounced or written offensive or defamatory statements against the emperor, having violated statues or images of the emperor, and having carried out magical practices in an attempt to find out the destiny of the empire".[18]

The *lex Iulia de vi*, was organized almost at the same time as the *lex de maiestate*[19] and in a certain sense represented a necessary addition. The Augustan law,[20] also in this respect, followed comparable Caesarian measures.[21] However, it cited a number of particular cases that encompassed various criminal actions, which were previously covered by crimes against the *maiestas*.[22] In the previous legislation[23] crimes that were contemplated included attacks against Roman magistrates, carried out in strategic sectors of the city and, in the case of the Sullan laws, included an intent to steal or kill. In titles 6 and 7 of the Digest 48 – those that according to Cloud can be attributed to the Augustan age – crimes were included such as the possession of arms, the attack of a property (and the expulsion of a proprietor/owner) with fire; and association in order to commit a crime.[24] The most important changes introduced by Augustus related above all to the limits imposed on the practice of *imperium* of a magistrate, who violating the right of *provocatio* kills, tortures or incarcerates a private citizen.[25] In addition, other *crimina* were contextualized and defined with more careful attention.[26]

It is probable that during the two-year period 18–17 BC, the *lex Iulia de collegiis*[27] was reintroduced. Also in this case there was a Caesarian precedent to which Suetonius refers.[28] Furthermore, in chapter 32 of the *Life of Augustus*, Suetonius dedicated a detailed discussion to the motives behind the Augustan measures and the way in which they were implemented.[29] Caesar dissolved all associations apart from the *collegia legitima*, i.e. the associations that were in line with the organization of the state and in particular, the relevant laws in effect at the time.[30] After his death, new associations were founded with no objective other than *facinus*. It is probable that Augustus re-established the previous status, dissolving all bodies with the exception of the long-standing (mostly sacerdotal) corporations, establishing decisive state control over the rights of association and making known its effects on all the imperial legislation in the field.[31]

Whoever wanted to form an association had to obtain permission from the senate in advance.[32] In light of this, the *lex Iulia de annona*[33] can be interpreted as a completion of the *lex de collegiis*, since it responded to questions over the instrumental use, particularly during the time of civil wars, of the system of provisions and rules relating to the management of foodstuffs set aside for the city and for Italy. The measure was designed to discipline instances of disturbance to public order, carried out by those who, through subversive association, impeded supplies.

However, Augustus did not limit his actions merely to the recovery and integration of the legislative activities of the two dictators, who, during the previous century of the Republic set up a new system of law courts and maintained or at times expanded the field of public security, limiting the rights of association, extending the list of recognized violent crimes both private and public against the Roman people and his representatives. New laws,[34] which regarded the behaviour of private individuals and its effects on the public sphere, at this point received a new and more restrictive formulation after the events of the first century BC.[35]

In order to make his action more incisive, Augustus established the foundations[36] of a new penal system, which for the whole first century would run in parallel to its traditional counterpart. It was characterized by the discretionality of the judicial entity: the *cognitio extra ordinem*. Alongside the *quaestiones* and the practice of law and order by the *praetors* and the *iudices*, repression began to be autonomously called for and therefore practised by the Senate, the emperor, his officials and appointed provincial governors.[37]

It is beyond the scope of this section to retrace the episodes that would lead to the completion of this process: the judicial activity of Augustus in the field of criminal legislation is described by the sources with relative richness of detail.[38] The case that has been the subject of most serious academic study is the series of edicts regarding Cyrene (all dated between 7 and 6 BC), and in particular the second.[39] Scholars have long argued over what could have been the constitutional foundation behind the discretion given to the emperor, which Dio defines as the legal right, conceded by a plebiscite, as far back as 30 BC, of *ekkleton dikazein* (to judge on request).[40] For many, the foundation can be traced back to the *auctoritas Principis*, that "meta-juridical and meta-constitutional power that assured the prince a sort of control over the entire *res publica*".[41]

The action against the violation of the *maiestas populi Romani et Principis*, a serious risk to the stability of the system of government,[42] is, in my opinion, the fundamental element of Augustan policies for security. The systematic revision, which Michael Peachin carried out of the evolution of the *crimen maiestatis*, and of the trials for high treason between the end of the Republic and the Augustan age clearly highlight the turning point, which was triggered by the last ten years of Augustus' reign. This moment appears to correspond with that when: "Augustus incrementally gained control of the State, and concurrently evolved as universally (though informally) perceived guarantor of the community safety and well-being, as his person came gradually to be identified with the state."[43]

Offensive behaviour against the emperor could be perceived as damage to the *res publica*, because it could represent a real threat to public stability, and should therefore be sanctioned. A turning point for this new and more profound understanding came about, imagines Peachin, in the year AD 6, a hypothesis to which I fully adhere. However, it is not a coincidence that the last and most decisive series of interventions of the *Princeps* for the safeguarding of internal safety, came about in that year or indeed those immediately following it: the establishment of the *cohortes vigilum* and their prefect; the creation of a military *aerarium*; the beginning of the distinction between the functions of the urban and praetorian cohorts; and strict repression for those who disturbed public order in difficult situations for the city.[44]

In addition to this, during the last ten years of Augustus' government we also saw the first signs of the new orientation in the field of the majesty of the Roman state, which began to manifest themselves. Here, it is worth mentioning the trials that involved Cornelius Gallus[45] and Cassius Severus,[46] Murena and Fannius Caepio[47]; even if, as is well documented, the examples would intensify with Tiberius, from Gnaeus Calpurnius Piso pater[48] to Cremutius Cordus,[49] to mention the most famous. The accusation of *crimen maiestatis domus Augustae*, which the latter cases can be characterized as, can be seen, in the view of many, as a silent revolution carried out by the *Princeps*.[50]

The relationship with the soldiers and the dawn of the security plan

The process of organizing the military and paramilitary forces and the interventions in the topography of Rome proceeded at the same pace as the legislative activity that regarded more directly the protection of the emperor and the spaces of the city. To give just one example, the praetorian cohorts and their urban counterparts were born at the same time as the ascent of the *Princeps*, while the creation of the cohorts of *vigiles* happened during a later period of the reign.

It seems appropriate, however, that a brief excursus of the relationship between Augustus and the army during the last phase of civil war precede the presentation and function of the armed and paramilitary forces: it allows us to clarify how this relationship changed after the year 27 BC and what results it had for policies regarding the security of Rome and Italy. From this picture, for obvious reasons, the reforms that regard the provincial armies will be excluded.

The relationship that Octavian had with the army in general and the soldiers was not idyllic.[51] The triumvir, in contrast to his adoptive father, and to Marcus Antonius, was not an *imperator* in harmony with his army. Suetonius, in particular, relates episodes that demonstrated, to use a cautious expression, a relationship that could be regarded as reciprocally diffident.

In approximately 40 BC, Octavian sent away from the spectacle of *Ludi* a soldier who was sitting in a row reserved for the equestrians. The word spread that the soldier had been sentenced to death and as a result there was a *turba militaris*. The *imperator* managed to save himself only because the soldier reappeared, unhurt.[52] This episode is indicative of the climate of suspicion that surrounded Octavian's attitude towards his soldiers.

Some years later, when the conflict with Antonius had began to deteriorate definitively, Octavian was accused of being a coward: he was not able to look at an army line up with a steady gaze; and the credit for the defeat of *Sextus Pompeius* and his allies could be attributed entirely to Agrippa.[53]

Even after the year 27, the attitude of Augustus towards the army did not undergo any significant change and there are no signs of an increased intimacy between the *Princeps* and his soldiers. However, despite his detachment and diffidence, from the perspective of representation, the unflattering profile of

Octavian as a poor man of arms underwent somewhat of a transformation. It was as the epithet 'Augustus' brought about a sort of profound mutation in the man Octavian, who was committed to the work of *restauratio/renovatio* of the *res publica*: from this moment, through artistic, epigraphic and literary representation, the picture of a close interaction and mutual praise between army and *Princeps* was built.[54] Numerous signs, often in the form of rituals, allow us to recreate the scenario: from the cult of images (and emblems); to the new military vow (*sacramentum*), the personification of the *Victoria Augusti*, the celebrated episode of the return of the insignia held by Parthians, and the ceremony of *adventus*.[55]

The literary sources limit themselves to registering measures and arrangements, creating a detached picture of the relationship between Augustus and his solders.[56] Suetonius tells us that the Princeps was always severe in maintaining discipline, in particular towards those who avoided service, undisciplined legionaries, those who persistently claimed advantage of the *commoda emeritorum praemiorum* and the cohorts who abandoned their positions.[57] He never called his soldiers *commilitones* (nor did he permit that other members of his family invested with military powers do so), and he avoided establishing strong camaraderie with his soldiers, which might evoke, in the eyes of a civilian population still traumatized by recent events, the strong personal bond between general and soldiers seen during the civil wars. He sparingly granted military rewards (*dona militaria*), being more generous with the soldiers than with their superiors.[58] Nevertheless he was ready to give honour to military value.[59]

Mindful of the episode of 40 BC, in which he had run a great risk of being subject to a false accusation, many years later Augustus established that in places where entertainments were watched by the public, soldiers were required to occupy places separate from the rest of the participants.[60] In this new phase, as we shall see,[61] the soldiers of Rome and in particular the praetorian guard, the *vigiles* and soldiers of the *cohortes urbanae* were responsible for the maintenance of order inside and outside the buildings and the different areas of the city on the occasion of entertainment, which included games, plays or sports; and were sometimes coordinated with civil guards.[62]

His cautious way of proceeding, in fear of an eventual rebellion, and a good strategic foresight aimed at influencing the action of his successors, led Augustus to introduce strict controls on wages and rewards. In 13 BC, returning from Gaul, he made the senate read a document that contained a report of his actions and a blueprint in which he acknowledged what had happened in a century of the professional army[63]: the number of years of service was now set to twelve for the praetorian guard and sixteen for the legionaries[64]; the procedures for discharge were also tightened. At this point therefore, the new statute for veterans created a well-defined situation. The resulting effect was twofold: the discharged soldier had rights and privileges and the small-scale proprietors were reassured.

With the same logic the creation of the military *aerarium* was established in AD 6: in the new treasury Augustus paid in one 170 million sesterces, committed himself to integrating further sums subsequently,[65] and organizing its management through the establishment of a new tax.

The sources suggest therefore that rather than a real Augustan policy for the soldiers,[66] there was a desire to construct a permanent army, attempting to contain the degeneration of troops caused by the professionalization. In this sense therefore, it is generally agreed that Augustus was not, and neither did he see himself as, a great reformer of the army. Rather he gave a clear signal that he wanted to close the brackets and then open new ones, which would then be the task of Flavians to close.[67]

At the end of his long period of government, in AD 14, rebellions broke out in Pannonian and Germanian provinces,[68] whose trigger was precisely the regulations introduced by Augustus (the duration of conscription, the sum of soldier's pay, the system of withholding pay, the conditions of discharge), in addition to the *militaris disciplina*, which was regarded as too strict. They undoubtedly are as a significant counterpoint both to the image of a perfectly tested machine (Suetonius, Cassius Dio) and to the idealized image of the Augustus as *Imperator Victor* in the sculpture of *Augustus* of Prima Porta.

It is always Suetonius and Cassius Dio, in addition to Appianus, (and not only them, as we shall see) to which it is necessary to turn in order to recognize, between the lines of military interventions of different sizes and others in fields apparently different, the signs of the slow development of Octavian Augustus' idea for the safety and security of Rome and Italy.

The first signs of the necessity of a plan of action (not yet systematic) emerge well before Actium:

> At this time Italy and Rome itself were openly infested with bands of robbers, whose doings were more like barefaced plunder than secret theft. Sabinus was chosen by Octavian to correct this disorder. He executed many of the captured brigands, and within one year brought about a condition of absolute security. At that time, they say, originated the custom and system of cohorts of night watchmen still in force.[69]

The words of Appianus provide the picture of the instability in Italy following the last terrible phase of civil war. A certain Sabinus (perhaps Caius Calvisius Sabinus) in the space of a year was able to restore peace. We do not know in detail what methods he used, but there is no doubt that, finding himself in an intense phase of civil war, the means and powers available to him were considerable. Perhaps a single year — such a short time compared to the purpose — is intended as a signal of the capability and above all, the effectiveness of the intervention promoted by the initiatives of the triumvir Octavian. No less interesting is the association made by Appian between the *vigiles* and security and the allusion to characteristic action of Augustus, which was based on his prevention and timely action, not only in the areas of the city.[70]

The measures on public security introduced in the fifteen years between 33 and 18 BC have a clear target in mind: those who practise foreign forms of worship and those who seek to tell the future. In 33 BC, the year of his *aedilitas*, Agrippa expelled astrologers and magicians from Rome. This initiative is registered by Cassius Dio, without further comment,[71] following descriptions of Agrippa's more grandiose building projects. Five years later, Octavian refused to consent to the celebration of the Egyptian rites inside the *pomerium*[72]; and in 21, Agrippa was called to Rome to contain a public disturbance "checked whatever other ailments he found still festering, and curtailed the Egyptian rites which were again invading the city, forbidding anyone to perform them even in the suburbs within one mile of the city". Another repressive intervention, in 18, involved the Jewish community. Two thousand Egyptians and Jews, if freedmen, were sent to Sardinia to fight banditry; if they refused to go, they were invited to leave Italy or at least abandon their forms of worship.[73]

It does not seem to be a coincidence that the removals from the city of who, foresaw the destiny of empire, or practised non-official religions, were concentrated in the years between the end of the civil wars and the introduction of the laws *de vi* and *de collegiis*. It would appear to be an inevitable measure in the politics of prevention.[74]

During the first period of the Augustan government, criminal incidents and episodes of banditry, such as traces of civilian anarchy, continued. The *Princeps* did not delegate, as he had done as triumvir with Sabinus, nor did he foresee rapid times of execution, but he started work on a series of preventative measures, which were to be much more systematic[75]: the dissolution of illicit associations went hand in hand with stricter control of the servile prisons of Italic ownership, and the creation of military stations in "considered strategic locations". Suetonius keeps his discourse generic, referring to *loca opportuna*: in the passage, his eye is as much on Italy as it is on the capital and its surroundings. In particular, the effectiveness of the stations, rather than being linked to the possibility of an intervention by the military, had value as a deterrent for the agents of disorder.[76] As to what happened with food supplies (*annona*),[77] the ban on illicit *coetus*, was interlinked with the establishment of numerous military placements in the territory in order to enhance prevention.[78]

Within this general context, a series of interventions for the security of Rome can be inserted. These because of their close interconnection, required lengthy times of execution. With the unscrupulousness that characterized much of his work, and in apparent respect for ancient traditions, Augustus created the basis for the violation of the principle that did not allow the stationing of soldiers within the city: military was in fact the status of the men charged with the protection of the public security and the private safety of the *Princeps* and the *domus Augusta*, and of many of those occupied in guaranteeing the safety of the streets and the public places of the capital.

The following paragraphs will explore the *securitas Augusti* of the *Princeps* and the man Augustus, this will be followed by an examination of the *securitas populi Romani*. In neither the first nor the second discussion will I make use of the general expression 'urban garrison': it seems inappropriate, as has been rightly noted elsewhere.[79] I will therefore not consider as a whole all the military and paramilitary corps that were created by Augustus and would later be granted a stable position in the capital.

My intention therefore in the following paragraphs of this Part II and in the subsequent Part III and Part IV is to present in a clear way the units responsible for the *maiestas Principis* and those responsible for the *maiestas populi Romani* with differing but concurrent spheres of competence. During the government of Augustus, as we shall see, such competences will be clearly shared between the praetorian and the urban cohorts,[80] while the cohorts of firefighters (*vigiles*) would coordinate their work with that of the district magistrates.[81] The desire of the *Princeps* was to create a synergy between authorities and forces, with the general aim to safeguard the *maiestas Principis et Rei Publicae*. For the whole course of the first century, the bodyguards of the *Princeps* remained private and the attempt to militarize them first by Caligula and then by Nero were ultimately destined to fail; the plan was doomed to be realized much later on.[82]

During the last period of the reign of Augustus, and more so following Tiberius and beyond, as we shall see, the autonomous action of the commander of the praetorian cohorts and that of the commander of the urban cohorts would become ever more distinct.[83] Starting from the second half of the first century, and even more so during the second century, the duties and sphere of influence of the so called 'big prefectures' would become more precise and would steadily get bigger.[84] The action of the urban prefect would gradually coordinate with that of the prefect of the *vigiles*, who obtained a progressive increase of responsibilities and had an ever more direct link with the emperor himself. The role of the praetorian prefect will be more and more extended to the control over other military forces, some of which not yet planned by Augustus, as, for example, the *equites singulares Augusti*, with their combined multiple functions (representation of power, personal protection, participation in military campaigns) since their birth.[85] It is not a coincidence that this birth was the result of a long gestation with its beginnings, in my view, in the period following the civil wars of 68–69. At the same time the *milites frumentari* were created, their barracks (the *castra peregrinorum*) were constructed in Rome, and the profile of the *statores* was transformed. It is the start of a new (and also the last) phase in the history of the *Securitas*, in which the military forces and the urban paramilitary originally expected work alongside the 'professionals' of war and control.

Starting with the advent of the Principate, the long history of the *securitas Principis et populi Romani* was destined to come to a close at the end of the second century. At this point, the creation of a legion at the gates of Rome (the *legio II Parthica*); the progressive militarization of forces that operated in the Capital, even those that until that moment had maintained a distinct profile (the *cohortes urbanae* and the *cohortes vigilum*); the ever greater involvement in theatres of war (of the Praetorian Guard, the urban cohorts and the *Equites singulares Augusti*) mark a definitive break in the equilibrium and of the project for parallel authorities and forces working in synergy, to which I have referred.

Notes

1 *Cuncta discordiis civilibus fessa*, to use the expression of Tac. *Ann.* 1.1.1.
2 Vitr. *Praef.* 2, *Res Gestae* 19–21, Suet. *Aug*, in particular 28–30. On religious architecture, Gros 1976. On the considerable role played by Agrippa, from the time of his
aedilitas in 33 BC, Roddaz 1984, de Klejin 2003. For general literature on the monuments and the images of Augustan Rome, starting with the classic Zanker 1988, see
the survey article of Patterson 1992, the effective synthesis of Wallace-Hadrill 1993,
pp. 50–62; and Galinsky 1996, part. chap. IV, Favro 1996, Hinard 2003; Wallace-
Hadril 2008, especially part II.
3 On the cultural commitment on the city as an Epiphenomenon of the transformations brought about by Augustus, refer to the recent Wallace-Hadrill 2007.
4 The cult of *Pax Augustae* is introduced during 13 BC when the decision is taken
to build the Ara Pacis. "Gli epiteti Σεβαστά e Augusta permettevano di qualificare
inequivocabilmente la pace come una divinità e ne delimitavano il campo d'azione
nell'ambito delle attività del princeps: la Pax era Augusta non tanto perché creata da
Augustus, ma in quanto inerente alle funzioni dell'imperatore ed ottenuta in virtù di
auspicia particolari mai concessi ad altro uomo" (Mastino, Ibba 2006).
5 See Part I, Chapter 3, pp. 40–46.
6 The *Princeps* himself refers to his legislative measure in *Res Gestae*. part. 8.5: *Legibus
novis me auctore latis multa exempla maiorum exolescentia iam ex nostro saeculo reduxi etipse
multarum rerum exaempla imitanda posteris tradidi.*
7 The expression *iustissimus auctor* is from Ovid. *Met.* 15.833–834: *Pace data terris animum ad civilia vertet/iura suum legesque feret iustissimus auctor.* The *Metamorphosis* are
from AD 2: a generation has passed since the end of the civil wars, and the world has
been completely transformed. That which in poetic fiction is only a prophecy is a
reality when, in AD 2, Ovid writes. The citation from Ovid returns in Mantovani
2008 (and now Mantovani 2015), p. 35. In this contribution (using ample compari-
sons of sources of a varying nature) there is a discussion of the nexus *iura et leges*,
which is found on an *aureus* of Octavian from 28 BC.
8 Many of the legislative measures introduced to restore the ancient *mores* of public
and private life of the citizens, the decorum of social behaviour, respect for the
family and faith in its capacity to be fundamental to the new order, can be indirectly associated with the category of reforms. On Augustus' legislation designed to
strengthen the family as the basic societal unit and to maintain moral standards, see
also Galinsky 1996, pp. 133–139, and Spagnuolo Vigorita 2010. On the progressivity of the implementation of the reforms, which according to Eck were conceived
by Augustus as a response to the emergencies and difficulties in a time span of four
decades, see Eck 1986, part. pp. 117 s.
9 During the second period "The Emperor had a better awareness of the distance that
separated him from the magistrates and the *res publica* and the stronger desire that his
position was amply recognised by every order of citizen" (Arangio-Ruiz 1938, p. 103).
10 For the text, refer to Riccobono 1945, pp. 142–151 no. 17. It was impossible to me
to read the monograph dedicated by F. Bertoldi, *La Lex Iulia iudiciorum privatorum*,
Giappichelli: Turin 2003. On the contents and the significance of the Augustan plan,
see Santalucia 1994, pp. 207–209 also n. 186, rich and abreast of the legislative activity.
11 The general picture is reconstructed with impeccable clarity by Pugliese 1982.
12 Cloud 1994, particularly pp. 505–530. "Le fait que l'on vienne à admettre, à fin
de la République, que des actes de violence perpétrés entre privés puissant nuire à
l'ensemble du corps social denote une prise de conscience de la valeur de la sûreté
collective " (Brélaz 2007, p. 229).

13 Paul. *Sent.* 5.23 8–9 and Crawford 1996 (ed.) II, pp. 749–753, n. 50. See also Cloud 1969 and Ferrary 1991, in particular pp. 420–423.

14 During Sulla's dictatorship, the *quaestiones de maiestate, de ambitu, de repetundis, de sicariis et veneficis* and *de falsis* undoubtedly existed (Santalucia 1994, pp. 197–198).

15 Santalucia 1994, p. 201, who also clearly illustrates the mechanism for the functioning of the criminal process (*iudicia publica*).

16 Cic. *Phil.* 1.9.21 (and followings): *Altera promulgata lex est, ut de vi et maiestatis damnati ad populum provocent, si velint.* In 44 BC Marcus Antonius had approved a law which conceded the *provocatio* to those condemned for *de vi* and *de maiestate.* Regarding the nature of *crimen maiestatis* see Gundel 1963, Levi 1969 and Ferrary 1983.

17 Dig. 48.4.1.1–2, Ulp. lib. VII de... *libro septimo de officio proconsulis: Maiestatis autem crimen illud est, quod adversus populum Romanum vel adversus securitatem eius committitur, quo tenetur is, cuius opera dolo malo consilio initum erit, quo obsides iniussu principis interciderent.* Riccobono 1945, pp. 156–160 no. 24: caput I. The translation of the text into English is that proposed by F. Del Vera, Quietis Publicae Perturbatio. Revolts in the political and legal treatises of the sixteenth and seventeenth centuries, in M. Griesse (ed.), *From Mutual Observation to Propaganda War*, Bielefeld 2014, p. 281 n. 28.

18 On the *lex Iulia maiestatis* see Cloud 1963; Pugliese 1982, p. 750 ff.; Santalucia 1998, p. 256 f. See also the bibliography cited *supra*, n. 15. On the implementation of the particular cases of *crimen maiestatis*, see also Russo Ruggeri 2006.

   On the *senatusconsultum de iniuriis et famosis libellis*, cf. Paul. *Sent.* 5.4.15 (*Qui carmen famosum in iniuriam alicuius vel alia quaelibet cantica, quo agnosci posset, composuerit, ex auctoritate amplissimi ordinis in insulam deportatur*); and Dig. 28.1.18.1, Ulp. lib. I ad Sab. (*Si quis ob carmen famosum damnetur, senatus consulto expressum est, ut intestabilis sit: ergo nec testamentum facere poterit nec ad testamentum adhiberi*). On Augustus and the *famosi libelli* of Cassius Severus, cf. Tac. *Ann.* 1.72.3: *Primus Augustus cognitionem de famosis libellis specie legis eius tractavit, commotus Cassii Severi libidine, qua viros feminasque inlustris procacibus scriptis diffamaverat.*

19 The period of time usually accepted by scholars is between 19 and 16 BC: one of the first to establish this was Coroi 1915. Dubious in regard is Santalucia 1998, p. 198 n. 37. For the sources, FIRA I 1909, pp. 111–112 no. 20; Riccobono 1945, pp. 129–140 no. 15: Crawford 1996 (ed.) II pp. 789–792 no. 62. A systematic study is available in Cloud 1988 and 1989.

20 A large part of which can be reconstructed thanks to Digest 48.6.7 and Pauli *Sent.* 5.26.1.

21 Cloud 1988; of a differing opinion is Arangio-Ruiz 1938.

22 According to Arangio-Ruiz 1938, p. 110: "An attack against the guarantee of liberty or against the functioning of the public powers, which encompasses abuses of authority on behalf of the magistrates ... such as the obstacles placed in the way of their activity, or the acts which determine a danger against the security of the state (appearing armed in public, collecting arms, preparing a sedition)."

23 The *lex Plautia de vi* of the post-Sullan age (whose date is now debated: see C. Renda, *La lex Plautia de vi: problem e ipotesi di recerca*, in Index 36, 2008, pp. 491–504, who proposes a chronology between 70 and 63) and a similar *lex* of the Cesarean age.

24 In particular, Dig. 48.6.1: *Qui quia arma tela domi suae agrove inve villa praeter usum venationis vel itineris vel navigationis coegerit*; 48.6.9: *Armatos non utique eos intellegere debemus, qui tela habuerunt, sed etiam quid aliud nocere potest*; and all the Chapter 48.6.11. Dig. 48.7.2: *In eadem causa sunt, qui pessimo exemplo convocatu seditione villas expugnaverint et cum telis et armis bona rapuerint*; and 6: *Eadem lege tenetur, qui hominibus armatis possessorem domo agrove suo aut navi sua deiecerit expugnaverit.*

25 Dig. 48.6.7: Dig. 48.6.7: *Qui, cum imperium potestatemve haberet, civem Romanum adver-*
sus provocationem necaverit verberaverit iusseritve quid fieri aut quid in collum iniecerit, ut
torqueatur.
Not acceptable, in this framework, is the opinion of C. Russo Ruggeri (2002 and
www.dirittoestoria.it/iusantiquum/articles/Russo-Ruggeri-Prove-per-Tormenta-
Leges-de-Maiestate.htm) that the most ample configuration assumed by the crime
against the *maiestas* gave space to recourse to the torture of *ingenui* in the first century AD. The A. states: "Torture is simply not used *contra legem*, but *secundum legem*:
indeed, more precisely, *secundum leges Iulias maiestatis et vis publica.*"

26 "In terms of the *vis publica*, are grouped together acts which, with the common
theme of violence (real or threatened), attack the regular and orderly carrying out of
the functions of the state, while at the same time not including the details of another
specific crime (e.g. *maiestas*)" (Pugliese 1939, p. 55).

27 Riccobono 1945, pp. 160–161 no. 25; De Robertis I, 1974, pp. 195–237; Id. 1987,
pp. 3–13 and 425–435 (both with previous bibliography).

28 Suet. Iul. 42: *Cuncta collegia praeter antiquitus constituta distraxit. Poenas facinorum auxit.*

29 Suet. Aug 32: *Pleraque pessimi exempli in perniciem publicam aut ex consuetudine licen-*
tiaque bellorum civilium duraverant aut per pacem etiam exstiterant: nam et grassatorum
plurimi palam se ferebant succincti ferro, quasi tuendi sui causa, et rapti per agros viatores sine
discrimine liberi servique ergastulis possessorum supprimebantur, et plurimae factiones titulo
collegii novi ad nullius non facinoris societatem coibant. Igitur grassaturas dispositis per oppor-
tuna loca stationibus inhibuit, ergastula recognovit, collegia praeter antiqua et legitima dissolvit.
See further, note 75 and Part IV, Introduction.

30 Jerzy Linderski (Linderski 1995) carried out an ample close examination of the two
passages by Suetonius in an attempt to better understand their meaning; in note 1
on p. 217 the previous rich bibliography is recalled.

31 Dig. 3.4.1, pr., Gai lib. III *ad edictum provinciale: Neque societas neque collegium neque*
huiusmodi corpus passim omnibus habere conceditur: nam et legibus et senatus consultis et
principalibus constitutionibus ea res coercetur. For the fundamental repercussions in terms
of organisation and detailed control of the territory and the commitment of personnel, both civilian and military in obtaining it, see *infra*, Part II, Chapter 4.

32 Riccobono 1945 p. 266 no. 60 = Dig. 47.22.1.1 (*Senatusconsultum de collegiis*). Vd.
CIL,VI 2193 cf. pp. 3304, 3826 = *CIL*,VI 4416, cf. p. 3416 = *ILS* 4966 = *FIRA*², III,
p. 111, no. 38: *Dis Manibus / collegio symphonia/corum qui sacris publi/cis praest[o] sunt*
quibus / senatus c(oire) c(onvocari) c(ogi) permisit e / lege Iulia ex auctoritate / Aug(usti)
ludorum causa. Cf. Manacorda 1999, pp. 254, 256, 261 no. 1. See also CIL, XIV 2112
I, 10–13 = ILS 7212.

33 Dig. 48.12.12 pr.; Riccobono 1945 pp. 200 s. no. 32: *Lege Iulia de annona poena sta-*
tuitur adversus eum qui contra annonam fecerit societatem coierit, qua annona carior fiat.

34 For the Augustan *leges* on the taking away or the appropriation of debts of money
or goods belonging to the state or destined for public worship; and against the
crime of electoral fraud, in my opinion, not relevant to relevant to the subject of
this research. I refer the reader to the sources and to the essential bibliography of
reference. On the *lex Iulia de peculatu*, Dig. 48.13; Coll. 9.28; Riccobono 1945,
pp. 161–165 no. 26 and the research of F. Gnoli from the 1970s. On the *lex Iulia de*
ambitu, of 18 BC I refer to Riccobono 1945, pp. 140–142 no. 16.

35 And the *lex Iulia de adulteriis coercendi* is the most significant example (Dig. 48.5;
Riccobono 1945 pp. 112–128 no. 14; Crawford 1996 (ed.) II, pp. 781–786 no. 60).

36 Even though "the introduction of the cognitio process did not constitute the realisation
of an informed political design, neither on behalf of Augustus nor by him immediate successors … the various cognitiones … take single concrete interventions

by the Emperor as a basis, which in practice end up being institutionalised" (Palazzolo 2015, p. 223).

37 Pugliese 1982, pp. 735–736. About the *cognito extra ordinem*, see the ten-year studies of Orestano, Luzzatto, Kaser and Buti, all amply remembered recently by Palazzo 2015.

38 Suetonius on several occasions underlines his *clementia* and *civilitas* in the administering law and order (*Aug.* 33.1–2 and 51.1–2. See also the words of Ovid, *Tristia* 2.131–132 (the latter in Nugent 1990).

39 The Edicts of Cyrene, all dated between 7 and 6 BC were published by Olivero in 1927. The second edict (Sherk 1970 pp. 174–175, was studied, among others, by Oliver 1949, Mantovani 2011 and more recently, Peachin 2015. We know from Pliny the Elder (*Epist.* 65.3; 79; 81) about analogous Augustan edicts regard Asia and Bithynia. Regarding Augustus' intervention in a private controversy that could have brought about problems of public order, refer to the episode recalled in Part I, Chapter 3 p. 41.

40 Cass. Dio 51.19.7. The bulk of reflections are considerable, starting with *Römisches Staatsrecht* of Mommsen. See also the review by Peachin 2015, pp. 499–511.

41 "Il potere metagiuridico e meta costituzionale che assicurava al principe una sorte di controllo su tutta la res publica." The definition is from Lovato, Puliatti and Solidoro Maruotti 2014, p. 121. See also Pugliese 1982 and Santalucia 1994, pp. 211–212. Regarding the *auctoritas* as 'staatsrechtlicher Begriff' see, for example, Volkmann 1935, p. 218; Santalucia 1999 p. 264. As a differently stated concept, depending on the different chosen forms of communication (literature, official art, coinage), see Galinsky 1996, pp. 10–41. As for the Senate, for the whole Augustan period, on *crimen maiestatis et repetundarum*, proceeded roughly using the rules in place for normal procedure.

42 Kelly 1957, pp. 38–46.

43 Peachin 2015, p. 535.

44 See Part II, Chapters 4 and 5.

45 The sources concerning the accusations and the trial of Cornelius Gallus (in particular, Suetonius, Cassius Dio and Ammianus Marcellinus) and the modern literature are collected and amply discussed by Arcaria 2015, who has repeatedly analysed the nature of senatory *cognitio* (e.g. Arcaria 1992, 2006, 2007, 2009). See also the brief paragraph in Peachin 2015, p. 519.

46 Tac. *Ann.* 1.72.3. See Peachin 2015, pp. 529–530 with extensive bibliography in notes 102–104.

47 On the conspiracy of Murena and Caepio (23–22 BC) there are numerous sources: Cass. Dio 54.3.4–5; Vell. 2.91.2, 2.93.1; Sen. *De Brev.* 4.5 and *De Clem.* 1.9.6; Suet. *Aug.* 19.1, 56.7, 66.6; *Tib.* 8.1; Tac. *Ann.* 1.10.3; Macr. *Sat.* 1.11.21.

48 On the *senatusconsultum de Gneo Pisone patre*, see Caballos, Eck and Fernandez 1996. On the theme of the *maiestas Domus Augustae*, see Corbier 1999, 2001. The *maiestas sua et domusque suae* is also mentioned in Suet. *Aug.* 25.1.

49 In Tac. *Ann.* 4.34 his speech on the defence opens with: "*Verba mea, patres conscripti, arguuntur: adeo factorum innocens sum. Sed neque haec in principem aut principis parentem, quos lex maiestatis amplectitur.*" This celebrated episode, object of the analysis by Theodor Mommsen in the Strafrecht, is connected by Bauman 1967, p. 275 ss. to Dig. 48.4.3. On the trial of Cremutius Cordus, see Canfora 1993.

50 Moreau 2005. Another form (?) assumed by crimes against the *maiestas* could be the resorting to torture of free men, which can be dated back to the first century AD. The acquisition of evidence through torture, regardless of the condition or social stautus of the suspects, could have been the outcome of provisions dating back in time. For further analysis see Russo Ruggeri (cit. *supra*, n. 18).

51 On the figure of *miles* at the end of the republic see Shaw 1984, p. 27 ff., who distinguishes between the *miles* ('state man of violence') and the *latro* ('private man of violence').

52 Suet. *Aug.* 14: *Nam cum spectaculo ludorum gregarium militem in quattuordecim ordinibus sedentem excitari per apparitorem iussisset, rumore ab obtrectatoribus dilato quasi eundem mox et discruciatum necasset, minimum afuit, quin periret concursu et indignatione turbae militaris. Saluti fuit, quod qui desiderabatur repente comparuit incolumis ac sine iniuria.*

53 Suet. *Aug.* 16: *Unde praebitam Antonio materiam putem exprobrandi, ne rectis quidem oculis eum aspicere potuisse instructam aciem, verum supinum, caelum intuentem, stupidum cubuisse, nec prius surrexisse ac militibus in conspectum venisse quam a M(arco) Agrippa fugatae sint hostium naves.*

54 Momigliano 1938.

55 For the relationship between Augustus and his soldiers and its representation in public monuments (and on the allegorical images of *Victoria Augusti*), just a few updating titles, in addition to the literature cited by Momigliano: Raaflaub 1987; Fears 1981 (on the Augustan developments in part. p. 804 ff.), Keppie 1984b; Hickson 1991; Rich 2003, p. 331; Keaveney 2007.

56 Suet. *Aug.* 49; *Res Gestae*, part. 16–17 and 30; Cass. Dio, in particular 52.27.

57 Suet. *Aug.* 24.

58 Suet. *Aug.* 25.

59 Suet, *Aug.* 38.

60 Suet, *Aug.* 44.

61 See Part II, Chapters 4 and 5 and Part III, Chapter 6.

62 According to Suetonius (*Aug.* 43) guards (*custodes*) were distributed around the city to contain the looting of goods from houses that were left unattended.

63 Cass. Dio 54.25.

64 The provisions for wills respect the same hierarchy of years of service and wages: 1,000 sesterces for each praetorian, 500 for each *miles urbanus*, 300 for each legionnaire (Suet. *Aug.* 101).

65 Suet. *Aug.* 49; *Res Gestae* 17 and Cass. Dio 55.25.2.

66 The creation of the three prefects (praetorian, urban and that for the *vigiles*) must be placed within the plans for "structural reforms", not therefore purely military: the first (Cass. Dio 55.10.10) was created in 2 BC. For the definitive arrangement of the second, the sources, since the role was with ancient origins, are not clear. However, the prevalent thinking (Vitucci 1963, Freis 1967 and, last, Nippel 1995, p. 92 and Keppie 1996 p. 110, footnote 81) assigns its formation to the end of the reign of Augustus; the prefect for the *vigiles* is born together with the cohorts in AD 6 (Cass. Dio 55.26).

Cosme 2012 p. 181 does not consider as a reform in the strict sense of the word the discussed interdiction of marriage for soldiers, which, among other things, but only in a hypothetical way, can be dated to the Augustan age. See the passage by Cass. Dio 60.24, relative to the reign of Claudius.

67 The idea of a long period of action, which goes from the demobbing post-Actium to the creation of the *vicesima hereditatium* to feed the *aerarium militare*, is of Fergus Millar (1964); recently it has been reaffirmed by Pierre Cosme (Cosme 2012). The literature on the Augustan reforms is abundant: I simply want to point out Eck 2009, Speidel 2009, p. 19 ff., both cited in the recent contribution of Eck 2016.

68 Tac. *Ann.* 1, respectively from pgf. 16 to pgf. 31.

69 The chapter by Appian *Bell. Civ.* 5.132 discusses Octavian 36 BC. A brief mention in Strabo 4.6.6 (C 204): "For in addition to his putting down the brigands Augustus Caesar built up the roads as much as he possibly could." On the systematic action

of control within the territory of Italy, Shaw (1984) observes the short duration the measures of the *Princeps* had, in terms of security (Tiberius within just a few decades had to take more numerous and rigid intervention). On the other hand, the same author underlines pointedly the double effect those measures had on political and ideological plans: Augustus, in this way, managed to present himself as the legitimate guardian of legality.

70 See Part II, Chapter 5, this volume. The conclusive observation of Appian seems to be an ill judged attempt to find a precedent for an institution that will be established forty years later. Sablayrolles 1996 does not comment on the passage, of which, perhaps, he is not aware.

71 Cass. Dio 49.43.5. On the policing operations of Agrippa, see Roddaz 1984, pp. 154 and f. who interprets such operations as a defence of traditional Roman thinking against the favour that some practices from the Orient enjoyed.

72 Cass. Dio 53.2.4

73 The passages are respectively Dio 54.6.6 and Tac. *Ann.* 2.85.

74 See the work of Rivière 2009 (which is inserted into a series of studies dedicated to exile and public order). The underlying idea is that the expulsion, and marginalization were carried out, in particular, against certain categories of people such as rhetoricians, philosophers, astrologists and the Jews as "were the most common means used by security forces to combat the troublemakers" (p. 56). For similar imperial interventions during the first century AD and for other interpretation of the phenomenon, see Part III Introduction.

75 Suet. *Aug.* 32: *Grassaturas dispositis per opportuna loca stationibus inhibuit, ergastula recognovit, collegia praeter antiqua et legitima dissoluit.*

76 Shaw 1984, p. 33

77 *Supra*, p. 76.

78 The state of the *stationes* at the gates of Rome and in the territory of Italy, will be discussed in parts III and IV.

79 The expression, which seems to refer to an unitary corps of soldiers, is rejected by Sablayrolles (2001, pp. 126–127) and by Rivière 2004 (p. 66) who also contests the idea that the concentration of military personnel could only be justified by "the management of the urban and human space".

80 See Part II, Chapter 4.

81 See Part II, Chapter 5.

82 See Part II, Chapter 4, p. 97 ff.

83 See Part II and Part III.

84 Perhaps the image of the prefects as an "energetic hinge (at times fiercely so) of the new constitution in his contrasting the different manifestations of the political opposition and common criminality" proposed by Cascione is a little exaggerated 2016, p. 190.

85 See Part III Introduction and Part IV.

4 The security of the *Princeps* in Rome

Military escorts and bodyguards

It is commonly understood by scholars that the praetorians were not a creation of Augustus. All the most important *imperatores* of the late Republican age had a special squad (sometimes a few hundred, sometimes some thousand men) called 'praetoriani', which was conceived and functioned as a mix of honour and bodyguard.[1] It was formed with the closest and most trustworthy soldiers; they probably also had requirements of strength and endurance, but we do not know the criteria by which the selection took place.

From 42 BC, Octavian and Antonius, had praetorian cohorts, which were fully functioning militarily.[2] Certainly this group of soldiers in the service of the *imperatores* was a precedent for the Augustan solution.[3] According to some scholars, the keeping of the praetorians in the *Urbs* is legitimated by the *imperium proconsulare* of 27 BC, thanks to which the provincial governors could have available, alongside a *consilium* of officers, a contingent of soldiers chosen to carry out personal protection duties. There was, however, a fair difference between the Republican bodyguards and the Praetorian Guard of Augustus. The continuity with is the idea of the need for protection of the person of Augustus as a magistrate with *imperium*, "in line with their Republican predecessors"[4]; the discontinuity is in the distinction between physical integrity and the image of power that Augustus (and his family) represented.

To prove this statement, I think the best way is to investigate the characteristics of this unit in the Augustan era, uncover where the praetorians lived[5] and, first of all, understand the functions they had, in strict connection with the prerogative that allowed Augustus to require the establishment of cohorts of soldiers at his disposal and at the disposal of the city. Some of these questions in recent decades have appropriately been emphasized,[6] and in addition to a discussion over the number and strength of the cohorts, have been analysed since the first decades of the twentieth century.[7]

In the chapter of the *Life of Augustus*, which reviews the emperor's behaviour with the survivors of the armies at the end of the civil wars, the new structure given to the troops in the provinces and the creation of military fleets, Suetonius states that the *Princeps* chooses some *milites in sui custodiam* (as well as others *in custodiam Urbis*), having dismissed the handful of *Calagurritani*, who had been in his service until Antonius' defeat, and that of the Germans, who until the *clades Variana* had been by his side.

It is not difficult to connect on the one hand, the *Calagurritani* to the *manus praetorianorum* at the disposal of the triumviri; and on the other hand the Germans with the future *corporis custodes*, which will be discussed shortly. It is difficult to chronologically frame the operation – which was probably executed in as early as 27, which is what Cassius Dio[8] appears to certify – but also to quantify the number of troops involved. The 'new praetorians' alternated turns so as not to leave the emperor or his residence unguarded. In order that their presence be accepted by the inhabitants of the capital, who were survivors of the last actions of the period of civil war and of the worsening situation of the previous ten years, Augustus did not allow the new cohorts to reside in the capital, deciding that the soldiers should be housed on the outskirts of the city.[9] Indeed, we know that their definitive installation at the gates of Rome was organized during the time of Tiberius, more than thirty years distant, when the nature of the regime had been largely unveiled.

Their salary was immediately different to that of the other soldiers. The creation alone of the trusted personal guard (*akribé phrourà*) and the expectation of a salary for it are signs, for Cassius Dio, of the plan to install a monarchical regime.[10] The instrumental reading of the measure by the Bithynian historian is naturally tied to the fact that during his time, after the Severan reforms, the praetorian cohorts, in addition to having been systematically strengthened and having changed composition and role, were not an insignificant item of spending for state funds.[11] These bare details do not provide further information on how and when the salaries for the praetorians were paid.[12]

Rather than when (of the creation) and how much (of the *stipendia*, *praemia* and so on), what interests this chapter is the profound sense of a completely new asset, which in my opinion, can be reconstructed through the functions assigned to the praetorians. My idea is that these functions were not defined from the beginning as a single block, but were articulated with time, in parallel to the legislative action of the emperors and the steady organization of the discourse on security which characterized the first century AD.[13]

In its initial phase, the organization of the praetorian cohorts was positioned between the *leges Iuliae*, which regarded public order (*lex Iulia maiestatis*, *de vi* and *de coetu*, etc.) and the moment in 6 BC when, as we have just seen, the control of the *Res Publica* on behalf of Augustus expanded and the image of his person, perceived as guarantor of security and community wellbeing, became progressively associated with the *Res Publica* itself.

It is not easy to follow the process of gradual enrichment of the functions of the praetorians. Clear references in the literary sources are missing and the archeological monuments of the praetorians are not datable with precision. Nevertheless, a careful examination can delineate the commitment of the praetorians during the first decades of the imperial age to a few, essential functions, which would progressively increase. These 'primigenial' functions were tied to the persona of the *Princeps*, to his family and to the protection of the *maiestas;* and these functions would be retained until the end of their history, even when, as we shall see, the more combatant character of the praetorians would be accentuated.

The primary role was that of guarding the residence of the *Princeps* and to escort him personally[14] in both public and official contexts. Cassius Dio, referring to AD 27 and explaining the etymology of the Palatium, talks about the residence of Augustus being near to his general quarters (*stratégion*).[15] It is impossible to determine much of what Dio intended to say precisely with this expression, if when referring to 'general quarters'; he is not referring to a reality to which he is witness, but rather to the beginnings of the Principate. However, it is certain that already in the time of Augustus, there was a steady contingent of praetorians (one or two cohorts) who took it in turns to guarantee the protection of Augustus and his residence.[16] Of this permanent presence and this concern for the protection of the emperor we have numerous confirmations: the slightest sign of a threat to his person would alert his defence team, who would jump to his aid.[17]

Equally well documented is the function of escort to the emperor in Rome[18] and outside the city.[19] The presence of the praetorians at the head of the cortège that accompanied the emperor in his movements was a sign and itself representation of imperial power.[20] It is not difficult to imagine that this second function came about at the same time as the first, of which, we may say, it was a direct consequence: at least a few soldiers who were by the Palatium accompanied the emperor outside. This duty also helps to explain why there were praetorians at the funeral of an emperor and of members of his family, and why they accompanied the remains of emperors who died elsewhere to Rome.[21]

Further duties were carried out by the praetorians occasionally and/or, starting with the late phase of the Principate of Augustus; other duties have, in my opinion, been inappropriately considered as specific to the praetorians.

There is no mention of the duties relating to the suppression of disturbances in Rome (as well as in Italy) before the age of Tiberius. A passage of Suetonius describes the preparation of troops in the city following the *clades Variana*

> He suffered but two severe and ignominious defeats, that of Lollius and Varus, both of which were in Germany. Of these the former was more humiliating than serious, but the latter was almost fatal, since three legions were cut to pieces with their general, his lieutenants, and all the auxiliaries. When the news of this came, he ordered that watch be kept by night throughout the city, to prevent any outbreak, and he prolonged the terms of the governors of the provinces, that the allies might be held to their allegiance by experienced men with whom they were acquainted.[22]

The measure clearly has an air of contingency and the use of the word *excubiae* does not necessarily mean that Suetonius intended to refer to praetorians rather than to *vigiles* or *milites urbani*, who more appropriately would expect to be tasked with the duties of surveillance and maintaining order on the streets.[23]

An indirect confirmation of the fact that, during the age of Augustus, the image of armed praetorians on the streets was not usual, can be found in the oldest example of a stela with a portrait of a veteran armed praetorian: it cannot be dated to earlier than AD 23, which corresponds with the height of the age of Tiberius; and moreover was not found in Rome but in Italy.[24]

In contrast, tensions were much greater under Nero, in particular in 63, as a consequence of the position taken by Trasea Paetus and his refusal to present himself to the senate:

> On the following morning, however, two praetorian cohorts in full equipment occupied the temple of Venus Genetrix; a body of men wearing the toga, but with sword unconcealed, had beset the approach to the senate; and companies of soldiers were scattered through the fora and basilicae. Under their menaces the senators entered their meeting-place, and listened to the emperor's speech, as read by the quaestor.[25]

The soldiers, it is said, were distributed throughout the fabric of the city: at the Temple of Venus Genetrix, at the entrance of the *Curia senatus*, inside the Fora and the basilicas. Properly armed praetorians (*duae praetoriae cohorts armatae*) are explicitly mentioned only in the first part of the passage. However, other soldiers are mentioned, even if they are not clearly identified (*globus togatorum* and *cunei militares*), perhaps belonged to the same troops or to the centuries of *speculatores*.

The tasks so referred to – the protection of the *Princeps* and his family inside their residence and during their movements around the city; and the prevention of uprisings – fall into the more general tasks for the protection of the *maiestas* of the sovereign. The slow modification of the profile of the praetorians in the course of the Julio-Claudian age, compared to what happened in the first years of the Augustan Principate, is a natural consequence of the progressive acquisition on behalf of the emperors of the role of guarantor of the security of the Roman State, which can be identified with them.

The role of bodyguard or personal protector did not therefore form part of the role of the praetorian cohorts, despite what many have said in regard to this.[26] This additional task would have been a superfluous duplicate of the duties of other military and civilian personnel. However, as we shall see, the principle reason lies in the fact that the role of bodyguard did not regard the *maiestas Principis*, but rather the physical protection of the man Augustus. It is no coincidence that all the examples proposed by scholars who support such a hypothesis[27] are after the Augustan age, and have, for the most part, their explanation in the function as military escort and guard of the palace, to which we have just referred.

The most cited passages are all taken from the period between the age of Nero and the years of civil war. The first of them refers to the moment of the acclamation of Nero, after the death of Claudius: "At last, at midday, on the thirteenth of October, the palace gates swung suddenly open, and Nero, with Burrus in attendance, passed out to the cohort, always on guard in conformity with the rules of service."[28] In this passage, there is no mention of bodyguards; the cohort is in fact posted outside (*Nero egreditur*) rather than inside the palace.[29] A comparable situation is presented in the *Historiae*, where reference is made to a cohort who "carried out the guarding of the Palatium"[30]; and in the *Life of Ortho* by Suetonius, where a "cohort which in that moment carried out the duty of guarding" is spoken of.[31]

Equally certain, in my view, is that among the tasks of the praetorians, there was not the function of political police.[32] If the existence of the *milites frumentarii*, to whom this duty would fall, it is not attested to before the second century,[33] in the first century there were other soldiers to whom this duty could be entrusted.[34]

Other tasks of an administrative and operative nature – all of which mentioned by the sources for a period beyond the Augustan age – are sometimes considered as characteristic of the praetorian cohorts. These duties are reviewed and commented upon by Sandra Bingham in her volume dedicated to the examination of aspects of the praetorian guard, which fell outside of their mandate, but was fundamental in order to guarantee the protection of the imperial household.[35] She highlights: "The Praetorian Guard of the Roman Empire developed into a multifaceted unit that not only looked after the personal safety of the emperor but also participated in the care of the state."[36] I believe, however, that some of the duties were entrusted to the praetorians only occasionally, in connection with their two primary functions (protection of the residence, escort and representation of the emperor).[37]

The first of these duties would have been to intervene in case of fire. According to Bingham, the praetorians who were called to intervene were those who, at the beginning of their careers, had been *vigiles*. However, this path does lead us to appreciable results, as Bingham herself recognizes.[38] In the same way, none of the episodes mentioned can be decisive in the demonstration of the task.[39]

To give some examples, on the occasion of the fire that started in Rome under Tiberius, Livia invited *populum et milites* to find solutions and Bingham speculates that "She [Livia] apparently has done the same sort of thing under Augustus."[40] In AD 15, the Tiberius's son Drusus went to the scene of the disaster accompanied by praetorians[41] and Bingham observes (emphasis added): "The aid of the praetorians is not commented on, but *it's probable* that they took part in the fight-firing process, rather than simply acting as protection to the emperor's son."[42]

To such episodes that occurred between the Tiberian and the Neronian ages it would be possible to add others; but without generous recourse to *argumenta ex silentio*, none of them is conclusive to the resolution of the matter. We can completely exclude that praetorians were ever involved in (fire) prevention. In addition, their presence was requested during emergencies, in case of particularly serious and expansive fires, but just as the intervention of all the other military, paramilitary and civilian corps present in the city was requested.

Another function considered by Bingham to be typical of the tasks of the praetorians was the maintenance of order in places where there was entertainment. To such activity we will return later, in relation to "Dangerous places".[43] For now it is enough to say that this activity was carried out by both the praetorians and the soldiers from the urban cohorts. The tasks of the praetorians, in such circumstances, were not far removed from their normal duties: the escort of the emperor who was present at the entertainment and who assisted at the event; the representation of the imperial power; and the protection of his *maiestas*. What can be excluded, as we shall see, is the presence of the praetorians in an official capacity at an entertainment, if the emperor or a member of his family was not participating in the event.

More careful consideration is required of the matter of the involvement of the praetorians in theatres of war. I believe that this more truly military role has changed over time.

The first account of participation in a military campaign is from 2 BC when, according to Cassius Dio, a tribune of praetorians was sent to aid Roman troops in the East. The measure appears to have impromptu character, besides the fact that Cassius speaks of an officer, but not of the cohorts in tow.[44]

There are two further reports of the presence of the praetorians in the provinces at the side of members of the imperial family at the end of the Augustus' Principate, or after: two cohorts and the prefect Seianus accompanied Drusus to Pannonia in AD 14[45]; two cohorts were with Germanicus on the Rhine in AD 16.[46] In neither case, it would appear to me, can we talk about a systematic participation of the praetorian cohorts in war.[47] However, a genuine combative function would appear to have been registered at the beginning of the reign of Claudius.[48] It is probable therefore, in line with what was imagined in his time by Lawrence Keppie, that the accompanying of the emperor or a member of his family to the battle front gradually lost its value as escort and instead accentuated the combative one.[49]

A widespread belief – as an inevitable outcome of the ancient authors' intent to return the negative image of the government of some emperors – is that the praetorians were the armed wing of the emperor. Tales of conspiracy in progress, or merely planned, and of related arrests had, as their protagonists, members of the imperial family or dissident senators who were are often assassinated, driven to suicide or exiled. The arrest of the senators or equestrians accused of conspiracy against the sovereign, fell within the duty to protect the *maiestas* of the emperor; and the examples, once again, are all from the Julio-Claudian age, but after the reign of Augustus.[50] In the passages of the authors (and in particular Tacitus) it is therefore difficult to identify who were the 'guards' referred to in generic terms such as *milites*, *excubiae* or *custodes*.

Sandra Bingham proposes a slightly different picture. In her opinion, the involvement of the praetorians in the arrest of members of the senatorian elite was a response to a (tacit) subdivision of their duties between the praetorians and the *milites urbani*. On the one hand, she believes that it would have been the task of the praetorians to prevent and oppose any kind of threat to public order posed by members of the high classes.[51] On the other hand, the urban cohorts would have been responsible for resolving similar problems brought about by members of lower social groups.

Apart from the fact that there are no accounts of such a division of tasks in the sources, other reasons lead me to exclude this appealing explanation.

The first substantial reason, in my opinion, brings chronology into play. As Keppie has convincingly demonstrated, until the last years of the reign of Augustus, there is no substantial distinction between praetorian and urban cohorts.[52] During this long phase, the twelve cohorts perform, as a whole, the task of protecting the *maiestas rei Publicae et populi Romani*, in which the emperor participates[53]; in other words, they act as guarantors to the *Securitas rei Publicae populique Romani* and that of the *princeps*.[54] Only following the late Augustan age and later with the advent of the successor and his new *pro securitate* measures (first of all was the construction of the *castra praetoria*), the *maiestas Principis* and the *populi Romani*, are entrusted to cohorts distinct in substance.

The memory of the strong tie between the praetorians and the *milites urbani* is clearly reflected in the episode of the deprivation of power from Seianus. At the moment in which it is decided to proceed with the arrest of the prefect of the praetorians, the *corporis custodes* and the *vigiles* are called in order to prevent or contain public turbulence and in fear of the reaction of the praetorians in defence of their Prefect. The *custodes* and the vigiles were in fact considered neutral and, unlike the *milites urbani*, without strong ties to the praetorians.[55] The year was AD 31, almost at the end of the era of Tiberius and the perceived closeness between the praetorian and urban cohorts (which also meant cohabitation in the same encampment) was still clearly perceived. It is therefore difficult to think that a net division of their duties could have been enacted by Augustus, as Bingham proposes.

Another reason better explains the apparent different 'social action' of the praetorians and the *milites urbani*. The arrest and punishment of members of the senatory elite was entrusted to the praetorians because their main duty was the protection of the security of the emperor and his family and not that of the *Urbs* and its inhabitants. The protection of the *Urbs* was the task of soldiers, who were commanded by the most important of the senators of Rome, the urban prefect. In the accounts of the pro-senator historians, the episodes that regard the accusations and the persecutions of senators receive, in contrast to the tasks of the soldiers of the urban cohorts, a more notable significance, thus providing a glaring example of the limitation of the *libertas* in the Julio-Claudian age as a direct consequence of the Caesarian and then the Augustan *lex maiestatis*.

In addition, when the praetorians found themselves outside the *Urbs* in order to contain the threats of uprisings or to quell them, the repression extended to all social strata, as Bingham herself must recognize.[56] The separation of the duties of the cohorts of Rome (praetorians and *milites urbani*) cannot be explained therefore with the responsibility assigned to them based on different strata of the urban society, but rather with the different ratio behind their creation and the different spheres of action of the prefects at their command.

Particularly close to the person *Princeps* were the *speculatores*. There are few accounts in our possession of this unit, nearly all of which are Augustan. However, like the praetorians, the *speculatores* already existed during the Republican ages, with Antonius and Caesar.[57]

Leaving to one side the hypothesis expressed recently that this unit constituted a sort of praetorians on horseback,[58] more valid, rather, is the idea of Otto Hirschfield, who viewed them as a group of people who carried out the role of informer for the emperor. The German scholar, who first makes a systematic collection of the evidence, begins his analysis with a passage from the *Life of Augustus* by Suetonius:

> While he was triumvir, Augustus incurred general detestation by many of his acts. For example, when he was addressing the soldiers (milites) and a throng of civilians (turba paganorum) had been admitted to the assembly, noticing that Pinarius, a Roman knight, was taking notes, he ordered that he be stabbed on the spot, thinking him an eavesdropper and a spy (curiosum ac speculatorem).[59]

Hirschhfield's idea is that, in this passage, the *curiosus* and the *speculator* was also placed in alternation (such as the *turba paganorum* and the *milites*): the *speculator* was a soldier in plain clothes set against the *curiosus*, a civilian spy.[60] In my view the doubt whether the expression *curiosus ac speculator* was used in this case as hendiadys remains; and therefore Suetonius had wanted to refer to a spy in the generic sense, also because the *speculatores* constituted a kind of special unit[61] and for the emperor they were a guarantee of safety and not a cause for fear. Indeed, it is always the *speculatores* who are remembered as being by his side, at the palace and during his movements: with Claudius, they participated in banquets[62]; and they accompanied Galba, when the emperor moved to the city.[63]

Another matter to consider is their placement in the military. Not an insignificant number of sources lead us to believe that this unit, not very numerous and carefully selected, referred to the praetorian cohorts. The problem is deciding when this began.[64]

A lot of the clues are linguistic: in the texts of the inscriptions, the *speculatores* are placed within the praetorian cohorts[65]; we find the expression *speculator Caesaris/Augusti*[66] or similar that link them to the emperor "who so expresses pride of command, not of property".[67] Suetonius states: "Augustus himself writes that he once entertained a man at whose villa he used to stop, who had been one of his bodyguards (speculator)."[68] In Pompei, next to the epitaphs of the praetorians who died there in service during the Julio-Claudian age, there are those of the *speculatores*.[69]

However, the most convincing evidence refers to the civil wars of 68/69 and the reign of Vespasianus. In a passage of the *Historiae*, there is a mention of Otho being accompanied by the selected units (*lecta corpora*) of *speculatores* and by "the rest of the praetorian cohorts"[70]; while in a military diploma from 76 BC "the names of *speculatores* who served in *praetorio meo* and of soldiers who entered into the nine praetorians cohorts and in the three urban cohorts"[71] are mentioned.

For this chosen unit (*corpus lectum* in Tacitus) it would appear reasonable to think of a precedent, inherited by Augustus from the triumviral era, which survived with a particular status or perhaps additional privileges, alongside the praetorian cohorts. Their public image became ever more unpopular because of their concealed role as informers and for the position they assumed during the events of late Claudian age and during the crises of 68/69.[72] With the advent of the new dynasty it is plausible that, as had happened with the *corporis custodes*, the *speculatores* were the object of a reform that would bring about their suppression.[73]

In the picture that Suetonius draws in Chapter 49 of his *Life of Augustus* of the distribution of troops in the province, of the creation of the military fleets and of the urban troops, the civilian personnel responsible for the safeguarding of the person *Princeps* is understandably absent.

However, we know that next to the praetorians and the *speculatores*, who carried out a representative function, there were also private bodyguards of peregrine or servile status, whom the emperor paid for with his own income: they were the *corporis custodes*.[74] In a few rare cases, they carried out military action, but for everyday duties and the manner of the service, they were conceived as personnel for the private security of the emperor.

After Augustus dismissed the Calagurritani, who he had kept for his protection until the defeat of Antonius (presumably for the decade between 42 and 31 BC), a new unit of bodyguards was recruited from the Germanies, and they were in service from AD 31 to AD 9, only to be disbanded following the *clades Variana*.[75]

Therefore, the last years of the reign of Augustus are left exposed.[76] A passage from Tacitus leads us to believe that, immediately after his death, the successors and the *domus Augusta* had begun to favour once more the service of the Germani:

> In spite of his secretiveness, always deepest when the news was blackest, Tiberius was driven by the reports from Pannonia to send out his son Drusus, with a staff of nobles and two praetorian cohorts . . . drafts of picked men raised the cohorts to abnormal strength. In addition, a large part of the praetorian horse was included, as well as the flower of the German troops who at that time formed the imperial bodyguard.[77]

In fact, according to Tacitus, among the forces employed by Germanicus to repress the revolt of the legions in Pannonia in AD 14 appear to have been selected Germanic guards. The same might have happened with the deployment of troops against the Cherusci.[78]

We also know of the presence of *corporis custodes* at the sides of Julio-Claudian principes until the age of Nero: Batavian bodyguards (*numerus Batavorum*) participated in the performance organized by Caligola[79]; while Nero took away from Agrippina his military escort and the *statio Germanorum*, to deprive her of every external sign of honour and *potestas*.[80]

I believe therefore, in contrast to Heinz Bellen, that "the guard in charge of the defense of Nero", which was removed in AD 68 by senate when attempting to eliminate the tyrant,[81] was the praetorian escort: Cassius Dio used the same expression (*phrourà peri ton Nerona*), when discussing the creation of the praetorian cohorts[82]; and the elimination of the escort preceded the entrance by the senators in the *castra praetoria*, to declare Nero public enemy and Galba the new emperor.

The definitive dissolution of the *custodes Germani* seems to date back to AD 69[83]; the reason was their support for Dolabella. Whatever the successive destiny of the Germans was,[84] it is a fact that, by the Flavian age, there are no more accounts of military *corporis custodes*.

The character of this unit is to be interpreted as a private force at the service of the Princep, based on the model of sovereigns of the Hellenistic monarchy and of the reign of Mauretania (Ptolemy and Juba II). In this way, they could be dismissed as wished and their tasks were limited, to recall the words of Tacitus, *ad custodiam corporis*, excluding, in contrast to the praetorians (and much of the *speculatores*), the representation and escort duties.

For this reason it is not easy to accept Bellen's hypothesis[85] that their organizational model, from the very beginning and later, was that of the decuries of praetorian *equites*. An acceptable idea is rather that their organization (probably as a consequence of their favourable and influential position in relation to the emperor) changed over time, as seems to be confirmed by the evidence.

Their denomination as a troop changed from Augustus to Nero: if they were called *manus* with Augustus, they became *numerus* with Caligula and *cohors* with Nero and Galba.[86] Finally, in an epitaph from the late Nero age, the *Germanus Nobilis*, defines himself as a soldier of Nero (*miles imperatoris nostri Neronis*).[87] It would appear therefore that those assigned to the person of the *Princeps*, with Caligula and even more so with Nero, assumed the duties of and behaved like the praetorians or the *speculatores*, who appropriately were given the task of protecting the *securitas Augusti* as an expansion of the *securitas Populi*. Their progressive militarization, tied to the strong link to both Nero and Galba, compromised their credibility and would constitute a strong push towards the dissolution of a unit that had become hated in the eyes of the people and the senate of Rome.

Ambiguous is finally the role of the *statores*, who in Rome, in the Julio-Claudian age, would progressively form a paramilitary unit that supported the sovereign and some members of his family and that perhaps, episodically, was responsible for security. They were a sort of orderly tasked with carrying out duties of different natures, which only during the second century took on a more distinctly military character.

We find *statores* in the retinue of the emperors from the first century, even outside of Rome and Italy, as demonstrated by the Egyptian inscription of the *viator* and *praefectus statorum* Lucius Publilius Labeo, in the service of the consuls, the praetorians and Tiberius Caesar.[88] This person, according to François Kayser, was part of the retinue that escorted Germanicus to Alexandria in AD 19 with the job of coordinating the *statores* in service to the imperial family. It can be no coincidence, if true, that as Tacitus said, Germanicus loved *sine milite incedere.*[89]

Notes

1 On the Pretorian guard before Augustus, see Passerini 1939, pp. 1–29; Bleicken 1990, p. 42; and Keppie 1996, pp. 102–107, to which we will often refer.
2 In addition to the numerous passages of Cicero (all recalled by Keppie 1996, starting with note 25), see also Appian *Bell.Civ.* part. 3.40 ss. and 5.3. See also Durry 1938, pp. 70–74, taken from Sablayrolles 2001, pp. 135–136.
3 Durry 1938, p. 67.
4 Keppie 1996, p. 119 who, for the relationship between the praetorians and Augustus in the role of proconsul, recalls Stein 1928, 54, who was not taken into consideration, if not openly criticized. See also Sablayrolles 2001, p. 135: "Ce corps aux functions de conseillers militaires et d'officiers de liaison s'ajoutait souvent une garde personelle, composée de soldats d'elite qui assurait, notamment en campagne, la sécurité rapprochée du promagistrate."
5 See Part III, Chapter 6.
6 In addition to Keppie 1996, Bingham 1997 (doctoral thesis) concentrates on the first period of life of the Cohorts, with particular attention given to the relationship between the emperor and the prefects of the praetorians as a litmus test for the evolution of the functions of the praetorians. To the thesis of 1997 and volume of 2015, we will repeatedly refer in this chapter.
7 Durry 1938 pp. 77–88; Passerini 1939, pp. 42–53; Angeli Bertinelli 1974; Kennedy 1978; Keppie 1996, pp. 107–11, n. 10. The situation is updated by Crimi 2009–2010.
8 *Infra*, n. 10.
9 Keppie 1996, part. p. 122. Too generic the affirmation by Nippel 1995, p. 91 that "the praetorians were the only troops regularly stationed in Italy"; so expressed, it gives the impression of them being an effect of an Augustan initiative.
10 Cass. Dio 53.11.5. The words of Dio are reflected in the affirmation: "The permanent military escort of those in power became a sign that Republic had given way to Principate" by Nippel 1995, p. 91, who cites Tac. *Ann.* 1.7.5 and *Hist.* 4.11.1. On the presence of soldiers in the city in the years of the civil wars see Rüpke 1990, pp. 56–57; and Nippel 1995, pp. 9–84.
11 Even more generic Suet. *Aug.* 49.1–5 when he discusses the measures relating to compensation for soldiers that were calibrated according to the rank and the length of service, to avoid the risk of sedition. Alongside the traditional funds, it is said, *nova vectigalia* and l'*aerarium militare* are instituted. This information does not specifically regard the praetorians and, as such, it is a pendant to the passage by Dio in which Maecenas gives advice on how to provide for the needs of the militias (Cass. Dio 52.28.1–6).
12 Durry 1938, p. 264–270 discusses in detail the payment fee for the military salary, its increase, the deductions made relating to equipment and the internal distinction between *deposita, viatica* and *seposita*. Lo Cascio 1989, pp. 119–120 pauses to reflect on the salaries of the legionnaires and the praetorians in the Augustan age, when an inflationary circumstance apparently brought about a raise.

13 See Part II Introduction and Chapter 6.

14 Flav. Jos. *Ant. Iud.* 19.38 (ἐπὶ φυλακῇ γε ... τοῦ αὐτοκράτορος).

15 Cass. Dio 53.16.5. Keppie 1996, p. 121 thinks of a *praetorium* on the Palatium (!), before the construction of the camp on the Viminal Hill. Royo 1999 pp. 245–268 proposes the translation "general quarter", which would correspond with the Latin *praetorium*.

16 Tac. *Ann*. 11.37; *Hist*. 1.29.2; Suet. *Otho* 6. On the Praetorian Guard and the Palatium, see Durry 1938 pp. 56 f., 275.

17 Two interesting testimonials refer to two successive periods. When the senator *Haterius*, who had gone to the palace to implore forgiveness, threw himself at the feet of Tiberius, perhaps causing him to fall, he is almost killed by soldiers (*milites*) who quickly intervene to protect their emperor (Tac. *Ann*. 1.13). With regard to the disturbance, which broke out following the repudiation of Octavia see Tacitus (*Ann*. 14.61) who describes soldiers made to go out (*emissi*) of the palace to disperse the crowd, with arms of various types.

    In the first case, soldiers inside the imperial palace are mentioned: it is not specified (which in truth, often happens), to which soldiers it refers, but it would not be surprising if it were to be the praetorians (and not *corporis custodes* or *speculatores*). Tiberius had a radically different relationship with the praetorian guard, in respect to that which Augustus had, and he would even obtain for these soldiers permission from the Senate to enter the Roman curia (see Part III Introduction, p. 122).

18 Cass. Dio 61.8.3, 62 (63) 4.3, both of which refer to the age of Nero. In fact the first passage refer to the ban on the praetorians assisting at gatherings of the people; it is not clear if their presence is meant as autonomous in respect of that of the emperor.

19 For this duty, see Part IV, Chapters 2–3. On the praetorian *chorographarii*, see the wonderful article by Nicolet 1988 (also Austin and Rankov 1995, p. 114). On the map of Agrippa (Plin. *Nat.Hist*. 3.16–17), completed by Augustus and exhibited in the *porticus Vipsania*, see Boatwright 2015. On the survey of Nero and a detachment of praetorians in Ethiopia, see Plin. *Nat.Hist*. 6.181, 184–186; and 12.18–19; cf. Sen. *Nat. Quaest*. 6.8.3.

20 From this viewpoint, the shock and indignation of the soldiers, who saw the wife and the son of Germanicus and the cortege of women who followed them without military escort, is understandable: *Feminas inlustres, non centurionem ad tutelam, non militem, nihil imperatoriae uxoris aut comitatus soliti* (Tac. *Ann*. 1.41.1–2).

21 Tac. *Ann*. 3.2.1: two cohorts of praetorians receive orders to carry on their shoulders an urn containing the ashes of Germanicus.

22 Suet. *Aug.* 23.1 (author emphasis): *Graves ignominias cladesque duas omnino nec alibi quam in Germania accepit, Lollianam et Varianam, sed Lollianam maioris infamiae quam detrimenti, Varianam paena exitiabilem, tribus legionibus cum duce legatisque et auxiliis omnibus caesis. Hac nuntiata excubias per urbem indixit, ne quis tumultus existeret, et praesidibus provinciarum propagavit imperium, ut a peritis et assuetis socii continerentur.*

23 In this Part II, Chapter 5.

24 *CIL*, XI 6125, *Forum Sempronii*. In regard, see Keppie 2000, p. 320.

25 Tac. *Ann*. 16.27: *At postera luce duae praetoriae cohortes armatae templum Genetricis Veneris insedere; aditum senatus globus togatorum obsederat non occultis gladiis, dispersique per fora ac basilicas cunei militares. Inter quorum aspectus et minas ingressi curiam senatores, et oratio principis per quaestorem eius audita est.* See also Tac. *Ann*. 14.61, just cited (n. 17) and 15.58, with reference to the conspiracy of Piso.

26 Nippel 1995, p. 92: "Charged in particular with protecting the Emperor's life, the praetorians were assisted during the Julio-Claudian period by a force of Germanic

bodyguards." Of our same idea is Keppie 1996, p. 118. The misinterpretation can be partly explained by the ambiguous term *somatophylakes* used by Cassius Dio 53.11.5; and partly by the frequent use of obscure periphrasis as "to be strictly protected".

27 See in particular Robinson 1992, p. 182 n. 59, who calls on all the examples indicated above to support the hypothesis that the function of bodyguard can be attributed to the praetorians.

28 Tac. *Ann.* 12.69.1: *Tunc medio diei tertium, ante Idus Octobris, foribus Palatii repente diductis, comitante Burro, Nero egreditur ad cohortem, quae more militiae excubiis adest.*

29 See also Mart. 10.48.1–2.

30 Tac. *Hist.* 1.29: *cohors, quae in Palatio stationem agebat.*

31 Suet. *Otho* 6: *cohors, quae tunc (Palatio) excubabat.* Similar considerations can be extended to the episodes that regard members of the imperial family (e.g. Tac. *Ann.* 13.18; Cass. Dio 57.4 and 61.8).

32 Sablayrolles 1996, p. 144.

33 Clauss 1973, p. 82 thinks of the Trajan era; Austin and Rankov 1995 on p. 136 talk generically of the first century AD and of "internal security agency throughout the empire"; while on p. 137, they specify that "we cannot trace the existence of a specific military intelligence-gathering agency based in Rome". Some bibliography on p. 259, no. 122. Not convinced of the idea of the presence of *frumentarii* in Rome at the very beginning of the imperial age, Sinnigen 1962, pp. 219–221; Clauss 1973, pp. 82–109. Much commented on has been the passage by Epictetus (*Diss.* 4.13.5) where he describes a "soldier disguised as a civilian" (a *frumentarius?*) who has the role of informer: "In this fashion the rash are ensnared by the soldiers in Rome. A soldier, dressed like a civilian, sits down by your side, and begins to speak ill of Caesar, and then you too, just as though you had received from home some guarantee of good faith in the fact that he began the abuse, tell likewise everything you think, and the next thing is – you are led off to prison in chains."

34 As the *speculatores* (*infra*, pp. 96–97).

35 Bingham 1997, Abstract, p. II

36 Bingham 1997, p. 233.

37 I would exclude the presumed task of the praetorians as tax collectors. Suet. *Cal.* 40: *Vectigalia nova atque inaudita primum per publicanos, deinde, quia lucrum exuberabat, per centuriones tribunosque praetorianos exercuit.* The only episode that this is referred to is from the age of Caius and regards officers and non-commissioned officers not men from the ranks; nothing brings us to think that this task continues to be entrusted to them in Rome; and that the taxes established ex novo had been maintained. For an in depth discussion of the taxes on prostitution, see McGinn 1998, pp. 256–264.

38 "Given the paucity of information of the movement of soldiers from the vigiles to the praetorians in the first century; then, it is impossible to determine whether members of the guard were employed to fight fires at this times because they had previous experience as vigiles" (Bingham 1997, p. 194).

39 In the sixty years between AD 6 and AD 68, recalls Bingham, Rome fell victim to at least ten fires of considerable dimensions. "Many of these would have involved the praetorians, though *the soldiers are not always mentioned*" (Bingham 1997, p. 182 n. 25; author emphasis). On the character and the motivations of the military presence in Ostia, see Caldelli 2012.

40 Suet. *Tib.* 50.3; Bingham 1997, p. 108.

41 Cass. Dio 57.14.10.

42 Bingham 1997, p. 183. In other cases there is disagreement in the sources over the category of soldiers involved. See, for example, those regarding the fire of the Theatre of Pompey in 22 (Tac. *Ann.* 3.72.3).

43 Part III 6. The sources almost exclusively describe *milites* which "to Roman ears undoubtedly would have been interpreted as designating the most visible and concentrated force in Rome, namely the praetorians" (Bingham 1997, p. 195).

44 Cass. Dio 55.10 A. Other examples in Cass. Dio 76.6–79.37. In addition a lot of inscriptions from the provinces.

45 Tac. *Ann.* 1.24.

46 Tac. *Ann.* 2.16.

47 Not to mention the staging of a theatre of war against the Germans by Caius in 39 (Suet. *Cal.* 43).

48 Other examples in Keppie 1996, pp. 120–121.

49 Perhaps they were also boosted numerically, which is what the episodes describing cohorts alongside Claudius in Britannia and Nero in Greece seems to suggest.

50 The arrest of Valerius Asiaticus (Tac. *Ann.* 11.1.3); the driving to suicide of Seneca (Tac. *Ann.* 15.60.4 and Iuv. 10.15–18); the order to assassinate the guilty by the Emperor (Tac. *Ann.* 1.53.5, where once again *milites* are generically spoken of!).

51 Bingham 1997, p. 174.

52 "The distinction under Augustus between the praetorian and urban cohorts may not have been as rigid as later sources indicate" (Keppie 1996, p. 109). His reasoning stems from the celebrated base of the statue by Lecce dei Marsi (*AE* 1978, 286). His arguments (the order in which the cohorts XI and IV are remembered, the epithet *praetoriae* referring to both, the lack of specifics about the cohort to which praetorians belong in the first century), are, in my opinion, thoroughly convincing (see also Crimi 2010 p. 140). Also successively, at least until the end of the first half of the second century, the praetorian and urban cohorts continued to share the military camp, burial sites, recruitment areas and had co-presence in dedications on *laterculi* (Keppie 1996, pp. 110–111 with n. 87). See also Part III, Chapters 6 and 7.

53 Dig. 48.4.1, pr. 1: *Maiestatis autem crimen illud est quod adversus populum Romanum vel adversus securitatem committitur.*

54 The *lex Iulia* affected the crimes of those who were found armed with swords and stones in the city; as those who were sought to meet with others in order to occupy public places or those who offended the imperial images (Dig. 48.4.6).

Liberati, Silverio (2010, p. 85) wrote thus: "The security of the *tribuni plebis* (as expression of the *maiestas populi Romani*) had to be protected . . . ; [in imperial times] attacking the *res Publica* was essentially the equivalent to attacking the powers, the functions, and the prerogatives of the *Princeps*."

55 Cass. Dio 58.9.2–6 and 12.2; Nippel 1995, p. 96. This episode will be returned to later, in regard to the *vigiles* (see Part III Introduction).

56 A discussion of the when and the how of this intervention by the praetorians in these episodes will be returned to in greater detail in Part IV Introduction and Chapter 3. The first interventions date back to Tiberius (Tac. *Ann.* 4.27). See also Nippel 1995, p. 91.

57 A collection and analysis of the predominantly literary documentation was carried out by Hirschfeld 1891, pp. 854–856. A useful overview of new documentation can be found in Crimi 2010.

58 The idea that the *speculatores* represented a kind of duplicate of the *equites praetoriani* has been expressed by Liberati, Silverio 2010, p. 94 and 117–118. The two authors argue that the *speculatores*, after the conspiracy of 97, were set aside by Nerva only to be substituted by Traianus with the *hastiliarii* (!).

59 Suet. *Aug.* 27: *In eadem hac potestate multiplici flagravit invidia. Nam et Pinarium equitem Romanum cum, contionante se admissa turba paganorum apud milites, subscribere quaedam animadvertisset, curiosum ac speculatorem ratus, coram confodi imperavit.*

60 Cf. Epictetus (*Diss.* 4.13.5), here remembered in n. 33.

61 Durry 1938, pp. 108–110 considers them a "special class" of praetorian, élite men in an élite corps. The idea that *speculatores* were a special unit, is also discussed by Passerini 1939, pp. 70–73.

62 Suet. *Cla.* 35: *Primiis imperii diebus quamquam, ut diximus, iactator civilitatis, neque convivia inire ausus est nisi ut speculatores cum lanceis circumstarent.*

63 Suet. *Gal.* 18: *ac descendentem speculator impulsu turbae lancea prope vulneravit.*

64 Clauss 1973, pp. 46–58 and 118 thinks that the link took place during the first half of the first century BC. The matter is taken up by Panciera 1974–1975, part. pp. 164–167.

65 The prevalent expression is *speculator cohortis.* In *CIL,* XI 6125 = *CLE* 986, from *Forum Petronii* (second half of the first century), a *speculator* [*in praet*]*orio* presents himself with a poetic language. To further evidence see Crimi 2009a, 2012: many *speculatores* are remembered on columbarium slabs from the Julio-Claudian age.

66 *Speculator Caesaris: CIL,* VI 1921a, cf. p. 3231 = 2782, cf. p. 3370 and 32661 = *ILS* 2014; *speculator Augusti: CIL,* 2755, cf. p. 3835 = *ILS* 2145.

67 Panciera 1968, p. 117

68 Suet. *Aug.* 74: *Augustus scribit, invitasse se quendam, in cuius villa maneret, qui* speculator suus *olim fuisset* (author emphasis).

69 For the reason of their presence in the Campanian town, I refer readers to Part IV Introduction.

70 Tac. *Hist.* 2.11: *Ipsum Othonem comitabantur speculatorum lecta corpora cum ceteris praetoriis cohortibus.*

71 *CIL,* XVI 16, 21 = *CIL,* III p. 853 = *ILS* 1993, Tomi; cf. RMD 1.1 = *AE* 1969/70, 420, from *Augusta Rauricorum* (Germania superior), incomplete and of uncertain age: *nomina speculatorum qui in praetorio meo militaverunt item militum qui in cohortibus novem praeto/riis et quattuor urbanis subieci.*

72 The *speculatores* were supposedly among the principle agents of Otho who managed to excite his colleagues – and the auxiliaries and legionnaires encamped in Rome – against Galba, "creating a climate of widespread terror and uncertainty". On 15 January 69 Otho was proclaimed emperor by twenty-three *speculatores.* The source is Tacitus (*Hist.* 1, from 24 to 35). For the disappearance between the end of the First and the beginning of the second century, see Millar 1977 and Crimi 2012, with the previous bibliography.

73 Suet. *Aug.* 49: *Ceterum numerum partim in urbis, partim in sui custodiam adlegit. Infra,* p. 166 n. 7.

74 On the role of the civilians *custodes* much is said by Echols 1958 (see Part I, Introduction).

75 A reliable source is Suet. *Aug.* 49: *Germanorum, quam usque ad cladem Varianam inter armigeros circa se habuerat.* Cass. Dio (56.23.4) remembers the departure from Rome of Gauls and Germans both unarmed and armed, the latter being exiled to the islands. To describe 'those carrying arms' Dio uses the word *doryphoroi,* the same often used to describe the praetorians.

76 Bellen's idea (1981, pp. 82–99), on the escort of Clauss, and also taken up by Nippel (1995, p. 92), is that after the *clades Variana,* Augustus substituted them with the *speculatores.* Interesting his consideration (at p. 33) that some powerful families (such as *gens Statilia*) had private bodyguards of a servile status, practising a kind of *aemulatio Principis,* carried on until the reign of Claudius.

77 Tac. *Ann.* 1.24: *Haec audita quamquam abstrusum et tristissima quaeque occultantem Tiberium perpulere, ut Drusum filium cum primoribus civitatis duabusque praetoriis cohortibus mitteret. . .et cohortes delecto milite supra solitum firmatae; additur magna pars praetoriani equitis et robora Germanorum, qui tum custodes imperatori aderant.*

78 Tac. *Ann.* 2.16: *Dein quattuor legiones et cum duabus praetoriis cohortibus ac delecto equite Caesar.* This second passage is less certain: it is more plausible that with *"eques delectus"* Tacitus is referring to the praetorian cavalry.

79 Suet. *Cal.* 43: *Militiam resque bellicas semel attigit neque ex destinato, sed cum ad visendum nemus flumenque Clitumni Mevaniam processisset, admonitus de supplendo numero Batavorum, quos circa se habebat, expeditionis Germanicae impetum cepit.*

80 Suet. *Nero* 34.1: *mox et honore omni et potestate privavit abductaque militum et Germanorum statione contubernio quoque ac Palatio expulit.*

81 Cass. Dio 63.27.2B.

82 Cass. Dio 53.11.5 (*supra*, nn. 10 and 26).

83 In the inscription *CIL*, VI 8806 = *ILS* 1727 a *corporis custos* by the name of *Nobilis* is defined as *miles* (soldier) *imperatoris Neronis Augusti*.

84 Grosso believes that the dissolution was the action of Nero, because he was betrayed. According to Bellen, once liquidated, the Germans were sent as reinforcements to the front (like, for example the *cohors Batavorum* of the rebellion of the years 69–70). See lastly the suggestive hypothesis made by Cosme 2011.

85 Bellen 1981, pp. 56–57. When Cassius Dio carried out the famous review of the armed services of the time of Augustus (55.24.7–8) he talks of *xenoi ippeis epilektoi*, of an indefinite number, to which was given the name *Batavi*. It remains uncertain whether the description was influenced by the *equites singulares Augusti* that existed during his time (see Part III, Introduction, p. 129).

86 The change, according to Grosso, took place in relation to the *expeditio Germanica* (AD 39–40), which apparently modified its status.

87 See the inscription *CIL*, VI 8806 = *ILS* 1727 of the *corporis custos Nobilis, miles Imperatoris Neronis Augusti* (*supra*, n. 83).

88 *CIL*, III 6589 = *ILS* 1920 = *IGRRP* I, 1075.

89 Tac. *Ann.* 2.59. Kayser 1990. See also Ricci 2008, with further bibliography on the subject.

5 The security of the urban area and its inhabitants

Civilian, paramilitary and military personnel

et Laribus tuum miscet numen
(Hor. *Carm*.5.4.39)

Mille Lares geniusque ducis qui tradidit
illos urbs habet et vici numina tria colunt
(Ovid. *Fasti* 5.145–146)

During the Augustan age, the plebeian associations were submitted to control and the magistrates of the plebs became "part of the cultural and administrative infrastructure of the new order".[1] The division of the *Urbs* into 14 regions also involved the creation of more than 200 *vici*[2] under the charge of *magistri* (mostly freedmen) and *ministri* (slaves). This reform was aimed at establishing a detailed control over the urban space and also of surveillance of all activities taking place in it. Apropos, evocative and entirely acceptable is the framework returned by Andrew Wallace-Hadrill:

> The city known and displayed, measured by professional surveyors, listed by census-officials, is a city under control; . . . it is not a mere exercise of rationality and administrative efficiency; it is part and parcel of a paradigm shift, from a knowledge conceived in traditional and ritual terms to the professionalized knowledge under imperial surveillance.[3]

In this section, after a survey of the construction of the 'security system' against the outbreak of *vis publica* behind Augustus' interventions, closer attention will be paid to the complementarity of forces patrolling the streets and key areas of the city. Considering their key role for the *Securitas Urbis et Populi Romani* and the change they underwent in the first two centuries of the imperial age, the functions of *milites urbani*, only treated in general terms in this chapter, will receive a systematic treatment in Part III. Chapter 7.

It is perhaps appropriate to specify once again that with the locution *Securitas populi Romani*, I mean safety in public places for inhabitants of the city, with security remaining a private matter, as in the Republic, not involving military personnel paid with public funds. Likewise, while still covering the physical safety of those living in Rome, the measures to prevent and remedy disasters (floods, collapse of buildings, etc.[4]) will be not treated in this volume. This is because these natural events had nothing to do with the will of human beings and, as the subject of civil administration, did not normally require intervention of military or paramilitary forces.

From the late-republican turbulence to Augustus' design

In the fourth chapter of the book *Roma e il Principe*, dedicated to the reorganization of the urban space in the Augustan age,[5] Augusto Fraschetti goes through the history of the most famous of the celebrations of the districts, the *Compitalia*, dating back, according to tradition, to King Servius Tullius.[6] In the Late Republican age, this celebration – considered to be a potential threat for public order because it involved all the quarters and was highly attended – was suspended for almost a decade. In a passage of the *Commentariolum petitionis* we have confirmation of the political relevance and the subversive potential of the *vici*[7]: the brother Quintus tells Marcus Cicero not to forget, in the electoral campaign, to establish a privileged rapport with the heads of the *collegia*, the *pagi* and the *vicinitates*. These represented, in the words of Quintus, the most efficient way to establish contact and therefore attract the *reliqua multitudo* to his cause.[8]

It seems today indisputable that the system of the *vici* of the Augustan age would take the place of the *monti* and *pagi* system, which had been in place until that moment. Inside each *regio*, as an ample framework, the *vici*, distributed like capillaries in the fabric of the city, would form the smaller articulations.[9] In the *Urbs*, their management was entrusted to the responsibility of the *vicomagistri*, *ingenui* or freedmen, whose epoch of creation remains controversial.[10] From the Augustan age these people, four elected every year, did not act autonomously, but were dependent on the State magistrates (*aediles*, *tribuni*, *praetores*); and in their turn they commanded slaves, called *ministri*, to whom they could entrust tasks of relative importance that were to be carried out with limited autonomy.

The oldest information we have about the division of the urban territory and the creation of the relative magistrates is that of Cassius Dio and Suetonius,[11] who refer (Cassius Dio and Suetonius) to the year 7 BC. The formalization of the intervention could, however, have been preceded by experiences of worship in the urban *vici*.[12] Such manifestations are to us well noted thanks to the recovery of a good number of altars that were destined to have been positioned in *aediculae* in locations relevant to the life of the districts, the *compita* (or crossroads), and were also for this reason called 'arae compitales'. The dedications, predominantly to *Lares Augusti* and to the *Genius* of the living emperor (but also to Mercury and to *Stata mater*), give us the names of the *magistri* and the *ministri*, with the indication of the year according to the calculation of the vican era, starting when the *magistri* entered into office.[13]

The role of the *vicomagistri*, according to a typically Augustan way of doing, put together a practical function and a symbolic message, joining tradition and innovation. The guardianship of the *vici* and of the altars at the *compita* and the celebration of rites that had been celebrated for centuries guaranteed the continuation of the tradition. To the ancient festivity of the *Compitalia* (18 May), was added a new one, the celebration of 1 August, the day that the *vicomagistri* (see Figure 5.1) entered into office. On this occasion the *Lares Augusti*, the ancestors of the imperial family, venerated together with the *Genius* of the living emperor were thought of as occupying the space of the city, making themselves personal guarantors and supervisors from those who were potentially dangerous.[14]

The *ratio* of the Augustan intervention for the organization of the territory of the city, even if it cannot be reconstructed in its entirety, can neverthe-less be sketched out in general lines. A first line can certainly be linked to the transmission of propaganda message of a new order of cohesion of all the components of the city around the figure of the *Princeps*.[15] The revised urban space would have contributed to a consolidation of the idea of the closeness of the *Princeps* – of his family and of his ancestors – to the *Urbs* and its inhabitants, around a common project of religious piety, of respect for traditional values and care for public affairs.[16]

The role of the *vicomagistri* did not, however, limit itself to religious celebra-tions and the care of the *aediculae compitales*, but stretched itself to the more concrete aspects of city administration and to the control of the urban fabric.[17] These other aspects are more directly interesting to investigate here, for the interaction that took place between the *vicomagistri*, the *vigils* and the *cohortes urbanae* with their prefects.

The division of the urban population according to the system of *vici* was first of all present in the Augustan plan, and aimed to ease the operations of

Figure 5.1 The so-called 'basis of the *vicomagistri*', AD 20–40
Source: open access, copyright free.

the *recensus*, and therefore of the procurement and the distribution of wheat. The organization of these activities was eased by the existence of a predisposed territorial framework and thanks to the collaboration of the *magistri* who worked on the territory as residents and provided a guarantee in terms of the prevention of unrest. Protests on the occasion of the distribution of rations, above all during periods of famine, were the most feared and represented an occasion in which the consensus might be coagulated around unelected popular leaders or demagogues that could unleash dissent, often manifested in a violent form.

In this sense, it is possible, without fear of exaggeration, to talk of a total rethink of the urban infrastructure, of scenarios of legitimate and 'legalized'[18] associative life, which Augustus developed, starting with – as it has often been reaffirmed recently – an idea that Caesar had already conceived but had not been able to put into practice. The transformation of the *vici* within the picture of the reforms of 7 BC[19] certainly constituted an occasion to completely revise the duties and the responsibilities of the small communities of neighbourhoods within the Metropolis.

The price to pay for the *magistri*, in exchange for the legitimization of their profile and official recognition of their role as protagonists in the life of the neighbourhoods, is expressed, without doubt, in economic terms, under the form of the direct contribution of expenses that their activity brought about. These expenses were not insignificant; indeed they required a quick – if not immediate – imperial subsidy.[20]

All the activities that were carried out in the *vici*, the distribution of rations as the management of the need for water, the disposal of rubbish and the commercial activity,[21] were under the supervision of the *magistri*, who naturally coordinated with the magistrates who oversaw their respective duties. For reasons of space and because of the strong connection with the maintenance of order,[22] we will pause here to examine in particular the fire service.

Octavian before 27 BC, and then for the entire duration of his government, constantly kept his eye on the problem of the prevention and the control of fires, adopting different solutions, arriving at that which would prove itself to be the most effective. We have already recalled the episode concerning Sabinus who was sent by Octavian to Italy in order to quell disorder in the period 36–35 BC, and the obscure expression of Appianus (the custom and system of cohorts of night watchmen).[23] Even if the comment made by Appianus sounds as backdating of the measure introduced in the late Augustan age, we cannot exclude the hypothesis that the initiative of 36–35 had a value of an experimentation that, in the space of at least fifty years, would bring about a structural reform.

In the fifteen-year period between 7 BC, year of the territorial reforms, and AD 6, year of the creation of the *vigiles*,[24] the prevention of fire and its extinguishment was entrusted to the *vicomagistri*. In the Augustan plan, they were used as intermediary figures, between the magistrates of the State who had to provide for the city in accordance with the traditional institutional plan (therefore the *aediles*) and the *familia publica*. This solution undoubtedly had positive attributes; in fact it rooted the preventive action and the interventions – entrusted to residents of individual neighbourhoods – in the fabric of the city, without increase in the number of personnel to employ.

However, the measure proved itself, once again, to be insufficient, opening the road to a more incisive and definitive intervention. In AD 6 the fire service and night policing was entrusted, in an experimental way, to freedmen, who were organized in cohorts, and were supervised by an equestrian officer, the *Praefectus vigilum*.[25]

The reasoning behind the choice of freedmen to carry out these duties is commonly explained as a conscious choice of Augustus.[26] Slaves were not chosen because, armed and concentrated in different points of the city, they would have been seen as a threat just a few years after the end of the civil wars. Neither were the *ingenui* chosen, who would have been seen as a *militia* imposed by the imperial power to control the *vici*.

According to Fraschetti, Augustus inserted, among the *vigiles*, freedmen of his *domus* so that others could not do it in his place. A clue to this behaviour can be found, continues Fraschetti, in the onomastics of the *vigils*: a lot of them, at the beginning of the first century, would have had the *nomen gentile* of the emperor, *Iulius*.[27] I cannot agree with this hypothesis, for a couple of reasons. First, it is too difficult to say with certainty which of the *vigiles* of Rome belonged to the first century: the vast majority of the epigraphic attestations that recall them are not prior to the following century[28]; those that are dated, are also predominantly from the period between the late Antonian and the Severian age, therefore, the end of the second century. Second, of the few *vigiles* called Iulii that we know, not all carried the *praenomen* Caius, which was typical of the freedmen of Augustus; and of these certainly none are from the first century BC.[29]

Those in charge of the *vici*: synergy between the new protagonists of security

From the moment of the creation of the *cohortes vigilum* and at least until the end of the Augustan age, *magistri vici* and *vigiles* continue to operate in strict synergy.[30] Augustus' desire not to exclude the *vici* from the fire-prevention activities, and therefore to promote collaboration between the *magistri* and the *vigiles* (civilian personnel, predominantly of the class of freedmen), can be detected between the lines of a passage by Suetonius, which concerns the year 7 BC and the division of the space of the city:

> He [Augustus] divided the area of the city into regions and wards, arranging that the former should be under the charge of magistrates selected each year by lot, and the latter under 'masters' elected by the inhabitants of the respective neighbourhoods. To guard against fires he devised a system of stations of night watchmen.[31]

From this passage it is possible to deduce clearly how, in this transitional phase between the Republican past and the following radical solution, fire prevention is tied to the territorial organisation of the city into (macro) regions and the smallest units, the *vici*. The last part of the passage is undoubtedly ambiguous: the *vigiles* are named, but the solution of creating a specialized paramilitary unit[32] is conceived by Augustus, as we have just recalled, only fifteen years later. If we exclude the idea that Suetonius commits an anachronism,[33] it is possible to conclude that the *vigiles*, as specialized figures (still not placed in cohorts and subordinate to a prefect), already existed before AD 6. Suetonius' phrase is therefore symptomatic of the intention of Augustus to entrust night patrols to "neighbourhood policemen", ready to collaborate with the specialists in the fight against fire, with the precise aim of increasing the forces already committed and to improve their effectiveness.[34]

Another symptomatic clue of the intent of Augustus is contained in an urban inscription (first years of the first century AD), which outlines the transitional phase between the preceding order and the new arrangement in relation to the patrol on urban security. The inscription is a dedication to Hercules, structured in two parts.[35] In the first, the *magistri vici*, in AD 4–5, put weights of gold and silver under the protection of the god; in the second, *invigulantes pro vicinia* of AD 12–13 unite with *magistri* to restore an aedicule. The expression *invigulantes pro vicinia* (a unicum), with all probability, alludes to a "group of people, tied to the neighbourhood by the duty of night watch, and united with the *magistri* of the year 19 to contribute, pro parte, to new expenses connected to the dedication". It could regard the "proto-vigils" of Augustus that were already active, even if they were not definitively put into place, before AD 6: people who, living in the same neighbourhood, organized themselves in order to help the magistrates who, a few years since the creation of the units of *vigiles*, continued to be involved in the watch of their neighbourhood.

There is no lack of other indirect clues relating to the close collaboration, at least in the Augustan age, between the magistrates of the neighbourhoods and the *vigiles*. One of them is made up of dedications to the *Stata mater* on the compitalian altars. *Stata mater*, sometimes mentioned with the epithet *Augusta*, is a divine figure who makes her appearance at the end of the Republican age: as the consort of Vulcanus, she is often venerated where a fire has broken out and then been quelled. The dedications that relate to this divine protector from fire and forefather to Saint Barbara, are all, with the exception of one, concentrated in the Augustan age, even after the creation of the unit of *vigiles*,[36] and represent a tangible sign of the proximity of the aims and the functions of the magistrates of the *vici* and the *vigiles*, each with their own responsibilities and their own times of intervention.

The loose definition of the discourse concerning fire safety is a confirmation of gradual imperial intervention: the organization of the *vigiles* in cohorts and the identification of the prefect as the person in charge, should not be interpreted as a brutal reduction of the duties of the neighbourhood magistrates and their ability to intervene. On the contrary: the juridical status of the *vigiles*; their profile that is clearly distinct from that of the more appropriately military units created in the same years; the highly probable lack of stable cantonment in the first decades of their existence[37] and, in a certain sense, the same 'transparency' of which they are victims,[38] are all clues that support the hypothesis that their action was thought of as a well-equipped, consistent and weighty support for the forces already available in the neighbourhoods.

Their involvement could be distinguished by the time and the type of intervention: a fire put out at its source did probably not require the intervention of the *vigiles*[39]; a fire of medium proportions could be singled out by the trained cohorts who intervened; for a devastating fire, the intervention of all the cohorts present in the city was required.[40]

It will be necessary to wait for Tiberius and then Claudius for the organization of the middle management, for the construction of the *excubitoria* and for the concession of citizenship after six years of service. In the second century, the age to which the first certain epigraphic attestations of the *vigiles* belong, the reform process can be described as complete: the measures introduced by Severus will bring about extreme consequences for the process of militarization of the units.[41]

The *ratio* of the Augustan interventions on the territorial arrangement of Rome for security was not limited to the containment of fires, and foresaw, in my opinion, the implementation of an integrated system with multifunctional supervisors. The collaboration did not only involve the *vigiles* and the *magistri vici*, but also the soldiers under the orders of the urban prefect.

The *milites urbani*, which according to Sablayrolles, constituted the most audacious and original construction of Augustus,[42] predated the *vigiles*, or at least their organization in cohorts, and therefore, the urban prefect (created between the end of the Augustan age and the beginning of the Tiberian age).[43] Like the *vigiles* and the praetorians, not being yet *castra* nor *excubitoria*, during the Augustan age and the first years of the reign of Tiberius, they had private lodgings. For at least a good part of those same years, as has been said, they were placed in cohorts together with the praetorians, with specific duties entrusted to them.[44]

A reference (rather pompous in nature) to the urban prefect – in this specific case Lucius Calpurnius Piso, in office between AD 13 and 32 – is made by Velleius who uses the expression: *Securitatis urbanae custos*.[45] His duties are listed in detail in the *Liber singularis* of Ulpianus[46] and are effectively synthesized by Tacitus:

> Then, upon his advent to power, as the population was large and legal remedies dilatory, he took from the body of ex-consuls an official to coerce the slaves as well as that class of the free-born community whose boldness renders it turbulent, unless it is overawed by force. Messala Corvinus was the first to receive those powers, only to forfeit them within a few days on the ground of his incapacity to exercise them. Next, Statilius Taurus

upheld the position admirably in spite of his advanced age; and finally Piso, after acquitting himself with equal credit for twenty years, was honoured by decree of the senate with a public funeral.[47]

Among the duties of his soldiers, which grew and became better defined with time, were a series of preventative actions: they evidently implied the necessity of being in the territory and of working, close to the *vigiles*.[48]

The action of the *vigiles* and the *milites urbani* coincided, first, in the patrolling of the streets. This duty was certainly one of the responsibilities of the *vigiles* at least during nighttime, and it was carried out together with the neighbourhood guards.[49] At night, we can imagine, the watch in the neighbourhoods had the aim of preventing fires and perhaps, in a complementary way, also aimed to prevent theft, which took place on the streets. During the day, we know, however, that the *milites urbani* carried out checks on areas considered to be at risk of disorder.

In this integration of action, the *modus operandi* of Augustus is evident. It maintained the pre-existing organization (the lower magistrates and their collaborators) without suppressing it, but at the same time substantially modifying it: the task of checking on the market zones and other public meeting places was not taken away from the *aediles* but, the development of surveillance on the streets as a task for urban soldiers and the *vigiles* limited, in a certain way, their function.

The integrated action was not only revealed in the action of soldiers, paramilitaries and civilian magistrates, but also in the substance of the intervention that sought to guarantee at least three essential functions of the economic life of Rome: the arrival and departure of imported and exported goods; the development of commercial activity; and the circulation of goods and products in the city. At the same time, it sought to limit the uncertainty caused by the non-secure hygienic conditions and the risk of disruptive action from delinquents, thieves, vagrants, etc. Indeed, we know that soldiers were also fundamental in checking the points of access to the city.[50]

A collateral effect of fire, which was particularly common and feared, was the malicious entry of houses affected by fire with the aim of looting[51]: it is no coincidence that among the responsibilities of the *praefecti vigilum* was the judgment of thieves (*fures*) in addition to that of those who started the fire (*incendiarii*); unless, as the jurist Paulus argues, the criminal in question was so brutal as to warrant the judgment of the urban prefect.[52] Therefore, the arrest and incarceration of the guilty could be carried out by both the *vigiles* and the *milites urbani*.

In the second and above all in the third century, the duties of the *vigiles* would progressively become greater, arriving at, for example, the handing out of corporal punishment for those who had shown signs of negligence in preventing the outbreak of fire; while the prefects were given the possibility to exercise a moderate *coercitio* against those criminals arrested by the *vigiles*.[53]

Another function brought the *vigiles* and the urban soldiers together. In Chapter 25 of the *Life of Augustus*, after having made clear the detached attitude, also in terms of language, of Augustus towards the soldiers in the aftermath of the civil wars,[54] the attention of Suetonius is moved to the employment of freedmen in the ranks of the army:

> Except as a fire-brigade at Rome, and when there was fear of riots in times of scarcity, he employed freedmen as soldiers only twice . . . even these he levied, when they where slaves, from men and women of means, and at once gave them freedom: and he kept them under their original standard, not mingling them with the soldiers of free birth or naming them in the same fashion.[55]

Evidently, since we are concerned with the rapport between the emperor and his soldiers (and fully aware of the perplexities that the ambiguous condition of the *vigiles* represented in the eyes of the Romans), the clarification was necessary: the *vigiles* were given the duty of intervening in case of fire, but also exceptionally took on a military function in the case of a legitimate fear that a particularly serious famine could provoke disorder. The vague reference made by Suetonius does not allow us to attribute the statement to a specific moment of the Augustan regime: with all probability his consideration has more generic value in judging the actions of Augustus. Rather than interpreting the passage in the sense that the *vigiles* should intervene when disorder brought about a fire, it would appear that, among their specific duties, was the need to intervene in case of turbulence, evidently alongside other forces engaged in the job of maintaining order, namely the *milites urbani*.

This function represented, in a certain way, a significant precedent. On one particularly delicate occasion, not long after (in AD 31), the *vigiles* would become preferred over the urban soldiers, as the guardians of public safety due to "their loyalty to the Emperor".[56]

Many of the duties of the *milites urbani* and the *vigiles* therefore, were superimposed, a fact that is unsurprising given that the two prefectures, that of the *vigiles* and that of the urban cohorts, both contributed in satisfying the needs of public order.[57] If, as we are told by Tacitus,[58] for at least twenty years the figure of the urban prefect was done without, maybe it is not so amiss to imagine that the soldiers of the urban cohorts, in addition to obeying the orders of their tribuns[59] and of the praetorian prefect, acted in this period of time also under the supervision of the prefect of the *vigiles*.

Notes

1 Nippel 1995, p. 86.
2 Plin. *Nat. Hist.* 3.66: *Ipsa dividatur in regione quattuordecim, compita Larum* CCLXV; Suet. *Aug.* 30: *Spatium urbis in regiones vicosque divisit.*
3 Wallace-Hadrill 2007, p. 77.
4 For the preventative measures of Augustus in this direction see Strab. *Geogr.* 5.3.7.

5 "Spazio urbano, ceti pericolosi e riorganizzazione augustea" (pp. 204–273).

6 Significant sources are: Dion. Hal. 4.14.3–4; Fest. 108 and 273 L; Macr. *Sat.* 1.7.34–35.

7 On the phenomenon of groups gatherings in associations in the Late Republic and its subversive potential, in addition to Fraschetti, see also Nippel 1995, pp. 72–73 and 86; Sablayrolles 2001, pp. 128–129; Lintott 2008. On the social composition of the *vici*, interesting observations are made by Tarpin 2008, pp. 49–52.

8 *De petitione consulatu* 8.30: *Deinde habeto rationem urbis totius: collegiorum omnium, pagorum, vicinitatum: ex iis principes ad amicitiam tuam si adiunxeris, per eos reliquam multitudinem facile tenebis.*

9 It is worth remembering, even if only cursorily, that the division in 'quarters/territorial districts' of the urban territory is not exclusive to Rome or Italy: for knowledge about the *vici* and the distribution of *compita* in the urban fabric, testimonies from Pompeii, for example, are fundamental (Tarpin 2008, pp. 38–40 brings various examples to our attention). If is difficult to say whether the organization of the city fabric in other areas of Italy and of the empire had the same character and a comparable significance as that in Rome.

10 The lack of attestations of *magistri vicorum* in the epigraphic documentation preceding the Augustan age leads Fraschetti (1990, pp. 249–255) to hypothesize that the *magistri*, before Augustus did not exist. His idea has been almost unanimously rejected, with the support of epigraphic testimonials that Fraschetti did not highlight: like the *vici* – argue, among others, Tarpin and Stek – even their magistrates existed since at least in the Late Republican age and the intervention by Augustus was intended to reorganize them rather than introduce a radical reform (Tarpin 2008, p. 36 and n. 6, Stek 2008, p. 125 and n. 8, with other bibliography and sources that refer not only to Rome). On the *vicinitates* and the urban division in *regiones* see Bert Lott 2004 and Kelly 2013.

11 Suet. *Aug.* 30.1: *Spatium urbis in regiones vicosque divisit instituitque ut illas annui magistratus sortito tuerentur, hos magistri e plebe cuiusque viciniae lecti*, cf. Cass. Dio 55.8.5–7.

12 Even if, according to Fraschetti, this date refers to the reorganization of the urban space, it is preceded by assumption of the cult of *Lares* and *Genius* in some of the *vici*. On the Lares of the *gentes* of the pre-Augustan *compita*, see Tarpin 2008, p. 60.

13 An epigraphic document, discovered in via Marmorata, contains the remains of a calendar and the almost complete list of consuls and censors from the year 43 BC to AD 3, in addition to a list of the *vicomagistri* who held office in a *vicus* of the XIII region (*Aventinus*), from the year of the foundation of the college to AD 21. The importance of the document also resides in the information it gives us about the change – which came about during the age of Tiberius (in the years between AD 16 and 18) – of the date of the entry into office of the *vicomagistri*: no longer 1 August of each year, but 1 January of each year like all the other magistrates.
    The architectonic plan of the monument was that of the aedicule (as that of the *lararium* in the domestic setting). The monumental model is considered to be the *compitum Acili* (AD 3–4), which arose at the *Tigillum sororium* and came to light in 1932 during works for the construction for the via dei Fori imperiali (*AE* 1964, 74a–b = *AE* 2004, 183. Cf. Pisani Sartorio 1993). On the altars of the *vicomagistri*, see Hölscher 1984, pp. 27–30; Hölscher 1988, pp. 398–400 no. 225 (more general, on the compitalian altars, pp. 390–398); Pisani Sartorio 1988; Wallace-Hadrill 2003, p. 203; Wallace-Hadrill 2008, pp. 276–290; Marcattili 2005.

14 Particularly evocative is the decorative language of the altar of Belvedere (12 BC) in which Augustus as *pontifex maximus* gives the Vestals his own *Lares*, which from *Lares gentis Iuliae* became *Lares Publici*. Image in Wallace-Hadrill 2008, fig. 6.5.

15 Marcattili 2005, p. 222. On the Augustan *vici* as an eventual 'top-down imposition', see Wallace-Hadrill 2003, p. 196, with interesting parallels drawn.

16 "Vici are deeply rooted in praxis, a grassroots neighbourhood formation which is recruited by Augustus to support his urban order" (Wallace-Hadrill 2003, p. 202).

17 On the administrative tasks of the *magistri vici*, see Gradel 2002, pp. 118 f., with a reference to Nicolet 1988, p. 209 ff.

18 The inscription of the Auriga and of the *collegium iuvenum*, studied by Panciera, is an interesting indication of the catalyst function of *regiones* and *vici* and of the new forms of non-aristocratic grouping (Panciera 1970). On the inscription and its significance, see also Lo Cascio 2008, pp. 74–75.

19 With the passage from non-codified districts to 'contrived communities', see Lo Cascio 2008, who goes further, stating that the start of the process cannot be back-dated to the Caesarian reforms. On the *cura regionum*, see Nasti 1999.

20 We do not have information on the direct financing of the *Magistri* from the *princi-pes*, but Gradel (2002, p. 119) notes that Tiberius inserts them in his will, therefore they evidently needed imperial support (Suet. *Tib.* 76).

21 Only a few late inscriptions connect the *regiones* with the *pistores* and the *popinarii* (of Rome and Ostia), as Lo Cascio 2008 observes.

22 It is no coincidence that the reform of the distribution of food supplies was the field, which, together with that of the prevention of fire, would occupy Augustus the most in terms of time (almost thirty years); indeed these last two were closely interlinked (Eck 2009, p. 242).

23 Regarding. *Bell. Civ.* 5.132 (*supra*, p. 80), see the comment by Fuhrmann 2012, pp. 101 f.

24 On the close coordination of soldiers, magistrates of the neighbourhood and *collegia* for the maintenance of order in the Capital, see Ménard 2004, pp. 31–33 (in reference to the temporal of which this author deals: see Part I, Chapter 1, pp. 13–14).

   For the events tied to the putting out of fires in the Republic and during the first period of the reign of Augustus, see Sablayrolles 1996, p. 13 (who identifies three weak points: lack of resources and personnel; ill-defined division of responsibilities; and the mix of official tasks and political interests of those involved) and Panciera 2006.

25 On the denomination of the prefect of the *vigiles* in Iohannes Lydus, see. R. Nicosia, *Il Praefectus Vigilum nella Nov. XIII di Giustiniano*, in *Mediterraneo antico* 6.1, 2003, pp. 469–510. On his competences, see Paul. Dig. 1.15; cf. Nippel 1995, p. 95.

   The creation of the seven cohorts is attested to by Dio 55.26.4; and by Paul. Dig. 1.15.1 (*Deinde divus Augustus maluit per se huic re consuli*); while Strabo limits himself to evoking the enrolment of a troop of freedmen (Strab. 5.3.7). On its precedents and the "rehearsals" before this date, see Panciera 1978.

26 Sablayrolles 1996, p. 25.

27 Fraschetti 2000, p. 726.

28 The oldest testimony of a member of the *vigiles* can be dated to AD 111 (*CIL*, VI 222 = ILS 2161): "Nous ignoronc donc absolument tous des vigiles du Ier siècle" (Sablayrolles 1996, p. 175).

29 My own verification reveals that the *Iulii vigiles* of Rome, certainly not from the third century (which naturally does not exclude that they were from the second century) are *Caius Iulius Secundus*, a centurion of the I cohort (*CIL*, VI 2961) and *Caius Iulius Peregrinus*, a soldier of the V cohort (*CIL*, VI 2980 cf. p. 3842). From the third century is the centurion of the V cohort *Iulius Rufus* (*CIL*, VI 2982 cf. *CIL*, VI 1057 = 31234, from 205); of the second half of the second century, or rather the third, is the soldier from the II cohort *Caius Iulius C.f. Cla. Sossianus* from *Iconium* (*CIL*, VI 2964). For the latter, see Sablayrolles 1996, p. 767 no. 2049.

30 For the coordination between *vigiles* e *milites urbani* it is interesting noting what happened at the ports of Ostia (see Part IV, pp. 126 and 206 n. 5).

31 Suet. *Aug.* 30.1: *Spatium urbis in regiones vicosque* [Augustus] *divisit instituitque, ut illas annui magistratus sortito tuerentur, hos magistri e plebe cuiusque viciniae lecti. Adversus incendia excubias nocturnas vigilesque commentus est.*

32 An indirect confirmation of their different status can be found in Tac. *Ann.* 4.5.3, who does not insert them in the review of the military forces of Rome. The *vigiles* could make a military will but did not count fully (?) as veterans, according to Robinson 1992, p. 107 n. 87 (who cites Dig. 27.1.8,4 Modest. 3 exc.; 37.13.11 Ulp. 45).

33 This hypothesis is not convincing, also because immediately after the draining and the widening of the riverbed of the Tiber are mentioned, which were carried out by the emperor in 7 BC (Fraschetti 1990, p. 255).

34 There are differing opinions on how the Svetonian passage should be interpreted. Not very clear is the contribution by Fraschetti (1990, pp. 254 f.), when he writes: "Regarding the fires [Suetonius] seems to not consider the new duties entrusted by Augustus in 7 BC to the *vicomagistri* and which the *vicomagistri* retained for a short period." Bert Lott 2004 is also vague (two distinct figures? Or rather the same but with differing functions?).

35 *CIL*,VI 282, cf. pp. 3004, 3756 = *ILS* 5615. *Sacrum Hercul(i) / mag(istri) vici anni XI. / A.A. Marcii, Athenodor(i) / lib(erti), Hilarus et Bello, / N(umerius) Lucius Hermeros / Aequitas, mag(ister)/ iter(um), / pondera auraria et / argentaria / viciniae posueruñt; / idem tuentur. / 'anno XIX' / 'pro parte, i' / 'vigul(antes) pro vicin(ia)' / 'una cum magisîr(is)' / 'contulerunt'.* On this inscription, see Panciera 1978.

36 According to Bert Lott (2004, pp. 167–168 and p. 182 with relative note) the dedications to the *Stata mater* attest the connection between *vici* and fires, at least until 7 BC (?). These dedications, in chronological order, are made by: a freedman in the late Republican age? (*CIL*,VI 762 cf. p. 3757 = *CIL*, I² 994 cf. p. 965 = *ILLRP* 259); a *magister vici* in the year 6 BC (*CIL*,VI 763, cf. pp. 3006 and 3757 = *ILS* 3307 and *AE* 2000, 132); the four *magistri* of the year 6-5 BC (*CIL*,VI 764 = *Suppl. It.* – Roma 2, no. 2935); the *magistri vici Armilustri* of 3-2 BC, who also address *Volcanus Quietus Augustus* (*CIL*,VI 802, cf. pp. 3007 and 3757 = *ILMN* 1, 18 = *ILS* 3306); the *magistri* of *vicus Sandaliarius* in AD 11–12 (*CIL*,VI 761); the *magistri* of *vicus Minervi*, in AD 42 (*CIL*,VI 766, cf. pp. 3006 and 3757 = *ILS* 3309, cf. CEMNR 2013 p. 73). A dedication to *Stata Fortuna Augusta* is made in AD 12 by a *magister vici* (*CIL*,VI 765 = *Epigraphica* 1978, p. 129).

It is not part of the group, but it is worth remembering the inscription *CIL*,VI 36809 = *ILS* 9250, dedicated to *Lares Augusti* from the *magistri* of *vicus Statae matris*, in 2 BC, cf. C. Buzzetti,Vicus Statae Matris, in LTUR V, 1999, p. 191.

37 See Part III, Chapter 6, pp. 144–146.

38 On the total lack of testimonies by *vigiles* from the first century, see *supra*, n. 28 (Sablayrolles 1996, pp. 114–115).

39 The underlining, which reaches comic levels, of the promptness of the interventions made by the *vigiles* can be found in Petronius (*Sat.* 78.6) and Tertullianus (*Apol.* 39). One of the latest testimonies can be found in *CIL*, VI 3744, from AD 362 (Sablayrolles 2001, p. 141).

40 Part II, Chapter 4, p. 93.

41 As we shall see in Part III Introduction and Chapter 6.

42 Sablayrolles 2001 p. 133: they were neither the bodyguard of the *Princeps* like the *corporis custodes*, nor the armed wing of the power like the praetorians, neither were they freedmen entrusted with putting out fires like the *vigiles*. Rather

they were created as a police force in service to a senator (cf. Durry 1938, p. 12). Vitucci (1956, pp. 26 and 43) believes that the restoration of the *comitia* and the creation of the cohorts that, until AD 13, remained without a prefect, occurred at the same time.

43 The lower limit for the creation of the cohorts is established by the passage by Cass. Dio 54.24.5, in which the military organization of the empire and the *militia Urbis* are described (AD 5). On the different dates, see Freis 1967 and Sablayrolles 2001, p. 133.

44 Echols 1958, Keppie 1996, p. 109 and Sablayrolles 2001, pp. 134 f.

45 Vell. 2.98. See Vitucci 1956, pp. 36–41.

46 Supervision of markets (Dig. 1.12.1.2 and 9) and of the places for entertainment (Dig. 1.12.1.12); surveillance of the *collegia* (Dig. 1.12.1.14), control on the slaves (Dig. 1.11.4, 1–2; and 12.1.1 and Tac. *Ann.* 6.11.2, see *infra*, n. 47).

The problem is deciding if and when, before the second century, these functions are formalized (Freis 1967, pp. 44–47). Caesar had already employed special guards for the markets as an exceptional measure, and they were supported by soldiers (?) and lictors, as Suetonius describes: *dispositis circa macellum custodibus . . . submissis non-numquam lictoribus atque militibus, qui, si qua custodes fefellissent, iam adposita e triclinio auferrent* (*Iul.* 43.2). See also Vitucci 1956, p. 51.

With regard to the supervision of brothels and taverns a passage by Tacitus, referring to the case of the matron *Vistilia*, informs us that, during the Tiberian age, this was still a task of the *aediles* (*Ann.* 2.85).

47 The passage by Tacitus (*Ann.* 6.11), after having efficiently synthesized the aims of the *praefectura*, present the first three urban prefects (author emphasis): *Mox rerum potitus ob magnitudinem populi ac tarda legum auxilia sumpsit e consularibus <u>qui coerceret servitia et quod civium audacia torbidum, nisi vim metuat</u>. primusque Messala Corvinus eam potestatem et paucos intra dies finem accepit quasi nescius exercendi; tum Taurus Statilius, quamquam provecta aetate, egregie toleravit; dein Piso viginti per annos pariter probatus publico funere ex decreto senatus celebratus est.*

On the idea of the *incivilis potestas*, see M. A. Levi, Incivilis potestas, in *Studi in onore di P. De Francisci* I, Milan 1954, p. 401 ff. (= Il tribunato della plebe, Milano 1978, p. 239 ff.); and S.A. Fusco, Insolentia parendi. *Messalla Corvino, la* praefectura urbi *e gli estremi aneliti della* libertas *repubblicana*, in Index 26, 1988, pp. 303–319.

48 Suitably, Sablayrolles 2001, p. 159 underlines the difference with the praetorians.

49 Suet. *Aug.* 30.1: *Excubiae nocturnae.*

50 Dig. 1.12.1.12 (see *infra*, p. 130 n. 94). On the *stationes* attested in proximity of the *Urbs* for the third century, like on key places for the security in the city, will be returned to in more detail in Part III, Chapter 6 of this volume.

51 Sen. *Controv.* 2.1.12: *Qui sive tectis iniectus est sive fortuitus, ruinae et incendia illa urbium excidia sunt; quippe non defendunt sua, sed in communi periculo ad praedandum ut hostes discurrunt appetuntque aliena, et in suis domini a validioribus, caeduntur, accenduntur alia ipsaque cum maxume flagrantia spolium ex alienis ruinis feruntur.*

52 Dig. 1.15.3, Paul. *De officio praefecti vigilum*: *Cognoscit praefectus vigilum de incendiariis effractoribus furibus raptoribus receptatoribus, nisi si qua tam atrox tamque famosa persona sit, ut praefecto urbi remittatur.*

53 The picture of the functions of the *praefectus vigilum*, also in the field of military jurisdiction, is pieced together by Paulus (see *supra*, n. 52).

Sablayrolles observes that in all the urban units (*vigiles, equites singulares Augusti* and praetorians) *optiones carceris* existed, an indirect confirmation of their "effective powers of the police" (Sablayrolles 2001, p. 144 and pp. 225–226). Something different is the *optio custodiarum*, the person in charge of the shift of the watch.

54 *Supra*, p. 73 ff.
55 Suet. *Aug.* 25.2: *Libertino milite, praeterquam Romae incendiorum causa et si tumultus in graviore annona metueretur, [Augustus] bis usus est. . . eosque, servos adhuc viris feminisque pecuniosioribus indictos ac sine mora manumissos, sub priore vexillo habuit, neque aut commixtos cum ingenuis aut eodem modo armatos.*
    On the interpretation of the passage, see Nippel 1995, p. 96.
56 For this episode, see the comment in Part III Introduction.
57 Fraschetti 2000, p. 728. For AD 69, see the significant passage by di Tac. *Hist.* 3.64: *At primores civitatis Flavium Sabinum praefectum urbis secretis sermonibus incitabant, victoriae famaeque partem capesseret: esse illi proprium militem cohortium urbanarum, nec defuturas vigilum cohortis.*
58 Tac. *Ann.* 6.11 (*supra*, n. 47).
59 As Vitucci 1956 suggests, it is of no help the passage by Cass. Dio 58.12.2 (AD 31), who, discussing officers of the *vigiles*, defines them in generic terms, such as "*pantes oi en archais ontes*".

Part III
Testing a dispositive

Part III

Introduction: the security of the *Princeps* and of urban spaces in Rome from Tiberius to the Severans

After outlining Augustus's 'security plan' in Part II, the focus now moves towards the changes – sometimes subtle, sometimes sweeping – of the picture.

In this second introduction, I would like to provide the fundamental lines of the emperors' policies on urban troops and on the security plan for the city. More or less, I will try to follow the same pattern of Part II, with minor adjustments: general imperial measures for security in Rome and Italy and targeted actions to urban troops and to the citizen security plan. In this framework, I have kept the urban cohorts on the margins, until the last paragraph of this part, specifically dedicated to their history; as well as quickly hinting to security events in Italy, referring to more specific information in Part IV.

In this introduction, time is a privileged point of view, being (here) the policies on the emperors' security considered in chronological order. In Part III, Chapter 6, the focus shifts on spaces: meeting places, where the population is denser and the security of people and the Emperors is increasingly threatened; checkpoints, where there is the crossing of people and various risks related to this; and buildings connected with political or administrative power, and the emperor's residence. An episode referring to 'dangerous places' (the circus, the imperial palace and the *castra praetorian*), told in more detail by our sources, will be treated more widely. Part III, Chapter 7, as I have already mentioned, is dedicated to the history of the *cohortes urbanae* in the first two centuries AD, and will give invaluable insight into the changes of the emperors' policies on the city and its inhabitants.

Overall, in this Chapter, I will continue in the effort to integrate different sources on control mechanisms in order to reconstruct both the circumstances in which military forces (separately or jointly) intervened, and the strategies of prevention that the authorities adopted against disorder. It would, however, be impossible and unproductive to refer in detail to single episodes: it was done specifically elsewhere[1] and my special interest, as often stressed, is to focus on the security dispositive during the first two centuries of the empire.

To reconstruct the evolution of the discourse on security from Tiberius to Severus[2] and to establish continuities and discontinuities, I will use certain indicators that can be considered either as a whole or only in part.

These indicators are: the construction of barracks; the modes of action and the protagonists involved in containing popular uprisings in public places; any changes occurring in the composition and in the roles of existing urban troops or in the command hierarchy; the creation of new troops or officials involved in the protection of the *Princeps* or in the safeguarding of the city; gradual extension or acquisition of skills of the new Augustan prefects.

The test of the dispositive of security designed by Augustus can also be done through other indicators, such as the estrangement of unwanted figures from city; or the setting up of military stations (*stationes*) in the territory of the city and its surroundings.

In reference to the first point, however, this was a recurrent phenomenon characterizing Rome in the first century AD, although not specific of a single *Princeps* or dynasty. A careful analysis of the incidents of estrangement from the city of Rome conducted recently by David Tacoma has clearly highlighted the differing levels, purely practical and functional or symbolic, in which the estrangement provision was claimed.[3] As to the second matter, regarding the places, it will be specifically addressed in Chapter 6.

The safety measures adopted during the Principate of Tiberius are a recurring motif in literary sources. Particularly eloquent is a passage from the beginning of the *Annales* by Tacitus:

> Yet, on the passing of Augustus he had given the watchword to the prae-torian cohorts as Imperator; he had the sentries, the men-at-arms, and the other appurtenances of a court; soldiers conducted him to the forum, soldiers to the curia.[4]

Tacitus often presents and comments on these measures as a signal of the emperor's uncertainty and obsessive fear of conspiracies or attacks. Tacitus' incisive remarks, read in conjunction with the passages of other authors, con-tribute in establishing the framework of a precise political plan.[5]

We have already focused on the passages by Velleius[6] and his wishes for perpetuated safety[7]; we shall now see what concrete measures were adopted and what contours were defined in terms of the security plan in the aftermath of Augustus' death.

Under Tiberius' reign there was a considerable number of cases of trials for lese majesty.[8] Rather than Tiberius' obsession on security, however, it seems more appropriate to speak of Tacitus' obsession, which becomes historiographical technique in attributing the drawbacks of the Principate primarily to the emper-or's bad intentions. The most interesting piece on this is included in the Annals:

> For he had resuscitated the *lex maiestatis*, a statute which in the old jurisprudence had carried the same name but covered a different type of offence – betrayal of an army; seditions incitement of the populace; any act in short of official maladministration diminishing the 'majesty of the Roman nation'. Deeds were changed, words went immune. The first to

take cognizance of written libel under the statute was Augustus; who had provoked to the step by the effrontery with which Cassius Severus had blackened the characters of men and women of repute in his scandalous effusions.[9]

In this passage, Tiberius is presented as the one who 'resurrects' a statute, formerly referred to certain cases that regarded the facts and not the words or intentions (*facta* not *dicta*), and which is significantly extended with Augustus' successor. However, as we have seen,[10] Augustus first had decided to identify and prosecute those who were responsible for *famosi libelli*.[11]

So the suggestion of the praetor Pompeius Macro to Tiberius, not surprisingly, is to resort to existing laws, namely to the previous Republican and to the Augustan *lex maiestatis*. The negative judgment that weighs on Tiberius thus appears to be linked to the implementation of the law, while how he operated does not seem to clash with what was already expected and acted by the founder of the Principate.[12]

In the city, many times, the new *Princeps* had to suppress or contain riots in places of entertainment. Suetonius and Velleius give an account of these episodes; the latter also takes the opportunity to celebrate the decision-making capacity of the *Princeps* and his desire to restore order and traditional values.[13]

Tacitus recalls one of these episodes by referring directly to the soldiers' prompt action in protecting the magistrates, and trying to prevent outbreaks of violence among the crowd.

> The disorderliness of the stage, which had become apparent the year before, now broke out on a more serious scale. Apart from casualties among the populace, several soldiers and a centurion were killed, and an officer of the Praetorian Guards wounded, in the attempt to repress the insults levelled at the magistracy and the dissension of the crowd. The riot was discussed in the Senate, and proposals were mooted that the praetors should be empowered to use the lash on actors.[14]

Thus, some soldiers and a few officers were victims of the riots or were injured; and the affair provoked such a discussion in the Senate that they wanted to decide on how to punish those who were considered in any way responsible for what had happened.[15] On another occasion, when the serious situation due to the *annona* prompted riots in the theatre, those spectators who had given proof of lack of self-control were punished; in the episode, however, there is no direct reference to the soldiers involved in the repression.[16]

As had already happened in the last years of the civil wars and then with Augustus,[17] Tiberius also prohibited Egyptian and Jewish religious practices and forced followers to burn vestments and ritual utensils[18]: young Jews were punished with exile and sent to the less salubrious provinces (*graviori caeli*) while the others, and 'those who professed similar cults' (*similia sectantes*) were evicted (as well as *mathematici*, if not repented), with the threat of slavery for life if they did not comply with the injunction.[19]

The most important Tiberian initiatives on security regarded the urban troops. The main and most renowned one is the construction of a permanent military camp on the Viminal Hill to house praetorians, *speculatores* and soldiers of the urban cohorts.[20] The praetorian prefect Seianus played a key role in the important decision taken in the early years of the reign of the Emperor. This measure, considered along with other distinctive episodes, which involved the praetorians in Rome and in Italy as the leading protagonists,[21] have often induced to identify Tiberius, through Seianus, as the true founder of the praetorian cohorts.[22]

In AD 33, Tiberius appealed to the Senate requesting that the praetorian prefect, some centurions and tribunes of the praetorian guard could attend the Senate with him every time he participated. The pretext was to avoid any enmities that could incur against the well being of the State (*offensiones ob rem publicam coeptas*), namely potential threats against the *maiestas populi Romani et Principis*. The Senate accepted the request without any restrictions in reference to the nature or the extent of the escort (*sine prescriptione generis aut numeri*).[23] It is legitimate to assume that, from this moment on, some praetorians and their officers would also accompany him to the *curia*, as confirmed by the above-cited passage of the *Annales*.[24] This information is an indirect confirmation of the fact that with Augustus it was not expected that the Praetorian Guard or their officers went to the Senate with him; and the temporary nature of the measure adopted by Tiberius is revealed by the facts that with Gaius we will assist to its reinstatement.[25]

The Tiberian interventions also affected the cohorts of the *vigiles* thus representing the first stage of a social promotion process that would continue the following years, and would have a turning point in the Flavian era; finally, in the last years of the second century, it ended with the *vigiles* being accorded full military rank. The *lex Visellia* of AD 24 assigned the status of *Latini Juniani* to freedmen (and therefore also to *vigiles*) and full citizenship after six years of service.[26] At the same time, perhaps, is dated the beginning of an adjustment of the command hierarchy of officers and centurions: a fragmentary inscription of Ostia of the Tiberian era attests most likely the existence of a tribune of *vigiles*; and the first notorious centurion of the *vigiles* (Rufellius Severus) was probably in service under the same emperor.[27] Thus, what begins to be defined is the classic circuit of equestrian careers, which will, however, be more fully defined more or less two decades later.

Even the episode describing Seianus' elimination, as reported by Cassius Dio,[28] indirectly testifies the emperor's consideration for and confidence in the *vigiles*. Tiberius spread the false news that he intended to grant the *tribunicia potestas* to Seianus. The Senate was informed of the maneuver implemented by Naevius Sertorius Macro, appointed secretly by the emperor at the head of the Praetorian Guard.[29] Macro entered Rome at night and gave to Graecinius Laco, prefect of the *vigiles*, the instructions he had received. After having removed, with a pretext, the praetorians who had accompanied Seianus to the Senate, he placed at the Temple of Apollo (meeting location), the *vigiles*, before reaching the *castra* in order to avoid the revolt of Praetorian Guard. The soldiers of the praetorian and urban cohorts (*stratiotai*), enraged for being considered less reliable than the *vigiles*, started to set fires and to loot.[30]

The whole episode is interesting for how the manoeuvre, which aimed at isolating Sejanus, was concocted. The last suggestion is particularly eloquent for another reason: the praetorian and urban cohorts, both housed at the *castra praetoria*, appear to (re)act collaboratively at a stage in which their functions were still closely linked.[31]

Despite the reign of Gaius is depicted as one in which the political role of the Praetorian Guard reached the acme, a closer analysis reveals that the signals of this do not appear so obvious. I will therefore try to consider them individually.

The increase in number of these soldiers, occurred in the decade between AD 37 and 47, seems rather to be linked to Claudius, whose rapport with them (even for the events that had brought to the accession to the throne) was certainly strong.[32] Moreover, in all the matters concerning murders commissioned by the *Princeps*, the praetorians are not explicitly indicated as the emissaries or perpetrators.[33] Apparently more revealing is the authorization obtained by the emperor in AD 40 to bring the armed guards into the *Curia*[34]: apparently due to the fact, as has been seen, that such permission is nothing more than the renewal of what was previously requested and obtained by Tiberius.

The bond between Gaius and the *corporis custodes* was certainly extremely strong. Perhaps in AD 14, during the revolt of the legions, both Germanicus' wife and son received their help.[35] Once he became *Princeps*, Gaius allowed the *custodes* to be actively involved during his leisure time[36]; and assigned Helikon, his beloved Egyptian slave, the post of commander of the bodyguards (*archisomatophylax*), as a token of his appreciation and trust.

> For he played ball with Gaius, practised gymnastics with him, bathed with him, dined with him and was with him when he was going to bed, as he held the post of chamberlain and Captain of the Guard of the House, a post greater than any that was given to anyone else, so that he alone had convenient and leisurely audiences to the emperor, where he could listen released from outside disturbances to what was most to his heart.[37]

In the tumultuous events that see the fall of Gaius and the rise of Claudius, in Flavius Josephus' account,[38] the military and paramilitary forces of Rome support one another and the *vigiles*, along with the fleet soldiers and the gladiators, were drafted to reinforce the Praetorian Guard. When the bodyguards learn about the emperor's death, they unleash and commit a massacre[39]; although, according to Flavius, the reason of their rage is to be grounded on interest rather than on a genuine personal attachment:

these men when they learned of the murder of Gaius were full of resentment, for they did not decide issues on their merits . . . but according their own interest. Gaius was especially popular with them because of the gifts of money by which he acquired their goodwill.[40]

Claudius gave ample proof of confidence in the soldiers who had sustained him. To his reign, as already mentioned, dates back the first adjustment of the command hierarchy of the Praetorian Guard and their participation in the campaign of Britannia[41]; along with the foregone *donativum* on the occasion of the first anniversary of his coronation.[42]

A passage by Suetonius, however, is symptomatic of a climate of suspicion and the permanence of old habits: Claudius banned soldiers (perhaps not only praetorians) from going to the senators' homes for the practice of *salutatio*,[43] which had spread and was considered a threat. In the same chapter, a little further on, there is no mention of the soldiers' involvement when referring to the deportation of Jews, followers of Christ[44]; nor with respect to the *senatusconsultum atrox* and *irritum* that in 52 provided for the eviction of mathematicians and astrologers.[45]

A year before, in 51, while Claudius was judging in court, he was attacked by the angry crowd, furious for the high price of grain. While Suetonius barely mentions how the *Princeps* managed to reach safety through the back entrance of the imperial villa on the Palatine Hill, Tacitus explicitly recalls the *globum militum* that drew him to safety.[46]

A particular problem arises with the soldiers of Rome in Ostia, the colony where Claudius built the famous port. It is known that, in the Augustan or perhaps early Tiberian age, the praetorians could be called by the city to fight the fires. The soldier who was drafted for this purpose, coming from Rome in the Augustan age,[47] rather than being specifically detached, as rightly pointed out by Maria Letizia Caldelli, could in fact have been part of those who were housed in neighbouring towns (*finitima oppida*), before the construction of the *castra praetoria*.[48]

It is more complex to define the nature of the measure that Suetonius attributed to Claudius in allocating in Pozzuoli and Ostia cohorts to contain the fires.[49] Until now, the cohort mentioned by Suetonius in Ostia has almost unanimously been considered as *cohors urbana*.[50] By no means it should be ruled out, however, that this was a cohort of *vigiles*, as some archaeological and epigraphic sources would seem to suggest. In particular, the analysis of archaeological remains recently unearthed in the *regio* II of the town allows us to consider possible the existence of a 'proto-barrack' for the *vigiles*.[51] If this hypothesis is true, like Augustus in Rome forty years before, Claudius was the first to take measures in Ostia to deal with the perennial problem of recurring fires: *vigiles* allocated here would provide the prevention of fires and assure a fire service throughout the city, and especially in the baths and entertainment buildings.[52]

Germani equites, quibus fidebat princeps quasi externis.[53] The bond that had linked Gaius to his bodyguards is restored with Nero. Most of the funerary steles of the *Germani corporis custodes* in Rome in fact belong to the period of the reign of this *Princeps*[54]; and, in the last days of life, the same bodyguards were the ones who protected him almost until the very end.[55]

Early in his reign (AD 55) is dated a measure, reported by Cassius Dio,[56] in which Nero would have prohibited the soldiers to be present, as they always had been, at all public gatherings. When trying to explain Nero's motives, Dio distinguishes between a pretext (the soldiers had to attend to their military duties, *ta stratiotikà*), and the real reason (the *Princeps* "wanted the greatest number of opportunities for those who caused public disturbance").[57] The passage is not easy to interpret, even due to Dio's use of the generic term of 'soldiers'. It is quite possible that the text is referring to the Praetorian Guard. Although nothing precludes that it may be alluding more generally to all the troops stationed in Rome, therefore, definitely also to the *milites urbani* and possibly the *vigiles*, who held the specific duty of containing and suppressing disorders in public spaces.

That Nero considered the praetorians trusted soldiers seems to be confirmed by some significant events. First, the *miles Cassius* (probably a praetorian) was commanded to arrest and put into chains the prefect Faenius Rufo.[58] Then, it was the Praetorian Guard who patrolled the path that led the participants in the Pisonian conspiracy of AD 65[59] to death. The *Princeps* himself, in 62, in front of the popular demonstration in favour of his wife Octavia, just repudiated, and against his new mistress Poppaea, reacted with the repression through the military corps (probably the Praetorian Guard), to reiterate the arbitrary exercise of power and his right to command. It was the year AD 62, and Tacitus gives us a detailed account of what happened:

> [After Octavia's return] at once exulting crowds scaled the Capitol and Heaven at last found itself blessed. They hurled down the effigies of Poppaea, they carried the statues of Octavia, strewed them with flowers, upraised them in the forum and the temples [. . .]. Already they (*multitudo*) were filling the Palace itself with their numbers and their cheers, when bands of soldiers (*militum globi*) emerged and scattered them in disorder with whipcuts and levelled weapons. All the change effected by the seditio were rectified and the honours of Poppaea were reinstated. [But Poppea told Nero that] those arms had been lifted against the sovereign; only a leader had been lacking, and, once the movement had begun (*motis rebus*), a leader was easily come by. The one thing necessary was an excursion from Campania, personal visit to the capital by her whose distant nod evoked the storm (*tumultus*)! [. . .]. A deserved castigation and lenient remedies had allayed the first commotion (*primos motus*); but let the mob once lose hope seeing Octavia Nero's wife, and they would soon provide her with an husband![60]

In this passage Tacitus makes use of a very varied vocabulary (the Latin terms are highlighted in brackets) and defines, through the mouth of Poppaea, the suppression of the riots a *iusta ultio* (a just revenge).

The rapport between the aspirants to the throne and the different troops of Rome in the crisis of AD 68 and 69, the conflict between the praetorians on the one hand and the *milites urbani* and *vigiles* on the other, and also the role of the *speculatores*[61] are a mirror of an exceptional situation, also indicated by the considerable increase in the number of praetorians and urban cohorts employed.[62] These years represent, from our point of view, the alteration of a delicate balance that in fact, over the course of nearly ninety years of reign of the Julio-Claudian dynasty, had led to a polarization of functions and roles between the troops and the private or paramilitary forces created by Augustus, producing a contrast between soldiers of the urban cohorts and *vigiles* on the one hand, and praetorians and Germans on the other.

If the extrovert Nero was opposed to philosophy because he considered it connected with magic,[63] in the very delicate year of AD 69, Vitellius issued an expulsion order of astrologers from Italy; and some of those who refused to comply, were sentenced to death.[64] Even under the Flavian dynasty there were orders of expulsions from the city. In a well-articulated passage by Dio, Mucianus tried to convince Vespasianus to hunt down all those people who, under the pretext of philosophy, spread inappropriate doctrines and thereby corrupted sympathizers.[65] Titus banished informers (*menutas*) from Rome,[66] while Domitian actually put to death some philosophers for the ideas they were promoting.[67]

When dealing with the reorganization of the army, Vespasianus also included the soldiers of Rome. The words of the general Gaius Licinius Mucianus,[68] when he entered the military camp (*castra praetoria*), shortly after the victory of Vespasianus, in order to nip in the bud the threat of mutiny, epitomize the basic principles that inspired the new emperor's future interventions. Mucianus calls the soldiers "children of the same oath, and soldiers of the emperor": in the aftermath of the civil war, the first appeal that was made could not but be to the sense of belonging of the Roman soldiers, united, beyond the differing specificities, at the service of both Rome and the emperor.

The following interventions of Vespasianus and of his sons, consistent with this principle, were aimed at returning to the past and its original equilibrium, consolidated through a few measures,[69] which meant to resize the predominant role of some *militiae* (Praetorian Guard, *speculatores*) and define the responsi-bilities of each; to ratify the dissolution of a seriously compromised corps (the *corporis custodes*) and start thinking about creating a new one (*equites singulares Augusti*). We have concrete evidence of some of these changes, and only indi-rect traces of others.

According to some, the advent of the Flavian dynasty sets a change for the *speculatores*.[70] The sinister role that they had played in the year of the four emperors induced Vespasianus to include them in the praetorian cohorts by containing their role and autonomy.[71] If one opts for such an interpreta-tion,[72] the intervention falls within the framework of the reorganization of the *cohortes praetoriae*, which, after the vicissitudes during the Julio-Claudian era, and the swinging numbers of the *longus et unus annus*, were restored to nine, as in the original plan. For the rest, the role of the Praetorian Guard did not undergo any substantial changes: with Domitianus their use as a fighting force persists and becomes more regular.[73] What is recovered is an intense relationship with the emperor, which they already had under Tiberius and, above all, under Nero.[74]

The *vigiles* who, during the contrast between Flavians and Vitellians, were considered by Tacitus among the urban forces upon which, in case of necessity, the *praefectus Urbi Flavius Sabinus*[75] could count on, had their permanent barracks in Ostia; and in this way they saw confirmed the extension of their range of action already experimented with Claudius.[76]

The creation of the *equites singulares Augusti* was the last piece of the organizational scheme of the *securitas Principis* and occurred a few decades after the elimination of the *corporis custodes*. The new soldiers on horseback, enrolled mostly in the Germanic, Danubian and Balkan provinces were structured after a phase in which the function of bodyguard was most likely carried out by civilians.[77]

In an early phase, as seen, for the *securitas Augusti*, the functions of the praetorians, as the emperor's security escort and guards of the Palatium (and only occasionally invested with operational tasks in military campaigns), and the *corporis custodes*, as bodyguards, were differentiated. The *equites* may seem as a replacement of the *corporis custodes*. In fact, modelled on the provincial *equites singulares*, they were a hybrid with characteristics of the vanished German guards, but also of the *equites praetoriani* and of the *speculatores*: right from the beginning of their creation, the emperor was escorted and accompanied by them, also in military campaigns.

The fact that this corps was not included in Augustus' original security plan suggests that it is a product of an ad hoc adaptation, required because considered indispensable after the alterations experienced during the years of reign of the Julio-Claudian dynasty and during the events of the civil war. If, on the one hand, Augustus – after having dismissed the *Calagurritani* and the *Germani*, legacy of the civil wars – promptly recruited new bodyguards; on the other hand, between the tormented phase of the years AD 68/69 and Traianus's reign, a new solution for the safety of *Princeps* was developed. The *numerus equitum singularium* appear to be the product of historical conditions that have changed, and of the emperors' need to conceive troops of a versatile and distinct profile.

Despite the relatively modest size of the corps, these soldiers are overrepresented due to the discovery of their burial ground. Numerous studies have focused on their uses, habits, social relations and burial customs.[78] Their loyalty to the emperor,[79] confirmed by numerous episodes, was also guaranteed by the isolation in which these soldiers lived, evidenced by a number of signs such as: the construction of a separate camp with stables and service buildings; the idea of a private burial place on the current Via Casilina[80]; the small percentage of family bonds with women from Rome; the persistence of ethnic traditions[81]; and finally the fact that they were promptly active on theatres of war. The oldest case confirming this dates back to the late first or early decades of the second century, at the same time of their creation: Flavius Proclus, an *eques singularis* native of Philadelphia who died in Mogontiacum at the age of twenty, may have been part of the cortège that accompanied Domitian[82]; and at Apamaea, the chosen equestrians Marcus Ulpius Severus and Ulpius Verecundus (the second deceased at the age of seventeen[83]), accompanied Traianus in the Parthian campaign.

The second century is commonly regarded, for the troops of Rome, as a period characterized by continuity,[84] if we exclude the presence of new protagonists in the *Securitas Augusti* such as the *equites singulares*[85]; and some changes in regard to the urban cohorts, which will be discussed in Chapter 7.

The possibility that, more or less around mid-century, the command of the urban cohorts was passed to the praetorian prefect seems incredible and without apparent reason.[86] The text of a dedication from Rome to Antoninus Pius, repeatedly recalled upon as evidence, commemorates in descending hierarchical order the two praetorian prefects, the tribunes and the centurions of the cohorts hosted in *castra praetoria*.[87] The urban prefect does not appear there simply because, as has been rightly observed, only the praetorian prefects were supreme commanders of the *castra* where the Praetorian Guard were quartered along with the urban soldiers.[88]

As of mid-second century, instead, substantial transformations regarded the *Securitas Urbis*. This was due to the emergence of new figures and the strengthening of the prerogatives already available to the existing ones; and to a greater coordination of the tasks of those involved in the management and control of urban spaces. If in fact, as we have previously mentioned, during the first century of the Principate, the civil jurisdiction continued to be entrusted mainly to the republican magistrates,[89] these find themselves gradually discharged of their responsibilities; while, at the same time, we witness the growing role of the prefects and in particular of the *praefectus Urbi* and *praefectus vigilum*.[90]

The territorial boundary of the jurisdiction of the *praefectus Urbi* existed, in all probability, since the age of Augustus,[91] although the information we have is not prior to the Severan period, when such boundary would be extended.[92] During the first two centuries of the Principate, his competence was related to the maintenance and supervision of order at places of entertainment and during gatherings,[93] in particular at the *collegia* and markets, with the aid of the urban cohorts.[94] Connected with this activity of safeguarding the city, was also the use of moderate *coercitio* on disturbance factors: only the most serious crimes were brought in court.[95] The first document that refers to his civil jurisdiction is the rescript of Hadrianus on the *argentarii*;[96] but the extension of jurisdiction in criminal matters, according to the doctrine, is not dated before the Severan era.[97]

To the age of Hadrian belongs a well-known document attesting the evolution of the administrative organization of the *vici*. We have already mentioned the Augustan regulations on the matter, which included providing support to traditional magistrates with the *magistri vicorum*, freedmen, and *vigiles*.[98] A dedication to the emperor Hadrianus by the heads of districts of AD 136 indicates *curatores regionum*, not attested before, beside them.[99] The exact function of these new figures and their hierarchical position compared to the others involved in the regions of the city is not clear. It can surely be considered that the new figures represented a connection between the emperor on the one side, and the urban prefect and *magistri vicorum* on the other.

Another inscription, from the time of Antoninus Pius, further expands this framework. It is a dedication to the Lares Augusti engraved on a slab covering the aedicule of a *compitum* of AD 149.[100] This document is an important source that complements the picture already outlined by the so-called 'Basis Capitolina'[101] and the literary evidence. In fact it informs us that, around mid-second century, it became the emperor's responsibility to approve renovations and interventions in the neighbourhoods. The emperor operated by delegation to the prefect of the *vigiles*, an official of modest importance even in the first century, who had gradually seen extended its authority, both in terms of policing, and in the civil and military jurisdiction sphere.[102] The prefect's role was facilitated by the 6,000 *vigiles*, distributed in all the regions, which ensured widespread surveillance on the territory of the city in order especially to contain social disorder.[103]

Other documents, of a later period, further support what has been said.[104] In the same direction of Hadrianus and Antoninus, moved Marcus Aurelius,[105] and perhaps even Commodus.[106] Still in 204, when arranging the secular games (*Ludi Saeculares*) Severus and his sons made an appeal to civilians to collaborate with the *vigiles* so that the nightly feasts did not degenerate, becoming a source of problems.[107]

What results from the cross examination of the sources was an increase, which took place in the second century, in the number of government officials of the city region and the synergy between figures with differing profiles and roles; and at the same time more space than in the past for the central power. The strategy for action in the field of administrative liability in the neighbourhoods had been, until then, more subtle, compared to the military forces, where the verticality in the command hierarchy and the delegation to prefects were explicit. At this time the same verticality and a strong centralization was applied to the government in urban regions with the creation of new figures (such as the *curatores regionum*) and the enhancement of the imperial prefects' responsibilities.

The apparently inexorable process of centralization was interrupted at the end of the Severan period. The *curatores*, which throughout the second century had been subjected to the prefect of the *vigiles*, with Severus Alexander were chosen among the senators of consular rank to collaborate directly with the urban prefect.[108] As it has been said, "it is possible that, after an initial phase in which the *Princeps* had left the situation unchanged, he considered more appropriate to reduce the high-ranking of commanders of the *vigiles*."[109]

The collaboration between the new consular *curatores* and the urban prefect can be read in the name of changes of the functions and the rank of the major prefectures, and with the important measure apropos the *annona*.[110] The measure introduced by Alexander also is part of what is known as the policy of openness in respect to the senators, adopted by this emperor and his entourage; and of his relationship with the praetorians, as we are referred to by the *Historia Augusta*. An indicative episode in this regard is the popular uprising, sparked by trivial reasons, in the early years of Alexander's reign, against the praetorians.[111] It seems strange, as duly noted by Fulvio Grosso,[112] that the best armed and best trained military force of the city is put in difficulty by civilians. The resistance may have been artfully orchestrated and supported by the senators, by tacit agreement with the *Princeps*.

It is no wonder that the last member of the dynasty of the Severans is the one who closed the circle of reforms, started from the thirties of the second century, on the safeguarding of the city.[113] A few decades earlier, his grandfather had given a significant change to the size and composition of the military forces in the city, which closed a parenthesis of substantial changes: the dissolution of the old praetorian cohorts and the replacement with new praetorian troops recruited from the legions; the growth in the numbers of these troops[114]; and the allocation of the legio II Parthica at the gates of Rome are the results of a process that had initiated since the mid-second century, certainly from the age of Lucius Verus and Marcus Aurelius.

In the period between the middle and the end of the second century the fighting role of the urban cohorts had become determined, through their increasingly intense involvement in military campaigns; and at the same time, with the strengthening of their prefect's responsibilities, even the territorial control functions had increased. It seems indicative, on this issue, the construction of a military camp reserved for the urban cohorts (*castra urbana*), strategically located on the Via Lata near the *forum suarium*, halfway between the Flaminia gate and the urban prefecture.[115]

The profile and the important role of the *vigiles* are substantially modified in the Severan era. A tangible example of the relationship that bound the *vigiles* to the Severans was the imposing basis with dedications to Caracalla between 205 and 210, unearthed in the quarters of Caelian Hill.[116] The names of more than 1,800 men inscribed on the basis, listed according to their hierarchy, representing, as Sablayrolles pointed out, 86 per cent of the known *vigiles*.[117] The randomness of the findings is not an adequate explanation for the disparity of the numbers respect to the previous eras. In addition to this, in the same period the growth of the *vigiles* of Rome; the enhancement of their prefect's responsibilities; the renovations or the construction of new barracks for them in Rome and in Ostia; and an increase in their salary are all clues of a closer bond that was established between the cohorts of *vigiles* and the new dynasty.

A different fate befell, in the same era, the Praetorian Guard. Over time, a larger area had been searched to recruit those soldiers, for several reasons: not least, a slack of interest towards the military profession that attracted less and less the inhabitants of Italy. With Severus, because of the position taken during the conflict between those aspiring to the imperial throne, the older cohorts, as said, were dissolved and reassembled with soldiers of different ethnic origin.

The forces, even in this case, increased, but their degree of Romanization decreased significantly: the praetorians no longer came from the regions of Italy surrounding the capital, but from the Danubian and Balkans; they could hardly speak Latin; they were rude and had no experience in close contact with the civilian population, especially that of a big city like Rome.[118] An episode such as that of 212, concerning the urban prefect Fabius Cilo, in which the urban cohorts and the crowd were opposed to the Praetorian Guard[119] is the sign of a change, of an incurable opposition, of a breaking of the balance of competencies and responsibilities that had been the guiding principle of Augustan policy.

The Augustan dispositive, which had proven to be fully effective for almost 200 years, albeit with subsequent adjustments, is about to become inadequate when its protagonists, at all levels (urban military forces, prefects, *magistri*), in form and/or in substance, change. In this new phase, the military forces present in Rome are opposed to one another, and the relationship with the civilian population always more often, as will be evident in some key episodes that follow one another between the end of the second and the first half of the third century, are the sign of a broken balance.

Notes

1 Refer to the authors reviewed in Part I, Chapter 1; on the riots in Rome and in Italy between the Triumvirate age and in the first Principate, see Benjamin Kelly (2007), who has closely examined the topic.

2 The evolution of praetorians and *milites urbani* after Augustus is delineated by Durry 1938, pp. 363–396 and Freis 1967, pp. 6–8.

3 Tacoma 2016, pp. 92–104. Somewhat approximate, although prior to the presentation and discussion of the episodes, the claim that "under the early Empire expulsion was no doubt the main mechanism used by the Roman state to control the population of Rome" (p. 93).

4 Tac. *Ann.* 1.7.5: *Defuncto Augusto signum praetoriis cohortibus ut imperator dederat; excubiae arma, cetera aulae; miles in forum, miles in curiam comitabatur.*

5 The famous episode of *Pollentia*, recalled within a series of measures *ad utilitates publicas* of the emperor (Suet. *Tib.* 37.3) will be more widely discussed in the chapter on Italy (Part IV, p. 194).

6 In particular, Vell. 2.89 and 98; 2.103.3.

7 See Part I, Chapter 3, p. 46.

8 Griffin 1995.

9 Tac. *Ann.* 1.72.2–4: *Nam legem maiestatis reduxerat, cui nomen apud veteres idem, sed alia in iudicium veniebant, si quis proditione exercitum aut plebem seditionibus, denique male gesta re publica maiestatem populi Romani minuisset: facta arguebantur, dicta inpune erant. primus Augustus cognitionem de famosis libellis specie legis eius tractavit. . .; mox Tiberius, consultante Pompeio Macro praetore an iudicia maiestatis redderentur, exercendas leges esse re spondit. hunc quoque asperavere carmina incertis auctoribus vulgata in saevitiam superbiamque eius et discordem cum matre animum.* See also 2.50.1.

10 See Part II Introduction, p. 76 and n. 18.

11 Griffin 1995, p. 52 indeed underlines that the use of the senatorial court for treason trials dates back to AD 8.

12 To which, for example, the creation of extra-urban *stationes* was linked (see Part IV, Chapter 6, Chapter 8).

13 Suet. *Tib.* 37.2; Vell. 2.126.2. See also Cass. Dio 57.14.4, discussed by Kelly 2007 and Le Roux 2002, p. 23 n. 18.

14 Tac. *Ann.* 1.77: *At theatri licentia, proximo priore anno coepta, gravius tum erupit, occisis non modo e plebe sed militibus et centurione, vulnerato tribuno praetoriae cohortis, dum probra in magistratus et dissensionem vulgi prohibent. actum de ea seditione apud patres dicebanturque sententiae, ut praetoribus ius virgarum in histriones esset.*

15 The *lex Iulia de vi publica* denied to those *qui artem ludicram faciunt* to recourse to the people's appeal (Pauli Sent. 5.26.1–2).

16 Tac. Ann. 6.13. 1–2: *Isdem consulibus gravitate annonae iuxta seditionem ventum multaque et pluris per dies in theatro licentius efflagitata quam solitum adversum imperatorem.* In response, the consuls enacted a measure and Tiberius' inaction was interpreted as a sign of haughty indifference, rather than clemency.

17 Perhaps Tiberius' intervention is more rigid, since Suetonius does not mention any territorial limitations that prohibited the practice the rituals, as was instead the case with Agrippa and then Augustus (see Part II Introduction).

18 Suet. *Tib.* 36.

19 The news is confirmed, among others, by Tacitus (*Ann.* 2.32) and by Cass. Dio 57.15.8: in AD 17 that *mathematici* and *harioli* (fortune-tellers), due to a Senate deliberation, if citizens would have been sentenced to *aqua et igni*, their property confiscated.

20 Suet. *Tib.* 37.1.

21 See Part IV.

22 Sablayrolles 2001, p. 138 includes a quote from Keppie 1996, p. 122 on the physical impact that the constant presence of soldiers and their camp had on the city.

23 Tac. Ann. 6.15.2–3: *Ad offensiones ob rem publicam coeptas, utque Macro praefectus tribunorumque et centurionum pauci secum introirent quoties curiam ingrederetur petivit. Factoque large et sine praescriptione generis aut numeri senatus consulto.*

24 Tac. *Ann.* 1.7.5, *supra*, n. 4.

25 According to Nippel 1995 p. 93, this does not mean that they were always present at meetings in the Senate.

26 Sablayrolles 2001, p. 133 rightly notes that the measure was in embryo in the Augustan regulations, in particular in the *lex Iunia* of 17 DC.

27 The inscription from Ostia is *CIL*, XIV 3947 = Sablayrolles 1996, p. 539 App. III no. 1; the one with the fragmentary cursus of *Lucius Rufellius Severus* is *CIL*, XI 6224, from *Fanum Fortunae* = Sablayrolles 1996, p. 579 App. IV no. 1. On both, see the same Sablayrolles 1996, p. 38.

28 Cass. Dio 58.9.2–3 and 6; and 12.2.

29 Here indicated as *somatophylakes* (cf. the same Cass. Dio 55.24, where the term clearly indicates praetorians), a definition more appropriate for the *corporis custodes*.

30 Cass. Dio 58.12.2.

31 See Part II, p. 95. On the episode, see Sablayrolles 1996, pp. 38, 42–43, who considers it indicative of the role of the *vigiles* as "professional troops", which could be drafted in case of a serious crisis.

32 Many consider the period of the reign of Gaius for the tyrannical aspects of his rule and for the role played by the Praetorian Guard in favour of his accession to the throne (Durry 1938, p. 79; Keppie 1996 p. 111). Naevius Macro, then prefect of Egypt, will be disgraced and will be driven to suicide (Suet. *Cal.* 12. 26; Dio 59.10.6); on the issue, see Durry 1938, p. 176 and Passerini 1939, pp. 278–279.

33 See, for example, Suet. *Cal.* 23 (where a generically defined *tribunus militum* is mentioned) 26, 28, 32, 35; Cass. Dio 59.8.2, 13.2, 13.4, 21.4.

34 Cass. Dio 59.26.3 (*en autò to bouleutèrio*).

35 Bellen 1981, part. pp. 82–99; Durry 1938, pp. 22–23; Speidel 1994b, pp. 21–24 (Gaius); Menéndez Argüín 2006, pp. 35–36; Liberati-Silverio 2013, pp. 91–92.

36 Suet. *Cal.* 45. On the episode, among the others, return both Speidel 1994b, p. 23; and Austin and Rankov 1995, p. 31.

37 Philo's *Legatio ad Gaium* 27.175. The translation is taken from Philo, *The Embassy to Gaius*, translated by F.H. Colston, vol. X, Harvard University Press, Cambridge, MA and London 1991.

38 *Ant.Iud.* 19.253. According to Robinson 1992, pp. 183–184 and Liberati-Silverio 2010, pp. 87–88 the episode of Claudius' access to the throne shows the attempt of the senators to use urban cohorts and *vigiles* against him and the praetorians (see Cass. Dio 60.1.1).

39 Suet. *Cal.* 58: *Ad primum tumultum lecticari cum asseribus in auxilium accucurrerunt, mox Germani corporis custodes, ac nonnullos ex percussoribus, quosdam etiam senatores innoxios interemerunt.* A little further on, Suetonius had reported the news that *Thraeces quosdam Germanis corporis custodibus praeposuit.*

40 Fla. Ios. *Ant.Iud.* 19.119–122.

41 Durry 1938, p. 367, with sources.

42 Cass. Dio 60.12.4 (in January 42).

43 Suet. *Cla.* 25: *Milites domus senatorias salutandi causa ingredi etiam patrum decreto prohibuit.*

44 Suet. *Cla.* 25: *Iudaeos impulsore Chresto assidue tumultuantes Roma expulit.*

45 Tac. Ann. 12.52.3: *De mathematicis Italia pellendis factum senatus consultum atrox et inritum* (referred to AD 52). See also Cass. Dio 61 (60).33.3b.

46 Suet. *Cla.* 18.2: *Artiore autem annona ob assiduas sterilitates detentus quondam medio foro a turba conviciisque et simul fragminibus panis ita infestatus, ut aegre nec nisi postico euadere in Palatium valuerit, nihil non excogitavit ad invehendos etiam tempore hiberno commeatus.* Tac. Ann. 12.43.1: *Nec occulti tantum questus, sed iura reddentem Claudium circumvasere clamoribus turbidis, pulsumque in extremam fori partem vi urgebant, donec militum globo infensos perrupit.*

47 A memorial stone in travertine from Ostia commemorates a soldier of the VI cohort that, in the Augustan age, distinguished himself for his commitment *in incendio restinguendo* (*CIL*, XIV 4494 = *ILS* 9494).

48 Caldelli, Petraccia and Ricci 2012, p. 292 also recalls, as an indication of the new situation after 23, the episode of Plin. *Nat.Hist.* 1.15.5.

49 Suet. *Cla.* 25: *Puteolis et Ostiae singulas cohortes ad incendiorum casus arcendos collocavit*, refers to the years between AD 42 and 47.

50 Bérard 1988 has already expressed himself against the *vulgata opinio*.

51 It should be said that of the out-and-out barracks for the firefighters in Ostia we have however a sure evidence only from the time of Vespasian (see *infra*, p. 129).

52 The issue, with extensive bibliography, from Mommsen to Sablayrolles, and discussion of the sources, is considered by Caldelli, Petraccia, and Ricci 2012 at p. 292, n. 39–41.

53 Tac. *Ann.* 15.58.2.

54 *CIL*, VI 4344 = *ILS* 1722 = *AE* 2012, 181; *CIL*, VI 8802 = *ILS* 1729; 8803 = *ILS* 1730; 8808 = *ILS* 1728; *AE* 1952, 145–149. All of them are later presented, with photos, and discussed in the series *Museo Nazionale Romano. Le sculture*, I.7, Roma 1984, pp. 115–119 (A. Ambrogi, S. Priuli).

55 "*Phrourà perì ton Nerona*" by Cass. Dio 67.27.2B that coincides, in my opinion, with the *statio militum* revoked by the Senate of Suet. *Nero* 47.3; Cass. Dio 67.27.3 refers about the removal.

56 Cass. Dio 61.8.3.

57 For a similar reason, Dio continues, the *Princeps* withdrew the protective measure in favour of his mother (Cass. Dio 61.8.4). A different motivation according to Tac.

Ann. 13.18.3: Agrippina, now hated by her son, is deprived of the protection of the Praetorian Guard, which she was entitled to as the wife and mother of the emperor.

58 Tac. Ann. 15.66.2: *Iussu imperatoris, a Cassio milite, qui ob insigne corporis robus adstabat, corripitur vinciturque.*

59 Tac. Ann. 14.45.2: *Tum Caesar populum edicto increpuit atque omne iter, quo damnati ad poenam ducebantur, militaribus praesidiis saepsit.*

60 Tac. *Ann.* 14.61.

61 Tac. *Hist.* 2.11 and 33 and 3.64.

62 These events have been often treated by specialists of the Roman army. I simply want to mention, with regard to the troops in Rome, Durry 1938, pp. 372–376 (who speaks about a "praetorian reign of Otho"); and Freis 1967, pp. 11–13. See also Liberati-Silverio 2010, pp. 88–89 and 92–93.

63 Phil. *Apoll.* 4.35: "Nero did not tolerate philosophy. Its practitioners he considered inquisitive creatures who concealed their practice of divination . . . To mention only one man, Musonius of Babylon a man second only to Apollonius, was put into chains because of his wisdom and while kept there was in danger of his life." See also 4.47; 5.19.

64 Suet. *Vit.* 14.

65 Cass. Dio 66.9.2 and especially 9.13, referred to the years 70 and 71.

66 Cass. Dio 66.19.3.

67 Suet. *Dom.* 10 and Cass. Dio 67.13.2–3 (murder of the philosopher *Arulenus Rusticus*). Many more died for their ideas and activities; and all the philosophers who were in Rome were expelled. It is the year 91.

68 Tac. Hist. 4.46.8: *Modo Mucianum, modo absentem principem, postremum caelum ac deos obtestari, donec Mucianus cunctos eiusdem sacramenti, eiusdem imperatoris milites appellans, falso timori obviam iret.*

69 Millar 1977. Also see Crimi 2012, with previous bibliography.

70 Durry 1938, pp. 108–110; Passerini 1939, pp. 70–73; Speidel 1994b, pp. 33–35; Menéndez Argüín 2006, pp. 32–33; Liberati-Silverio 2013, pp. 89–91.

71 See the sources that have been quoted in Part II, Chapter 4, pp. 96–97.

72 Rather than for an integrated presence in the praetorian cohorts since the time of their creation (see Part II, Chapter 4, p. 97).

73 Suet. Dom. 6: *In Dacos [expeditiones] duas, primam Oppio Sabino consulari oppresso, secundam Cornelio Fusco, praefecto cohortium praetorianarum, cui belli summam commiserat.*

74 See also Part IV, Chapter 10.

75 Tac. *Hist.* 3.64.

76 "Les soldats des vigiles «descendaient» à Rome aux ides de décembre, avril et août et remontaient à Rome quatre mois plus tard quand arrivait la relève," according to Sablayrolles 2001, pp. 383–384.

77 According to Busch 2012, after the dissolution of the *corporis custodes*, "next to the regular Praetorian Guard there was a bodyguard made up of his freedmen controlled by one of them named *Parthenius* (Suet. *Dom.* 16–17; Cass. Dio 67.15.1)". In the texts indicated (*Parthenius* is also mentioned by Martial, in particular 4.45 and 5.6.2) there are not, however, elements that lead us to believe that among the tasks of the *praepositus Parthenius* there was also the one to organize Caesar's bodyguard.

78 See the monographs by Speidel 1994a and 1994b and Panciera 1974.

79 For example, towards Commodus (Cass. Dio 73.9.2, referring to the years AD 185–186; and Herod. 2.5.3).

80 Unlike what happened for more than a century to the soldiers of the urban cohorts and the Praetorian Guard, buried along with the civilians in the main cemeteries of the city (Panciera 1993 and Busch 2005 and 2007).

81 Sablayrolles 2001, p. 142; Roymans 2009. On the *castra priora* and *nova* built respectively in 113 and in 193, see Speidel 1994b, pp. 126–129 (with extensive bibliography and sources at p. 194 nn. 162 and 163); C. Buzzetti, *Castra Equitum Singularium* in LTUR I, 1997, pp. 246–248.

82 *AE* 1962, 289; Schillinger-Hafele 1977, no. 72; Speidel 1994a, no. 684: *Flavius Proclus, / eq(ues) sing(ularis) Aug(usti), domo / [P](h)il<a>delp(h)ia an(norum) XXI, / [h(eres) f(aciendum)] c(uravit)*. Speidel suggests that *Proclus*, during the revolt of *Saturninus* of 89, was a member of the cavalry of Domitian, which would represent an evolution of the *singulares* on horseback that accompanied Vespasianus in the Jewish campaign (Ios. *Bell.Iud.* 3.97 and 120).

83 Speidel 1994a, no. 689 and no. 690.

84 To the age of Traianus are dated the verses of the third satire (*Sat.* 3. 302–308) in which Juvenal attacks the ineffectiveness of the measures for the safety of Rome and its immediate surroundings. The attack is certainly aiming at the system that took shape in the first century of the imperial age.

85 Which, however, beyond the precise period of their creation, are related, considering the ideation and the occasion of its birth, to a context of late first century.

86 The idea of a joint command is supported by Durry 1938, pp. 166–167 and rejected by Vitucci 1956, pp. 46–48 and 86–87 (the last followed by Freis 1967, p. 43). For a synthesis, see Rucinski 2009, p 165.

87 *CIL*, VI 1009, cf. pp. 3070, 3777, 4315 = *ILS* 2012 = *AE* 2004, 43 (AD 140): *M. Aurelio Caesari / Imp(eratoris) Caesaris T(iti) Aeli Hadriani /Antonini Aug(usti) Pii fil(io), Divi / Hadriani nep(oti), Divi Traiani Parthici / pronep(oti), Divi Nervae abnep(oti), co(n)s(uli). / Petronius Mamertinus et Gavius Maximus pr(aefecti) pr(aetorio), / tribuni cohortium praetoriarum decem et / urbanarum trium, centuriones cohortium / praetoriarum et urbanarum et statorum, /evocati cohortes praetoriae decem et / urbanae X, XII, XIIII centuriae statorum / optimo ac piissimo.*

88 Robinson (1992, pp. 187–188) observes that "the *equites singulares* . . . are not listed in this inscription for the simple reason that they were quartered elsewhere . . . ; it is simply inconceivable that [the command] should have been taken away from him [the *praefectus Urbi*] and given to the praetorian prefect without adverse comment from the literary sources which are predominantly senatorial and anti-praetorian". On the praetorian prefects under Marcus Aurelius, see Rossignol 2007; on those of Severan era, see Coriat 2007 and Christol 2012, with extensive bibliography and sources.

89 On the jurisdiction of the urban prefect in the Severan age, see Solidoro Maruotti 1993 (civil jurisdiction) and Mantovani 1988 (criminal jurisdiction). According to Rucinski 2009, pp. 127–129, at the beginning of the Principate, the civil jurisdiction is not attributed to the urban prefect, but still to the Republican magistrates.

90 For the birth of the *praefectus Urbi*, the reference texts are still today Vigneaux 1896 and Vitucci 1956, which, however, do not devote specific attention to the first years of existence of this role. An article by Della Corte 1980 analyses the profile and the functions of the first *praefectus Urbi*. Systematic are the recent works by Mantovani (already mentioned, previous note), Rivière 2009 and Rucinski 2009.

91 Dig. 1.12.1.4, Ulp. De off. praef. Urbi: *Quidquid igitur intra urbem admittitur, ad praefectum urbi videtur pertinere. Sed et si quid intra centesimum miliarium admissum sit, ad praefectum urbi pertinet; si ultra ipsum lapidem, egressum est praefecti urbi notionem.* The problem, scarcely addressed after the Patsch's study of the late nineteenth century, was recently the subject of a systematic study by Rivière 2009 (part. pp. 235–250), with extensive discussion of the evidence and the secondary bibliography. Rivière also dealt with the means of coercitio available to the Late Roman Empire urban prefect.

92 Mantovani 1988.

93 As confirmed by the literary sources, in addition to the fundamental letter of Septimius Severus to the prefect Cilo. One only needs to remember the words of eulogy addressed to Piso by Seneca (Sen. *Ep.* 83.14: *Officium suum quo tutela urbis continebatur diligentissime administravit*); or those of Velleius Paterculus addressed to the same personality (Vell. 2.98: *Diligentissimum atque eundem lenissimum securitatis urbis custodem*), already discussed in Part I, Chapter 3 of this volume. See also Tac. *Ann.* 6.10.3.

94 Dig. 1.12.1.12.

95 Dig. 1.12.1.10: *Praefectus urbi adiri solet et pro modo querellae corrigere eum. Aut comminari aut fustibus castigare aut ulterius procedere in poena eius solet..etiam metalli poena in eum statui debet.*

96 Dig. 1.12.2, Paulus *liber singularis de off. praef. Urb.: Adiri etiam ab argentariis vel adversus eos ex epistula divi Hadriani et in pecuniariis causis potest.*

97 Dig. 1.12.1: *praefatio: Omnia omnino crimina praefectura urbis sibi vendicavit.* See Pugliese 1982, pp. 722–789; Mantovani 1988; Robinson 2000, p. 6; Rivière 2009; and, recently, Palazzolo 2015.

98 See Part II, Chapter 5.

99 *CIL*, VI 975, cf. pp. 3070, 3777, 4312, 4340 = 31218 e *ILS* 6073. SupplIt Imagines – Roma 1, 169 (this monument is known as the *Basis Capitolina*). The *curatores* are mentioned in the col. 1, lines 2 and 52; col. 2, line 27; col. 3, line 71. On the monument, see Boatwright 1987, pp. 26–27; and Ménard 2004, pp. 32–33.

100 Panciera 1970. The text, according to the integration proposed by the editor, contains a dedication to the Lares Augusti and the Genii of the Caesars, "*permittente Imperatore Caesare T(ito) Aelio Hadriano Antonino Augusto Pio*", and closes with the indication of the name of the prefect of the *vigiles* and the date.

101 *Supra*, n. 99.

102 Dig. 1.15, in particular 15.1 and 15.3. Baillie Reynolds 1926, pp. 30–42; De Robertis 1937, part. pp. 35–41; Sablayrolles 1996, part. pp. 51–55 and 124–129; Nippel (1995, pp. 95–97) however, declares to be sceptical that the Severus' reform on the judicial competences of the prefect of the *vigiles* would foresee an implementation of this function over time.

103 Lo Cascio 2007, p. 157. With time the status and the rank of *curatores* would change and "the same evolution appears to correspond to a change and, perhaps, to a widening of the tasks (control of the population, maintenance of public order, exercise of jurisdictional powers) of those to whom the care of the regiones was entrusted" (Lo Cascio 2007, p. 157). Although, due to the current state of the documentation, it is impossible to specify in detail.

104 *CIL*, VI 30960 = *ILS* 3621 (AD 223) and 30961 (222–235); *RAC* 17, 1940, p. 24 (205–208).

105 Scriptores Historiae Augustae, *Vita Marci* 11.9: *Dedit praeterea curatoribus regionum ac viarum potestatem, ut vel punirent vel ad praefectum urbi puniendos remitterent eos, qui ultra vectigalia quicquam ab aliquo exegissent.*

106 An inscription (*CIL*, VI 31420 = Sabbatini Tumolesi 1988, pp. 48–51, no. 44), perhaps from the Forum of Caesar, containing an *excusatio magisterii* by a *magister vici* of AD 186, in relation to the organization of a *venatio*; the competence of the judgment seems transferred from the praetors to the emperor, represented by the prefect of the *vigiles*. The document has aroused much debate, recalled by the same Sabbatini.

107 *CIL*, VI 32327, p. 3824 = *ILS* 5050a (AD 204) contains the fragment of a letter of recommendation of the emperors to the college of *quindecemviri* (in particular

to the lines 20–22): *admonemus Quirites dominos urbano[s. . . et eos quo]que qui mercede habitant in noctibu[s feriarum illarum ut una cum mili]tibus nostris circumeuntibus [reg]ionum tutelam diligenter administrarent.* See Panciera 1978, p. 268 and note 21; Sablayrolles 1996, pp. 377–378.

108 The main source is, once again, the Historia Augusta (*Vita Alex. Sev.* 33.1): *Fecit Romae curatores urbis quattuordecim, sed ex consulibus viros, quos audire negotia urbana cum praefecto urbis iussit, ita ut omnes aut[em] magna pars adessent, cum acta fierent.* It has been recently treated by Nasti 2006, pp. 214–226, with discussion of other sources and extensive bibliography.

109 Nasti 2006, p. 220.

110 Coarelli 1987, recalled by Nasti. If Rome and the district within a radius of 100 miles was under the praefectus Urbi, outside that radius the praetorian prefect had jurisdiction; while the prefects of the *annona* and the prefect of the *vigiles* were given the power to proceed by way of justice for those who committed crimes within their powers (Dig. 1.15.3.1, Paulus *lib. sing. de off. praef. vig.*: and 1.2.2.33, Pomp. *lib. sing. ench.*). Pavis d'Escurac (1976, p. 270 and passim) excludes the possibility that the *praefectus annonae* exercised criminal jurisdiction.

111 Dio 80.2.2–4.

112 Fulvio Grosso (also quoted by Coarelli 1987), for the external support provided to the people, recalls other episodes (Grosso 1968, especially pp. 431–432).

113 Rucinski 2009, pp. 62–66 and 151. On the Severan legislation about the *collegia* and the *crimen maiestatis*, see Daguet-Gagey 2003, p. 503 and n. 11

114 Cass. Dio 74.1.1–2, 2.4–6; Herod. 2.13.1–12, Scriptores Historiae Augustae *Vita Severi* 6.22.

115 On the evolution of the urban cohorts, see Part III, Chapter 7.

116 *CIL,*VI 1056, 1057 and 1058 = 31234.

117 Sablayrolles 1996, pp. 176–178. Of all the dedications of *vigiles* that remain, as many as seventeen belong to the time of Severus, while sixteen regard the other emperors all together.

118 Durry 1938, pp. 383–385.

119 Cass. Dio 77.4.2–5 and Scriptores Historiae Augustae *Vita Car.* 4.6.

6 A topography of security and dangerous places

With an episode

The focus moves, in this chapter, to the places of security, namely buildings or areas potentially dangerous because of substantial gatherings[1]; or because our sources provide us with reports on public disturbances; or naturally for both reasons.

Certainly neither barracks nor prefectures can be defined, strictly speaking, as 'dangerous places'. The presence of military personnel in the offices of the prefectures, the impenetrability of the walls and doors, and the existence of stocks of arms (*armamentaria*) in the barracks[2] made them overall safe places. The only episode of attack on the barracks of the praetorian guard occurs – not surprisingly – in AD 238.[3]

The issue on the location of urban barracks and of the headquarters of the prefecture is, however, of particular importance, and in recent years has received adequate attention. Apart from the description and the examination of these structures – beyond the scope of this chapter and already conducted exhaustively elsewhere – the interest here is that of the possibility of identifying a plan and a coherence in the implementation of what can be called 'a preventive topography of security'.

The dangerous places that will be considered are enclosed public places. As in the Republican era, also in imperial times the security of the aristocratic residences remained a private matter.[4] To move through the streets of Rome was certainly a risk, but wealthy Romans could be accompanied by a security escort, even large in number, composed mostly of domestic slaves.[5] Whoever was traveling in a litter would have robust men who, besides from carrying this human-powered transport, also deflected any potential attacker with their impressively built body. Interesting symptom of this is the inscription from an urban sepulchral monument constructed by the powerful *gens Statilia* that tells us of Iucundus, a Statilius Taurus' *lecticarius*, who "as long as he lived, he was a man and defended himself and others".[6]

The buildings or complexes, of which I will discuss in the following sections, are important and prestigious and, as such, they have been the subject of numerous studies. I will try, on this occasion, to look at them from a different perspective, not directed at their structure or capacity: rather I will focus on their strategic position in the framework of the imperial security (for prefectures, military camps and stations); or on what happened when, with respect to the primary purpose for which they were built, they become the scene of riots and clashes (for imperial palaces, places of entertainment, baths). What matters, ultimately, is to highlight the causes and the occasions of implementation of the security dispositive in such circumstances, and the specific role of personnel, mainly military, involved to ensure or to restore order.

In Rome there were other sites of course, besides those considered here, where public gatherings took place (inns and taverns, venues for association meetings, banquet halls, etc.), in addition to spaces where assemblies (*contiones*) met. For all these places, we have little evidence from the imperial era, and none mention instances of disturbance to public order.[7]

Castra praetoria

The camp of the Praetorian Guard – a giant rounded rectangle with the short sides measuring 380 metres and the long ones of about 450 metres – occupied a large area between the present Via Tiburtina and Via Nomentana (Via del Policlinico, Via Gaeta and Via San Martino della Battaglia) (see Figure 6.1). The complex is portrayed on coins of the Julio-Claudian age and reliefs. From the

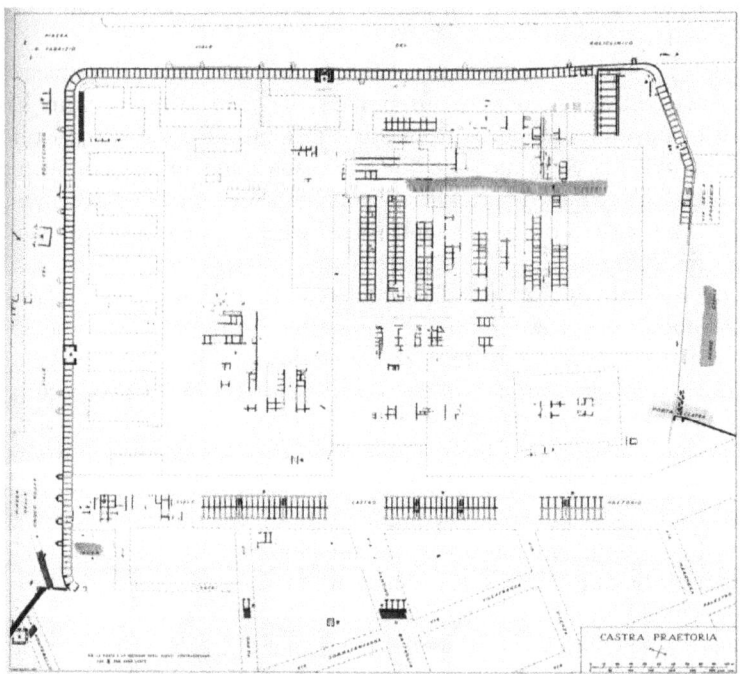

Figure 6.1 Map of the *Castra Praetoria*

Source: drawing from Lissi Caronna 1993, *s.v.* '*Castra Praetoria*', in *Lexicon Topographicum Urbis Romae*, I: 251–254.

scarce archaeological remains we can identify the interior space that was composed of cells tightly placed against the wall, with chemin de ronde around the top. Other buildings, in addition to the *principia*, certainly present because attested by inscriptions or ancient authors, were a temple of Mars, a *tribunal*, a *schola* and an *armamentarium*. For the soldiers' drills a nearby campus, which stretched towards the west, was available.

Whole sections of the complex cannot be reconstructed, and there is no possibility of systematic investigation due to the presence, on the above ground, of structures, including military ones, or heavy-traffic roads. In the 1930s, Marcel Durry's monograph devoted ample space to the birth and the construction phases of the barracks of the Praetorian Guard on the Viminal Hill[8]; while archaeological investigations, carried out since the 1960s, have helped to better define some details of the complex.[9]

Lastly, Alexandra Busch's monograph on the barracks in Rome adopts, for the *castra praetoria*, the same scheme of analysis later applied also to the *castra* (*vetera* and *nova*) *equitum singularium*, the *castra peregrinorum*, the stations of the *vigiles* and the great legionary camp of Albanum. This scheme provides a general historical context, history of the studies and description of the topographical situation and that of the external and internal structures of each complex with the furnishings. Apart from the new evidences relating to the architectural survey and new excavations that have been carried out at the *castra* of Albanum, for the camps in Rome the author makes use of secondary literature, without bringing new elements.[10]

Most studies concerning the praetorian cohorts and, more generally, the military presence in Rome, have highlighted both the physical impact of their cantonment,[11] and the concentration of different forces that were quartered there. The barracks on the Viminal Hill, in addition to the praetorians and the urban soldiers, also housed the *statores* (perhaps after their creation) and the praetorian *speculatores*, the chosen body integrated in the praetorian cohorts at least since the end of the first century AD.[12]

Marco Maiuro has recently been intrigued by a specific aspect of the military camp of the Praetorian Guard: namely the legal nature of the plot of land where it was built. Starting from a passage of Ulpianus,[13] the author reveals how, among the imperial properties, there were some that could be easily sold, although they were practically inalienable; and to this end he cites the Albanum (*fundus Albanus*) and the Sallustius Gardens (*horti Sallustiani*). It is the latter, of which Tiberius took possession at the end of the year AD 20, chosen for the construction of *castra praetoria*, completed within three years. Maiuro believes that the topographical location of the *castra*, at a scarcely urbanized but strategic locality outside the Servian wall, was possible because of the recent imperial appropriation. He also thinks, by reversing the perspective, that the *horti* were destined to become the place for the imperial suburban residence par excellence because of its position near the huge military complex. This way of reasoning allows Maiuro to draw a parallel with what, after two centuries, will occur with Septimius Severus and the *castra legionis*

II Parthicae at Albanum. In this case, the appropriation of the property was not necessary, since it had already been in imperial hands for some time; and compared to Tiberius, Severus could act with the nonchalance characteristic of the winner of a bitter civil conflict.

The only episode of attack on the barracks of the Praetorian Guard to which we have referred occurs after the time frame that is being considered here. It takes place after emperor Gordianus' death, when the Senate proclaims Pupienus and Balbinus as his successors. The furious people, ably provoked by the senators, rage against the *castra praetoria* – where at that time there were only soldiers waiting to be discharged – as a symbol. It is part of a far more wide-ranging clash, in a year of great transitions, which has scenarios like Rome, Italy and some provincial areas; and as protagonists, the Senate and the people of Rome, the soldiers (legionaries and urban troops) and emperors. Our sources are the Scriptores Historiae Augustae[14] and Herodianus.[15] The version of the latter is particularly interesting for the reconstruction of the protagonists and of the responsibilities of the narrated events. The conflict took place in different places of the city: initially at the Curia, seat of the Senate, then at the barracks of the Praetorian Guard, to end, during the most intense phase, at other buildings and spaces of the city, with a quite heavy death toll.[16]

Only a few words may be added on the seat of the praetorian prefecture. We have no information of its location. It is more credible to think, in the light of recent suppositions in this regard that it the praetorian prefecture was located on the Palatine Hill, also because, as we have seen, a cohort stationed there permanently.[17]

Stationes vigilum, praefectura vigilum

The topography of the barracks and of the guard posts of the *vigiles* in the fabric of Rome were gradually defined, with subsequent interventions during the first two centuries of the Imperial age, which can only be partly reconstructed. The location of some of the barracks seems today aberrant compared to a theoretical model of rationality and can be actually explained, as suggested by Robert Sablayrolles, because of the profound changes taking place between the first and third centuries (see Figure 6.2). Still in the Julio-Claudian period, for example, the cohort VI, housed in the regio X, changed its location because of the construction of the imperial palace, where security could be however guaranteed by the *familia Caesaris*, the Praetorian Guard and the *corporis custodes*.

The biggest problem actually occurred with the initial stage of the construction of the complex. It is questionable whether, from the moment in which the corps was created, the *vigiles* were housed in quarters created especially for them; or if, instead, like for the praetorian and urban cohorts, in a first phase (Augustan) a temporary solution was followed by a second more coherent planning to deploy permanent stations in different regions of the city. The two opposing views, in this respect, collide.

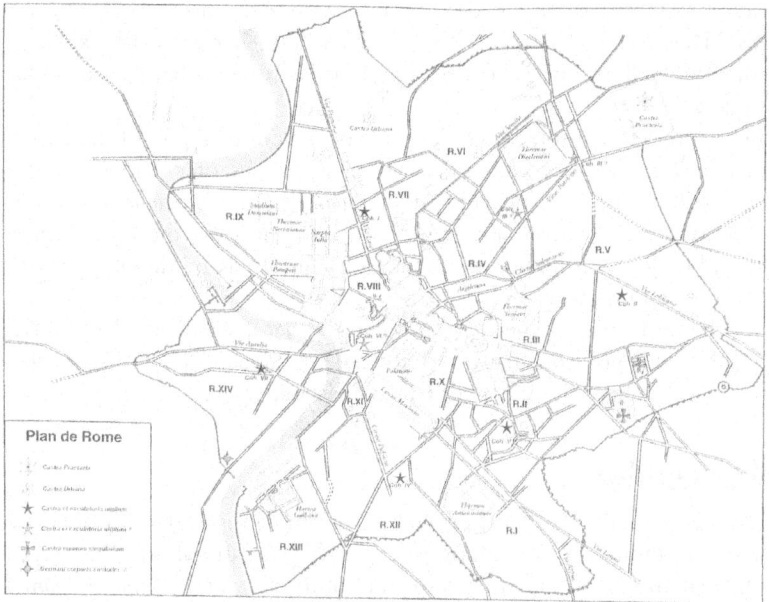

Figure 6.2 Map of Rome with *Castra* and *Stationes militum*

Source: drawing from Sablayrolles 2001, p. 153. Courtesy of PUM (Presses Universitaires du Midi) Editions.

The dedication to Hercules with a reference to the *invigulantes pro vicinia*[18] makes Silvio Panciera (and those who share this view) think that, even in AD 12–13, the *vigiles* did not have *stationes*[19]; and also makes Panciera wonder when and to what extent the barracks began to be built. In fact, we do not even know if the men at the prefect's disposal were, from the outset, more than 3,000.

Besides, Robert Sablayrolles who has dedicated a reference monograph to the *libertini milites* in 1996, does not rule out that the construction plan of the barracks of the *vigiles* dates back to the Augustan age.[20] Some issues may, in his view, be regarded as an indirect indication of such an event: the potential danger of the *vigiles*, in the first century, was – for their number, for the assigned functions and the type of arms at their disposal – incomparably lower compared to that of the Praetorian Guard and could not be considered as an obstacle to integrate them, from the initial moment of their existence, into the urban fabric. There was, in Sablayrolles's view, the actual need to concentrate the *vigiles* in the heart of Rome in order to ensure maximum efficiency and timeliness of intervention. Finally, the rationality adopted in the distribution of the barracks would be compatible with the idea of an Augustan plan.

However, we consider the matter, it is a fact that only at a later time, in the period between Tiberius and Claudius, the *vigiles* obtained a more defined hierarchical organization and the possibility to be granted citizenship after six years of service. Moreover, the first known prefect of the *vigiles* (Quintus Naevius Macro) dates back to Tiberius. Therefore, it cannot be excluded a priori that the well-articulated construction project of their barracks dates back to this phase.

Other issues, in addition to whether or not the construction of the barracks had been built gradually, concern the possible existence of a centralized camp; and the possibility to identify with sufficient precision *excubitoria* and *stationes*, before the time of Severus. Sablayrolles is the one to offer the most comprehensive and debated review on the structures identified and divided into sure, probable and still unknown.[21]

As for the prefecture of the *vigiles* it is certain that it was located at the Balbi crypt,[22] in central position, halfway between the area of the Forum and the *porticus Minucia*, where the wheat distributions took place.

Praefectura urbana

In the next section (Part III, Chapter 7), I will examine the time and place where the *castra* were built, and which hosted exclusively urban cohorts. This section is devoted exclusively to the prefecture and not to the *castra militum urbanorum*.

The headquarters of the prefecture most likely changed location various times, adapting to the changing reality of the buildings of the central area of the city.[23]

The original location of the prefecture, according to Filippo Coarelli, was in a basilica built by Augustus[24]; then it was transferred, during the Flavian period, to the *templum Pacis*, where it remained until Late Antiquity, when some sources place it at the *aedes* in the Carinae.[25]

Somehow divergent are the opinions of Giuseppina Caruso, Rita Volpe and Eugenio La Rocca who think that the building unearthed during the recent excavations of the Baths of Trajan (mid-first century AD) housed at least part of the offices of the prefecture.[26] Their opinion seems to be confirmed by the concentration of epigraphic attestations in the border area of regions III and IV, in the zone between Via della Polveriera, the former route of San Pietro in Vincoli and via degli Annibaldi.[27]

Finally, the fascinating hypothesis of a polycentric prefecture, with multiple offices in different areas of the city, functional to the multiplicity of tasks that its officers had to carry out, should not be rejected. One of such offices could have been, for example, near the *castra urbana*.[28]

Castra corporis custodium (?), castra equitum singularium

Some scholars have suggested that the *corporis custodes*, at least during the reign of Nero, had their own camp; and this one would have been located along the Portuense road (Trastevere).[29] If the first hypothesis is not totally far-fetched, the second, in my view, is built upon very weak foundations.

When the *corporis custodes* were created, as slaves and part of the *familia Caesaris*, they were buried in the columbarium of the imperial family and their eldest funerary monuments come from the Appian Way.[30]

The character of this paramilitary corps, however, has changed in the course of time until obtaining a more distinctive military profile.[31] From the reign of Claudius, they maybe received permission by the emperor to be buried in the necropolis of Rome: one could be that on the Via Aurelia[32]; another in the 'Pozzo Pantaleo' area, on the Portuense road.[33] The latter has been identified as the site of the barracks of the *Germani*. We know, however, that the place of burial of the soldiers of Rome has not always been close to their camp[34]; in addition, the five steles discovered here cannot be considered tangible evidence that confirms the theory.

A discussed passage of Suetonius alludes to a camp of the Germani:

> He also disbanded a cohort of Germans, whom the previous Caesars had made their bodyguard and had found absolutely faithful in many emergencies, and sent them back to their native country without any rewards, alleging that they were more favourably inclined towards Gnaeus Dolabella, near whose gardens they had their camp.[35]

The expression "*iuxta cuius hortos tendebat*" means "near whose gardens (the cohort of Germans) has its camp". The only clear information we receive from Suetonius is therefore that the *horti* of Dolabella were near the bodyguards' camp. Unfortunately, we do not know the location of neither. That is why if the hypothesis of a camp of the *corporis custodes* on the Portuense road is not based on solid foundations, nor the location of the Dolabella Park, in absence of decisive clues, can at the moment be determined.

The emperor's new chosen guards on horse, the *equites singulares Augusti*, were initially quartered in special barracks (*castra priora*) in today's Via Tasso (Via Labicana), identified thanks to the inscriptions brought to light during the excavations between the end of the nineteenth century and the 1930s; a new and more spacious barracks (*castra nova*) was built by Septimius Severus in the Lateran area.[36] Of the first only a few vestiges remain and is mostly epigraphic material that enriches our knowledge; the recent work by Paolo Liverani and the new excavation project of a systematic study of the whole Lateran area[37] is intended to clarify and implement considerably our knowledge on the orientation and layout of the new complex beneath the great basilica of Constantine.[38]

At the gates of the city: the *stationes militum*

The sources do not lack to inform us of the many risks wayfarers incurred even at the gate of Rome, where there were numerous hidden dangers and frequent ambushes.[39] It has been seen[40] that, by order of Augustus, repeated and extended by Tiberius, military posts to deal with public order were placed throughout Italy. The famous and often quoted passage of Ulpianus,[41] recalling as a priority task of the urban prefect to keep the peace among the citizens and order during performances, stated that it was up to him to prepare *stationes* in order to ensure the control and peace.

We unfortunately have little information on the composition of the corps of guards of these garrisons outside Rome: only a few urban inscriptions of the Severan age help us locate a few. In fact, at this time a military *statio* existed at the third mile of the Appian Road, not far from the tomb of Caecilia Metella, as testified by two dedications. The first is engraved on a column probably intended to support a statue, offered to the Genius of the *castra peregrina* for the salvation of Alexander Severus; the devotees are two *frumentari*,[42] detached respectively from the Legion VIII Augusta (Quintus Haterius Valerianus) and the Legion XIII Gemina (Marcus Aurelius Sophaenitus). The two claimed to have built the *statio* 'for members of the association' at their own expense.[43] Of the second dedication, where another *miles frumentarius* says to have worked on a restoration of the building, and is therefore next in order of time compared to the previous one, we unfortunately have only a fragment.[44]

A second guard post certainly existed on the Via Latina, by the catacomb of Praetextatus: here, in 1933, an inscribed slab was found, re-used in the catacomb to close a tomb.[45] Again this is a dedication addressed to the emperor, Septimius Severus and to the *principes* Geta and Caracalla, by a soldier recalled to duty, *Iulius Primianus*, who on the Via Latina in addition to a *statio* also built a *schola*; while *Septimius Proculinus* (probably one of his colleagues) took care of having the inscription engraved. We are not able to establish to which corps the soldiers belonged, but according to Géza Alföldy (*CIL*) probably they were praetorians or urban soldiers: the scholar excludes the possibility that there was a second *statio frumentariorum* at such a short distance (about a kilometre) from the third mile of the Appian road, I would not rule out in principle that also in this case they are *frumentarii* and that the two *stationes*, rather than co-exist, may have occurred one after the other within a few decades.

To complete the picture, another three testimonies can be linked to the previous two[46]: none of them explicitly appoint a station, but their content reasonably induces to think that they were referring to a structure of this kind.

The first is a dedication, once again, from the Appian Road, and concerns legionaries of the II Parthica. The dedication is carved on a large marble slab reused in a mausoleum built over the catacombs of San Callisto and is dated 24 July 242[47]: a Genius, in this case that of the Legion II Parthica, together with Fortuna Pacifera and Redux, is called to protect Emperor Gordianus III, and his wife Sabinia Tranquillina. The curators of the dedication are two officials: Valerius Valens, acting on behalf of the praetorian prefect Timesitheus (engaged in the Persian campaign, perhaps with the emperor), and Pomponius Iulianus, at the command of the legionnaires who were in *Albanum*.[48]

Not only does the large slab (which probably covered a base) show the loyalty of the legionaries of *Albanum* towards the imperial family, but it is also a significant indication of the division of powers among legionaries in control of the most important Roman roads. If, in fact, outside the city up to the third mile, the *frumentarii* of the *castra peregrina* were the ones to have control responsibilities; beyond that limit the legionaries, camped in the nearby town of Latium, were in charge.[49]

The comparison with the testimonies of Ostia of *stationes frumentariorum* offers insight that these stations had different character and purposes, related to the nature and the needs of the urban centers where they were placed.

Between Ostia and Porto quite a few testimonies, dated between the second and third centuries, reveal the presence of the *frumentarii*, connected to the port activities of the town and in particular to the organization of the *annona*.[50] In the text reported on a marble slab from the area of the port of Claudius (AD 210),[51] the link between annona and *frumentarii* is very clear, as confirmed in a dedication from Porto of ten years later,[52] which speaks of the allocation of a space for a *statio frumentationum* by three different figures, a *procurator* of libertine status and two centurions, one of which explicitly connected to the *annona*.

Next to these texts, full of details are epitaphs and some dedications (to the Genius of the *castra peregrina*, to Isis and to Genius *cenaculi*) of soldiers and their officers found respectively in the necropolis of Isola Sacra in Porto and in various areas of the city (near the theatre, the baths and so on). All these testimonies are a sure sign of the presence, perhaps stable[53] of a *statio frumentariorum* at Ostia between the second and third centuries. The exact location of such station is unknown; in all likelihood one, and perhaps not the only one, of the reasons that determined the installation was linked to the administration of the grain supply.

In Rome, there were stations near the south gates of the city (Appia, Latina); and, perhaps, at the imperial palace[54] or its vicinity. It seems reasonable to assume that the function performed was that of controlling the access to the city; even considering the fact that the roads which led to the north and east – the Tiburtina, the Nomentana and the Flaminia – were guaranteed by the presence of large encampments (*castra urbana* and *castra praetoria*). At Ostia and Porto the *milites frumentarii* were not placed outside the walls: they were fully active at ports and in the city center. The control function was linked to the commercial activity of the port of Rome; and the soldiers were used in close collaboration with other authorities, such as the prefect of the annona and perhaps the prefect of the *vigiles*. One cannot but emphasize how that integration of action between administrative officials and military officers that, at the beginning of the Principate, was typical for the city of Rome (*vigiles*, urban soldiers and magistrates of the *vici*) was implemented, at a different time and with different protagonists, at its most important port on the Tyrrhenian Sea.

To the diversity of places where the *stationes* arose corresponds a substantial chronological uniformity. The dating of documents, both urban and ostiense, testify a particular concentration of initiatives, such as the construction of buildings and the erection of dedications, in the third century and in particular during the Severan era.

Palatium

Studies related to the complex of the *domus imperatoriae* on the Palatine Hill have known, in the last two decades, a real surge in connection with excavation activities. However, it is still very difficult to be able to both distinguish the different building phases and to identify the functionality of the single rooms in the different complexes.

Some scholars have not failed to consider the matter, which concerns us here, of the protection of the *Princeps* and his family in their home. In this section I will try to resume such reflections by putting them into relation with those that have been carried out in the previous chapters. Valuable insights, for a parallel reflection, are also suggested by what in Part IV will be discussed in relation to the security service of the emperor during his travels in Italy. As hitherto, the discussion will try to follow a chronological order.

We know that different forces were committed to ensure the emperor's safety in the residence where he exercised his public and private activities. The monitoring of the different solutions adopted is anything but easy: the Palatine residence was the place of representation par excellence.[55] The protection and representation of the emperor at his home in fact is entrusted to a number of figures, which alternate throughout the day and at different times, depending on the emperors and on the events that mark its end. The various solutions are further subdivided between internal spaces and external spaces, to ensure both the emperor's safety and security at the same time.

The advances in research regard the space of action of the praetorians (particularly in the second half of the first century AD) outside the Palatine palace; while it still is difficult to follow the track of *speculatores* and *corporis custodes*,[56] for the first imperial age; or of the *equites singulares Augusti*, from the second century on.

The issue regarding the Praetorian's Guard service is not an easy one for two reasons. Although the presence of a praetorian cohort that alternated daily at the palace is certain since the age of Augustus,[57] the details on the service are expected to remain obscure. Only one passage by Martialis seems to allude to the fact that there was an on-call duty service also during the night, with a shift system.[58] However, we have no explicit indications on where the guard post was placed.

A recent contribution has actually considered this last aspect: namely, the distinction between the internal security guaranteed to the *Princeps* in his own home, against accidents that could occur to him; and the one offered against external threats.[59]

The *fores Palatii* and the *aditus domus* are two elements that cannot be mistaken, and are directly related to the distinction made between the public and private part of the imperial palace. If on the one hand the *fores* are real double doors that open when the *Princeps* and his entourage enter, and are locked behind him; on the other hand, the *aditus* is the threshold, the entrance to the inner vestibule. Despite the changes and adaptations that the imperial residences undergo (however difficult to prove), the distinction between the different entrances must still be considered; and as Yves Perrin[60] acutely observes, sources clearly distinguish between the men who watch the *aditus domus* and the secret passages from the inside; and those who instead watch the *fores Palatii* from the outside. To assign to each of the entrances the respective figures of reference, one could indicate for the first the *speculatores*[61] and the praetorians for the second.

The examples on the relationship between the *fores* and praetorians are numerous: the seventeen-year-old Nero, once having found out about Claudius' death, approached the sentinels (*excubitores*) who acclaimed him emperor on the steps of the palace (*proque Palati gradibus imperator consalutatus*) before taking him to the *castra* and then to the Senate.[62] When his time occurred, Messalina was at the park of the palace, lying on the ground, when suddenly the palace gates from the outside opened wide (*pulsae fores*) for the impetus of those who were coming, and a silent tribune (belonging to the Praetorian Guard)[63] appeared. The building gates also opened from the inside towards the outside (*foribus palatii repente diductis*), for instance, when Nero, accompanied by Burrus, moved towards the cohort on guard, located externally.[64] Another example of integrated control, can be identified in the 'Life of Vitellius': when the emperor, after a brief attempt to escape, returned to the palace that was now devoid of the Praetorian Guard, took refuge in the custodian's cubicle, after tying to the front gate which led to the complex a dog (*religato pro foribus cane lectoque et culcita obiectis*), as a weak protection.[65]

There are various passages that recall the Praetorian Guard at the staircase of the building (*gradus domus*) in the years 68 and 69.[66] Today, unlike in the early last century, it seems that, in the Flavian era, one of the entrances to the palace could be identified as the highest point of the Via Sacra,[67] where there must have been a guard post; another, equally if not more important, was on the *clivus Victoriae*, where rooms, in the east and west side of the road, elsewhere interpreted as shops (*tabernae*), are now thought of as reserved to the security personnel.[68]

Some structures arranged around the upper level of the peristyle of the *Domus Augustana* were also interpreted as positions for the guards. It is reasonable to think, despite the lack of concrete evidence, that such structures represent viewpoints, but what has been previously said makes it difficult to understand if they were occupied by the praetorians; or not, for example, by the *speculatores* or by domestic staff of servile status and freedmen.[69]

In the second century some changes occurred in the forces manning the person of the *Princeps*: outside the palace, the service of the praetorian cohort at the *fores* did not undergo any significant changes inside the palace, while inside the palace the *corporis custodes* and the *speculatores* were replaced by the *equites singulares*. The complexity of the system of protection of the *Princeps* at this time is emphasized by Fulvio Grosso,[70] who recalls the episode in which Pertinax was killed by a Tungrian *eques singularis*.[71]

In the Severan period, the system of access to the Palatium from the side of the Clivus Palatinus did not change, according at least to what has been suggested by the excavation and study of the Vigna Barberini.[72] There are also no significant elements related to the possible accesses on the south-east corner or on the north-west side of the hill.[73]

A dedication to Alexander Severus, Julia Mamaea, Julia Maesa and the Genius of the *castra peregrina* of the late Severan period[74] may, however, be an indirect indication of a change in the security system near the residence of the emperor. It is the fulfillment of a vow made during the initial stage of his career by a *princeps peregrinorum*, Titus Flavius Domitianus, at the *Atrium Vestae* in the Forum, where perhaps there was a *statio peregrinorum*. The dedications to the Genius of the *castra peregrina*, in Rome and in Ostia, were in fact located within the stations where these legionnaires, detached for service, were operating.[75] It is natural to put into relation the dedication of the officer Domitianus to those of the *stationes* of Via Appia and Latina; and to interpret the placement of a *statio peregrinorum* in the Forum as a significant sign of the changes introduced by the Severi dynasty as in the composition of the urban troops and their functions, also in the urban security system.[76]

If also before the Severans we have some evidence of the progressive loss of the praetorians' privileged role as guard and escort of the *Princeps*[77]; the picture that emerges from the events from the years 20 and 30 of the third century on, is that of different forces, with changing roles, sometimes involved occasionally in the emperor's personal protection, because of the situation of political fluidity and the alternation to the summit of power. A particularly clear example is successive to the time span considered here and regards the events that led to the deposition of Maximus and Balbinus after only three months of regency in 238, when the praetorians, hostile to the two *Princeps* appointed by the Senate, feared to be replaced by the Germani (this term, at this moment, designated the elite guard composed of auxiliaries who had accompanied Maximus in Rome).[78]

Places of entertainment (theatre, amphitheatre, circus)

The numerous and spacious buildings in Rome dedicated to games and shows, even for the symbolic value of the events hosted here, were frequently the context of episodes of unrest. Historians and archaeologists have dealt extensively with the mechanisms that loosened the viewers' inhibitions in order to better interpret the inconvenience hidden behind each episode.[79] Also in this case, our interest is drawn upon the protagonists charged with controlling the riots and upon the functioning of the security dispositive.

In 2 BC, the people sent an embassy to Augustus, who was at Antium, asking him to accept the title of *pater patriae*. When the *Princeps* came back to Rome, the plebs approached him, and greeted him with laurel branches and, as he was making his entrance to the show, they repeated their request.[80] This is the first instance where the people use a public building for entertainment to obtain something from the emperor. In the same way, with inverted roles, the performances during the imperial age could be used by the *Princeps* to test the popularity of an initiative, to answer the people's questions, to express his paternal kindness or severity,[81] or even to gain approval for the administration of the criminal justice, adopting sometimes spectacular modes.[82]

The performance venues were nevertheless a privileged scenario where the people could express their discontent against a measure considered contrary to their own interests; or to give their vibrant support to their heroes, actors, gladiators or charioteers. Excessive vitality of these demonstrations or clashes between supporters of different parties could lead to real disorders,[83] in which the protagonists were sometimes punished in an exemplary manner.[84]

All these circumstances imposed that, in places of great popular concentration, as a preventive measure, soldiers and other personnel with supervisory duties had to be present. The ancient authors, referring to this presence, are divided between those who see in it a potential risk of unrest; and those who consider it not inappropriate.

The first evidence of the presence of the praetorians in a theatre dates back to AD 15.[85] In the passage by Tacitus, the security force (a tribune, a centurion and some praetorians) are among the victims; and it seems clear that the cohort of which they were part was following the *Princeps*.[86]

The praetorians are banned by Nero in 55 from attending any places with popular gathering.[87] A wide range of explanations are proposed by Tacitus on this fact: the *Princeps* maybe intended to provide the audience with the illusion of a greater freedom of action; or gave them a chance to test their self-control once the guards were removed (*amotis custodibus*); but also distracting the soldiers from the typical atmosphere of license of the theatre. Less than a year later, however, the measure is withdrawn, coinciding with an expulsion of the actors.[88] Of a resumed participation of the praetorians to the shows further evidenced is by Tacitus, when he refers to an entire cohors, with officers and centurions and the same praetorian prefect, applauding and cheering in 59 the emperor, while mingling with the crowd of flatterers.[89]

When the sources specify which soldiers were in charge of prevention or suppression of disturbances at the theatre or at the stadium, it is always the praetorians. The tasks and the areas of control were divided up between praetorians, on one hand, and the *milites urbani* (and the *vigiles*), on the other: if the praetorians were on duty within the entertainment places, the *milites urbani*, and the *vigiles* were charged with what was happening in the rest of the city, with the patrolling of streets and homes to avoid riots and thefts.[90]

The praetorians are then, inevitably, also the protagonists of the repression. When in AD 40 the people rush to the circus to protest against fiscal measures, Gaius orders soldiers to contain manifestation by sheding blood.[91] Some time later, the same soldiers, distributed among the various sectors (*per cuneos*) of the Taurus's amphitheatre, obeying to Nero's order, should batter those spectators who were not showing interest by applauding.[92]

After a century of peace and freedom, at least in appearance, from the end of the second century there are again signals of gratuitous violence on the part of the emperors but now, as an instrument of repression, they also employ other figures in addition to the Praetorian Guard. An episode that insists on Commodus' ruthlessness seems traced on that of Gaius' model: the only difference is that this time the performers are the soldiers of the fleet of Misenum:

> And although the people regularly applauded him in his frequent combats as though he were a god, he became convinced that he was being laughed at, and gave orders that the Roman people should be slain in the Amphitheatre by the marines who spread the awnings.[93]

Caracalla was responsible for a massacre when, during a chariot race, the crowd teases one of his favourite charioteers. Although the order given to the troops (*to strateuma, oi stratiotoi*) was to make arrests and carry out summary executions, however, it was only partly obeyed. Soldiers in fact, according to Herodianus' account, demonstrated a disorderly and uneven conduct: while some, failing to identify those who were responsible, carried out killings at random; others instead did not interfere, and, unseen, tried to save the people.[94]

Thermae

The baths were another potentially risky place, due to the promiscuity of contacts and the dense concentration of people. Thermal baths were a social space and an ideal context for meeting and exchanging views or information; members of different associations could also meet here. The danger was represented by the presence of thieves who could easily take advantage of the guests' personal belongings left unattended; and also by ease with which, due to the heating systems, fires could erupt.

Among the *vigiles'* duties was the supervision of the thermal baths.[95] The chapter of the Digest on the responsibilities of the prefect of the *vigiles* provides details on the widespread coverage that its jurisdiction had with possible interventions on officers guarding the clothing (*capsarii*) and dishonest people who would sneak in the dressing rooms.[96] And it is always the prefect of *vigiles* who, ultimately, resolves, in mid-third century, a dispute that lasted nearly two decades on the non-payment of the amount due by cloth-launders[97]: in fact, the *fullones* required a large quantity of water for their work, which was mostly concentrated in the thermal complexes.

However, we do not know from when the *vigiles* began to exercise their duty of prevention and intervention at the thermal baths. A graffito on the walls of the *statio* of the VII cohort in Trastevere recalls a vigil charged with the Baths of Nero.[98] It is difficult to specify the time when his service took place: the complex, built by Nero in the Campus Martius,[99] kept its original name until the third century, when Alexander Severus build on the same site the *thermae Alexandrianae*.

An episode

A famous episode of popular insurrection that took place first at entertainment buildings and then outside of the imperial villa dates back to Commodus. I will retract its origins and development here, at the end of this chapter, for the details we know about it and because some of the most important 'dangerous places' that have been discussed so far are directly involved.

The episode I am referring to is narrated by Herodianus and Cassius Dio.[100] The pretext for the unleashing of the conflict is given by the rich and powerful freedman Cleander, Commodus' favourite, who, after being his chamberlain and bodyguard,[101] had become praetorian prefect. Cleander's rapid success and arrogant attitude made him so unpopular with the Senate and the people of Rome that the crowd began to show signs of impatience, holding the prefect responsible for the evils that were happening, because of his insatiable greed[102] and of the influence he exerted on the emperor.

During a competition of races at the Circus Maximus,[103] a group of youngsters began to launch invectives against the praetorian prefect who was absent (Dio), and the crowd, getting involved, started to move in procession towards Commodus' villa,[104] to call for the head of the prefect. Both Cassius Dio and Herodianus agree that Cleander asked the equestrians (praetorian *equites* or *equites singulares Augusti*, both under the command of the praetorian prefect) to lead the charge against the crowd that, despite significant losses, tried to resist and backtracked. The Romans, those who until then had remained passive, barricaded themselves in their homes and from the roofs they hit the equestrians with rocks and bricks ("the mob . . . attacked in total safety").

The guardians of city order, the *milites urbani*, "for their hatred towards the *equites*", attacked them to show solidarity with the population.[105] "In this civil war, nobody wanted to inform the emperor of what was happening, so much was believed to be the power to Cleander."[106] Only Commodus' sister, Fadilla, had the courage to do so, accusing the prefect of arming against her brother both the soldiers and the people. The interference of the urban soldiers precipitated the situation: Commodus resolved by convening his protégé and putting him to death.

This is not a matter of little importance, as also attested by the consequences (victims and devastation), which would not have been such to induce Commodus to sacrifice Cleander, if it were just a brawl between praetorians and people.[107]

The dynamics of the matter are quite clear: the audience at the competition responds to the provocation that comes from one of the spectators apparently in a random way, actually orchestrated by the senators, and moves to turn its protest directly to Commodus. The equestrians are ordered to stop the march by attacking (with sticks and swords) the crowd, and they execute the order before the protesters had time to have their say, directly or through spokesmen, to the emperor. When, with the first wounded, some backtrack, a third protagonist, namely the soldiers of the urban cohorts, intervenes at the demonstrators' side, as the guarantor of public order in the city. It would seem that, behind the *milites urbani* and even before, behind the provocation leading to the whole episode, were the senators: through the urban prefect, they gave its soldiers the order to intervene, not so much to take the defenses of the helpless population, but to express all their dissent against the praetorian prefect. The clash between demonstrators and the security forces is, so to speak, remote controlled from the top. The praetorians (and/or the *equites singulares Augusti*) obey to the order to attack, the 'city soldiers' perform the task of containing the clash in the interest of social peace, not least to reaffirm especially tasks and areas of responsibility.

Commodus' act, the sacrifice of his own man, is the lesser evil, which gives satisfaction to popular discontent and does not respond to the challenge of his most dangerous enemy, the Senate. The feeling is that the emperor is losing control of the security forces in Rome, as when in the Colosseum he called the soldiers of the fleet and not the praetorians to quell the riot.[108]

It is no coincidence that the clash between the Praetorian Guard and the urban soldiers takes place during the reign of Commodus, when the relations between the two military units were undergoing a change.[109] The soldiers of the urban cohorts will be the protagonists of the last paragraph of this third part, in which I will try to follow their progressive evolution. The idea I'm going to propose, partly discordant compared to recent hypothesis, is that in fact the substantial change in the profile of these soldiers occurred just in the second half of the second century AD.

While indeed in the age of Hadrian we have no signs of changes, with Marcus Aurelius and his son Commodus the signals of a shift compared to the past become evident. With regard to the *milites urbani* the most significant are: the construction of a new military camp on the Via Lata, close to the Forum Suarium, therefore a definitive separation from the Praetorian Guard; their new orientation in the choice of burial areas[110]; the involvement in the Marcomannic Wars, which determines the substantial alteration of their profile until then specifically linked to the city. In light of this, the deployment of urban cohorts and their prefect at the side of the people in the episode of Cleander acquires a deeper meaning.

Notes

1 "Le lieu ou le moment n'est pas considéré comme criminogène en lui-même, mais parce qu'il permet l'établissement d'une sociabilité – de la rue aux cabarets en passant par les thermes" (Ménard 2004, p. 27).

2 Weapons were supplied not only to the soldiers of the praetorian and urban cohorts, but also to the *corporis custodes* and *vigiles*. On the use of the *fustis* (Dig. 1.12.10.21, for the urban cohorts, and 1.15.3.12, for the *vigiles*), see M.P. Speidel, *The* fustis *as a soldier's weapon*, in *Ant.Afr.* 29, 1993, pp. 137–149; and Rivière 2004, pp. 76–77. Interestingly, in this regard, as for the episode of AD 238 (*infra*, p. 152), there is a

frequent reference in the texts to the weapons used in the conflict. It seems to me that Herodianus' insistence on this point is not random and follows almost a kind of crescendo, the progressive degeneration of the conflict. There is everything: illegally held weapons (the swords with which the senators kill the praetorians, spears, axes and even the swords that the crowd runs to gather from homes), various objects uses as weapons (tiles, bricks, pottery of the people), weapons recovered for the occasion (those used during parades hosted in the *armamentarium*), professional weapons (those of the gladiators, but also arrows or spears and others types of which the arsenal of the Praetorian Guard was equipped).

3 *Infra*, p. 152.
4 For some descriptions of structures inaccessible to thieves, see the passages of the authors reported by Kelly 2013, pp. 222–223 (with his remarks on the archaeological remains of the houses of Ostia and Pompeii). The same author also cites two passages (Mart. *Epigr.* 5.22 and Sen. *De constantia sapientis* 14.1–2), which allude to doorkeepers.
5 Gell. *Noct. Att.* 2.13.4, referring to Tiberius Gracchus; Juv. *Sat.* 3.282–285; Prop. 2.29 A.
6 *CIL*,VI 6308, cf. pp. 3419 and 3851 = ILS 7408d; Caldelli, Ricci 1999, p. 92 no. 93, with photo: *[I]ucundus Tauri [l(ibertus)], [le]çticarius. Quandi/us vixit vir fuit et se et / alio[s] vindicavi(t). Quan/dius (!) vixit honeste vixit. Callista et Philologus dant.*
7 Millar 1992, pp. 368 f.
8 Durry 1938, pp. 43–63 (pages devoted to what is called the 'praetorian topography'. The literary reference sources are Suet. *Tib.* 37.1–2; Tac. *Ann.* 4.2; Cass. Dio 57.19.6; Juv. 10.94–95 (who speaks of *castra domestica*).
9 On the *castra praetoria*, see the relative entry of E. Lissi Caronna (1993) in the Lexicon Topographicum Urbis Romae. Interesting is the architect Giacomo Cavillier's (2007) point of view on the construction technique and the information obtained from it.
   For an updated overview on the new discoveries related to the excavations of the 1980s in the access path of the National Italian Library by Castro Pretorio and in subsequent text excavations in neighbouring areas, see the work by Morretta 2007.
10 In addition to the monograph of 2011 (pp. 29–109), see also the article by Busch 2007.
11 "Préfectures et castra marquaient donc la topographie urbaine et imposaient physiquement la présence des soldats et des préfets" (Sablayrolles 2001, p. 146).
12 See Part II, Chapter 4, p. 96–97. Not quite free from the confusion between the accommodation of the praetorian *speculatores* and that of the legionary *speculatores* Rankov 2006, pp. 133 f. and Gex 2013, p. 117 n. 22.
13 Dig. 30.1.39.7–10, Ulp. 21 ad Sab., put in parallel with Papin. 10 *quaest.* (Dig. 18.1.72 *praef.*).
14 *Vita Maxim. et Balb.* 10.4–6.
15 Herodian. 7.11–12.
16 It is a much-quoted episode. Recently it has been analysed by Rivière 2004, pp. 82–84; Ménard 2004, pp. 87–89; Kelly 2007.
17 Sablayrolles 2001, p. 146 quoted by Palombi 2013, p. 40. The same hypothesis is formulated by Yves Perrin who, stressing the symbolic value of the imperial palace on the Palatine Hill, highlights how it focused on the "technical services of the direction of the Empire", among which Perrin also cites the praetorian prefecture (Perrin 2003a; see also Perrin 200b, non vidi).
18 See Part II, Chapter 5, p. 110.

19 For the *castra Ostiensia* (defined in this way in two inscriptions of AD 207: *CIL*, XIV 4381 and 4387), see the Roman lead pipe inscription *AE* 1954, 170: *[I]mperatoris Domitiani Caesaris Aug. / [- - -] quae ducunt in castris.*

20 Sablayrolles 1996, p. 33 and 247.

21 Sablayrolles 1996, pp. 249–273.

22 Sablayrolles 1996, pp. 274–275 and 2001, p. 146; Coarelli 2001, p. 341.

23 The bibliography on the seat of the urban prefect (or the various offices of the prefecture) has been greatly enriched in the past decade. I will just refer to the major contributions on this issue. The starting point can only be an article by André Chastagnol released in 1997, which was followed by Coarelli 1999 and 2010 (the latter in particular built on the assumption of the link between the courts of the late antique prefecture and the Basilica of Maxentius); Palombi 1997, part. pp. 149–153; Färber 2012. Amoroso 2007 and Marchese 2007 reveal the particular concentration in the area of S. Pietro in Vincoli of inscriptions mentioning urban prefects.

24 Generally identified with the *basilica Iulia*; but Coarelli 1999 considers most likely the *basilica Aemilia* (or *Paulli*): in addition to Iohannes Lydus (Lyd. *Mag.* 1.34), a passage taken from a lost work of Suetonius, Coarelli recalls Martialis 2.17 who refers to the offices of the prefecture, or some of them, placed at the entrance of the Subura (Coarelli 2009).

25 In detail, two views are contrasted: that of Coarelli (1986, 2010) and Amoroso 2007 on the one hand, who identify the Temple of *Tellus* in a cement structure west of the *compitum Acilii*; and on the other hand that of Palombi 1997 (part. p. 150 note 47), followed by Babliz 2007, pp. 39–43, who prefers to place it in the south-west of today's Church of St. Peter in Chains. The situation has recently been effectively synthesized by Marchese 2007, already cited; and extensive bibliography reported by Orlandi in 2013, part. p. 51 n. 2, which I complete with the reference to Daguet-Gagey 2000, part. pp. 73–77.
    For the positioning of the prefecture under the south-western portico of the Baths of Trajan, where the fresco called 'the Painted city' was painted, see Volpe 2000, pp. 519–520; Caruso-Volpe 2000, pp. 54–56; Carnabuci 2006, pp. 182–192.

26 Caruso, Volpe 2000; La Rocca 2000, the last, Carnabuci 2006 (with extensive earlier literature).

27 Marchese 2007.

28 For this hypothesis, see Färber 2012, p. 59, who refers to Coarelli 1993.

29 Bellen 1981, pp. 56–57, recalled by Speidel 1994b, pp. 25–26; Eck 1996b, p. 114. The hypothesis is also found in Cosme 2011, pp. 306 f.; Busch 2011, p. 94. The scholars commonly refer to two (the first archeological, the second literary) sources none of which, as we will see, makes explicit reference to barracks.

30 From the via Appia: *CIL*, VI 4340, 4342, 4243 cf. pp. 3416 and 3850, between (late) Augustus and Gaius; *CIL*, VI 4437 and 4716, cf. p. 3416, from the second Codini's Columbarium, between Augustus and Gaius. Unknown the origin of *CIL*, VI 8810 and 8812, between Tiberius and Nero.

31 See Part II, Chapter 4, pp. 97–98.

32 The monuments from the ages of Claudius and Nero come from the via Aurelia (*AE* 1968, 32; 8802, 8803, 8804, 8806, 8807, 8808, 8809; 37754 and 37754 A, Claudian era); from the Magliana area (*AE* 1983, 58, Nero).

33 *Not. Sc.* 1950, p. 88 = *AE* 1952, 148; *Not. Sc.* 1950, p. 87 = *AE* 1952, 145; *Not. Sc.* 1950, p. 88 = *AE* 1952, 147; *Not. Sc.* 1950, p. 87 = *AE* 1952, 146; *Not. Sc.* 1950, p. 89 = *AE* 1952, 149.

34 See for instance the *equites singulares Augusti, infra*, note 38.
35 Suet. Galba 12.2: *Item Germanorum cohortem a Caesaribus olim ad custodiam corporis institutam . . . dissolvit ac sine commodo ullo remisit in patriam, quasi Cn. Dolabellae, iuxta cuius hortos tendebat, proniorem.*
36 For the bibliography, see *infra* notes 37 and 38.
37 Liverani 1988. The Lateran Project, involving the University of Florence, the Newcastle University, the KNIR and the University of Amsterdam, has been recently presented in occasion of the Conference 'The Lateran Basilica', Rome, 19–21 September 2016).

In her book, Busch 2011 (pp. 72–83) does not provide any new information, especially for the *castra vetera* where she is substantially indebted of the reports on the excavations of the end of the late nineteenth century.
38 On the burial ground of the *equites singulares Augusti* along the via Labicana, in the locality called *ad duas lauros* and the underlying catacombs of Marcellinus and Peter, see W. Henzen, in *Ann.Inst* 22, 1850, pp. 5–53; F. Grosso, in *Latomus* 25, 1966, pp. 900–909 and in *RendAccLinc* s.VIII, 21, 1966, pp. 140–150; J. Guyon *Dal praedium imperiale al santuario dei martiri. Il territorio "ad duas lauros"*, in A. Giardina, A. Schiavone (eds), *Società romana e impero tardoantico*, II, Roma-Bari 1986, pp. 299–332. Other bibliography in Panciera 1974, p. 292 n. 2 and in Speidel 1994a.
39 In addition to the literary sources, even some inscriptions. See, as an example, the *carmen* of the Julio-Claudian age *CIL*,VI 5302 = *ILS* 8513 and *CLE* 1037 line 6 = EDR 135953 (V. Di Cola, 2–3 2014), with extensive bibliography; or the later *CIL*, VI 20307A = *ILS* 8505 (III–IV century?).
40 Part II, p. 41 and, in this chapter, Part III Introduction, p. 122.
41 Dig. 1.12.1.12, Ulp. *De officio praefecti Urbi.* The existence of *stationes* is also explained by the customs service that the urban troops certainly carried out, as testified for the fourth century by the Codex Theodosianus (Cod. Theod. 4.13.3 = Codex Iustinianus 4.61.5). "We know that in fourth century at least they had some role in the collection of customs duties at the gates of the city" (Kelly 2013, who also cites Freis 1967, p. 46).
42 On them, see Part II, p. 93 with note 33.
43 *CIL*,VI 230 = 36748 cf. p. 3004 = *ILS* 2216: *Pro salute Imp(eratoris) Caes(aris) M(arci) Aur(eli) Severi / Alexandri Aug(usti) Genio sancto kast(rorum) per(egrinorum) / totiusque exercituus Q(uintus) Haterius Valeria/nus frum(entarius) leg(ionis) VIII Aug(ustae) et M(arcus) Aurelius / Sophaenitus frum(entarius) leg(ionis) XIII Gem(inae) Severi/anarum stationem collegiis suis / impendi(i)s fecerunt* (AD 222–235).
44 *CIL*, VI 3329 cf. p. 3844 = *ILS* 2222: *[- - - V]ictor, subprinc(eps) peregrinor(um), / [stationem ad mil(liarium)] III vi(a)e Appi(a)e frumentaris de / suo refecit* (post AD 235).
45 *CIL*, VI 40623; A. Ferrua, 'Le iscrizioni pagane della catacomba di Pretestato', *RendLinc* 28 (1973), p. 67, no. 8 =*AE* 1973, 75: *Felicissimo saeculo do[minorum nostrorum L(uci) Septimi] / Severi et M(arci) Aureli Antonini et [P(ubli) Septimi Getae Auggg.], / T(itus) Iul(ius) Primianus, evocat(us) ex [coh(orte) - - - numini] / eorum dicatissim(us), sch[olam - - -] / et stat(ionem) viae Latinae a solo sua [pecunia ? - - -], / Septimius Proculinu[s - - -] / titulum station̦e [- - -]* [between AD 209 and 211]. Cf. C. Ricci, Latina via. Statio, in LTURS III, 2007, p. 144.
46 Of a fourth dedication, unearthed near the *Atrium Vestae* in the Forum, there will be mention later (p. 152) in regards to the *Palatium*.
47 A. Ferrua, *Cimitero di San Callisto*, in RAC 57, 1981, 17–21, no. 18, fig. 10 (B) = *AE* 1981, 134; D.E. Trout, Victoria Redux *and the First Year of the Reign of Philip the Arab*, in Chiron 19, 1989, pp. 223–224 = *AE* 1989, 62. See also Ricci 2000, p. 403: *Genio*

leg(ionis) II Parth(icae) Gordianae et Fortunae Reduci / Paciferae conservatoribus d(omini) n(ostri) / Imp(eratoris) Caes(aris) M(arci) Antoni Gordiani Pii Felicis Invicti Aug(usti) et / Sabiniae Tranquillinae Aug(ustae) coniugi(s) Aug(usti) n(ostri). /Milites leg(ionis) II Parth(icae) Gordianae p(iae) f(elicis) f(idelis) aeternae, / qui militare coeperunt Sabino II et Anullino co(n)s(ulibus), / quorum nomina cum tribus et patrias (!) duobus tabulis aereis / incisa continentur devoti numini maiestatique eorum / sub cura Valeri Valentis v(iri) p(erfectissimi) vice praef(ecti) praet(orio) agentis / et Pomponi Iuliani p(rimi)p(ili) praep(ositi) reliquationis. / Dedic(averunt) VIIII Kal(endas) Aug(ustas) Attico / et Praetextato [v(iris) c(larissimis) co(n)s(ulibus)]. A similar dedication was erected in Ostia, in honor of Furia Sabinia Tranquillina, the Gordianus' wife, of the prefect and the subprefect of the *vigiles* and the subprefect of the annona, between 241 and 254, on behalf of the seven cohorts of *vigiles*: *CIL*, XIV 4398 = *ILS* 2159. See also *Epigrafia latina. Ostia: cento iscrizioni in contesto*, Roma 2010, pp. 241–242 no. 70.2, with photo (M. Cébeillac-Gervasoni).

48 On the meaning of the word 'reliquatio', see G. Migliorati, *Proposta d'interpretazione del termine* reliquatio, in M. A. Bertinelli, A. Donati (eds), *Opinione pubblica e forme di comunicazione a Roma: il linguaggio dell'epigrafia* (Epigrafia e Antichità 27). Atti del Colloquio AIEGL – Borghesi 2007, Faenza 2009, pp. 309–317.

49 On the changed ethnic composition of the praetorians before and after Severus, see Durry 1938, pp. 247–257 (chapter on the recruiting from 193 onwards) and 383–389 (chapter on the Praetorian Guard and Severans).

50 All testimonials are now collected and commented by Caldelli 2014.

51 Cébeillac-Gervasoni 1979, pp. 267–277 (AD 210).

52 *CIL*, XIV 125: Imp(eratori) Caesari M(arco) [A]urelio / Severo Alexandro / Felic<i> Aug(usto) et Iuli(a)e Mameae / matri domini n(ostri) et castror(um) / totiusq(ue) d(omus) d(ivinae). Statio n(umeri) fr[u]mentariorum, / locus adsignatus ab Agricola Aug(usti) lib(erto) proc(uratore) p(ortus) u(triusque) / et Petronio Maxsimo (!), | (centurione) ann(onae) et Fabio Maronae (centurione) / operum. Dedicatum III Non(as) Aug(ustas) Appio Cl(audio) Iuliano et Brutt(io) / Crispino co(n)s(ulibus), patrono Q(uinto) Turranio Masila, cura(m) / agente P(ublio) Flavio Fl(avi) filio Iuniore / et Valerio Donato cur(antibus).

53 The list of frumentarii in Ostia is also found in Clauss (1973, p. 90 and n. 62), according to whom they were detached from the camp on the Caelian Hill.

54 *Infra*, pp. 150–152.

55 A key moment is the opening of the construction site for the Domus Aurea, in the aftermath of the 64; the *communis opinio* wants the Palatine to remain the official pole of power, while the Esquiline pole would always maintained a private character.

56 The *Germani*, in all likelihood were initially part of the *familia publica* with whom they shared burial places (*supra*, p. 98).

57 Various are the episodes: see for example, Tac. *Ann.* 12.37.2 and *Hist.* 1.38.4.

58 Mart. 10.48.1–2: *Nuntiat octavam Phariae sua turba iuvencae / et pilata redit iamque subitque cohors* (in this and in the following cited passages, emphasis is added by the author).

59 Perrin 2003. By scrutinizing carefully the archaeological and literary sources, the researcher focuses on the accesses to *Palatium* as a place of power between Nero's and the beginning of the Flavian era. His starting point is a lexical research, the difference between the words *ianua, porta, fores (thuriae) Palatii* and *aditus domus*, terms or expressions that, in his view, refer to distinct realities (that the excavations, at least in part, help to identify), with a different nature, a different position and, what interests us, differently kept.

60 Perrin 2003, p. 361.

61 For them, however, we have no sources about the duties performed within the Palatine palace. However see Suet. *Cla.* 35 (at the side of Claudius at banquets, although not in Rome).

62 Suet. *Ner.* 8. On the passage and, generally, on the relation between the *Princeps* and the Palatine, see M.A. Tomei, *Nerone sul Palatino*, in Tomei, Rea (eds) 2011, pp. 118–135. On the passage, also see Royo 1999, p. 295.

63 Tac. *Ann.* 11.37.3–4: *Isque (the freedman Euodus) raptim in hortos praegressus repperit (Messalinam fusam humi) . . ., cum impetu venientium pulsae fores adstititque tribunus per silentium.*

64 Tac. *Ann.* 12.68–69: *Tunc medio diei tertium ante Idus Octobris, foribus palatii repente diductis, comitante Burro, Nero egreditur ad cohortem, quae more militiae excubiis adest.*

65 Suet. *Vit.* 16.

66 Tac. *Hist.* 1.29.2 (Piso, appointed as successor by Galba, tries to convince the praetorians *pro gradibus domus (Palatinae)*; Tac. *Hist.* 3.74.5 *(stantem pro gradibus Palatii Vitellium et preces parantem pervicere . . .)* and Suet. *Vit.* 15.2 (Vitellius, *pro gradibus Palati apud frequentes milites*, announced his resignation in power).

67 Arce, Mar 1990 and Perrin 2003 A, p. 369, with previous bibliography. Very general Mar 2009, on p. 252 suggests that the Domitian building of Santa Maria Antiqua is to be defined as a complex (vestibule plus ramp and guardroom) for public access to the annexes of the Domus Tiberiana.

68 Perrin 2003 and Wulf-Rheidt 2012, p. 99 and fig. 3 at p. 104 ("before the big garden in front of the entrance of the complex"), who takes into account, I believe, Villedieu, Andrè 2003 who have identified the substructures system that, during the Domitian age, connected the area of the Vigna Barberini to the rest of the Domitian palace). Also see Villedieu 2006.

69 These facilities are located inside the building, at the entrance to an area with high representative function. Wulf-Rheidt cites Zanker 2004 and Mar 2009.

70 Grosso 1968, pp. 213–214: "the Palatine was obviously well defended and controlled".

71 Scriptores Historiae *Augustae.* 11.9: *Sed cum Tausius quidam, unus e Tungris, in iram et in timorem milites loquendo adduxisset, hastam in pectus Pertinacis obiecit.* Partly different is Herodian's account (2.5). Cass. Dio (73.9) highlights the complicity between the palace guards and the imperial freedmen.

72 Villedieu 2001, 2003, 2013. When Geta and Caracalla came back to Rome after their father's death, they barricaded themselves in the Palatium, closing all accesses, including secret passages. Only the outdoor entrances remain open, guarded by 'particular guards' for each of them (Herod. 1.5).

73 For the south-east area, see the investigation of Wulf-Reidt and his working group (Hoffmann, Wulf 2004, 2013), some specifically devoted to the Severan buildings; for the northwest corner, see Meylan Krause 2002 and above all, Tomei, Filetici 2011.

74 *CIL*, VI 36775 = ILS 484 (222–226 d.C.): *Pro salute domini / nostri Imperator(is) / Severi [[Alex[an]dri]] / Augusti et / Iuliae [[Maesae et]] / Iuliae Avitae / [[M[ameae]]] sanctissimarum / Augustarum / Genio Sancto castror(um) / peregrinorum. / T. Flavius) Domitianus, / domo Nicomedia, quod / speculator leg(ionis) III Parth(icae) / Severianae vovit has/tatus leg(ionis) X Fretensis / princeps peregrinorum / reddedit (!).*

75 Panciera 1989, p. 380 no. 11 and then p. 382, who agrees with Henzen: "(the dedication is) to be linked with a detachment to the service of the imperial palace on the Palatine".

76 The studies about the constructions made by Severans on the Via Nova have not been published yet (courteous reporting by Francoise Villedieu).

77 Our information primarily concerns the security escort of the emperors and the guard service at the imperial villas outside the city (see Part IV).

78 Herodian. 8.7.8. To the soldiers of different formations, and the conflict of roles between them, that Herodianus' account refers to, is to be added the staff "engaged in defense services of the doors" (Herodian. 8.8.6); it is probably personnel of servile condition or freedmen, helpless because disarmed and untrained and because numerically inferior. Both Eutropius (9.2: *itaque cum Romam venissent, Balbinus et Pupienus in Palatio interfecti sunt*) and Aurelius Victor (*De Caesaribus* 27.6, with a generic reference to the *tumultus militarium* and to the murder *intra Palatium*) are laconic in reporting the episode. Far richer in detail is the account of the Scriptores Historiae Augustae, *Max. Balb.* 14.2–8.

79 Yavetz 1984, pp. 44 ss.; Rivière 2004; Nippel 1995.

80 Suet. Aug. 58: *Patris patriae cognomen universi repentino maximoque consensu detulerunt ei: prima plebs legatione Antium missa; dein, quia non recipiebat, ineunti Romae spectacula frequens et laureata; mox in curia senatus, neque decreto neque adclamatione, sed per Valerium Messalam.* It remains uncertain which was, in this case, the building for the shows (Circus Maximus? Statilius Taurus's Amphitheatre? Pompeius's Theatre?) referring to; Nippel 1995, p. 87 believes it is the circus (he speaks of games); J.C. Edmondson, Dynamic arenas: Gladiatorial presentations in the city of Rome and the construction of Roman society during the Early Empire, in W.J. Slater (ed.) *Roman Theater and Society.* E. Togo Salmon Papers I, Ann Arbor 1996, p. 76 gives no indication.

81 Suet. *Aug.* 42 and *Dom.* 13.1.

82 Suet. *Tit.* 8.5; Mart. *Spect.* 4 ("the vast arena did not have room enough for the guilty"); Plin. *Pan.* 34.1–4.

83 On disputes between supporters, surprisingly often of pantomimes more than of the beloved charioteers, see Jory 1984 and Wistrand 1992.

84 The punishments consisted: for the actors in whipping or, predominantly, in orders of exile (Suet. *Aug.* 45.3–4; *Tib.* 37.2, *Dom.* 7.1; Tac. *Ann.* 4. 14.3 and 13.25.4; Plin. *Pan.* 46. 2, all recalled by Leppin 1992, pp. 64–67); for spectators, in arrests or banishments by the magistrates responsible for the conduct during games (Cass. Dio 54.2.3).

85 Tac. *Ann.* 1.77.1 (of which has already been discussed in the Part III Introduction); also see Rich 1994.

86 Eight years after (AD 23), yet in Tiberian age, the actors were banished (Cass. Dio 57.21.3: "He [Tiberius] banished the actors from Rome and would allow them no place in which to practise their profession, because they kept debauching the women and stirring up tumults", cf. Tac. *Ann.* 4.14.3 and Suet. *Tib.* 37.2.

87 Cassius Dio's explanation (61.8.3) on the possible reasons for it has been previously described (in this Part III Introduction, p. 127).

88 The two passages by Tacitus are, respectively, *Ann.* 13.24 (AD 55): *Fine anni statio cohortis adsidere ludis solita demovetur, quo maior species libertatis esset, utque miles theatrali licentiae non permixtus incorruptior ageret et plebes daret experimentum, an amotis custodibus modestiam retineret*; and *Ann.* 13.25.3–4 (AD 56): *Non aliud remedium repertum est quam ut histriones Italia pellerentur milesque in theatro rursum adsideret.*

89 Tac. *Ann.* 14.15.4: *Accesserat cohors militum, centuriones tribunique et maerens Burrus ac laudans. tuncque primum conscripti sunt equites Romani cognomento Augustianorum, aetate ac robore conspicui, et pars ingenio procaces, alii in spe[m] potentiae.* On the passage,

cf. Nippel 1995, p. 94 and Le Roux 2002, p. 23 and n. 19. On the Augustiani (Suet. *Nero* 25 calls them *milites* and Cass. Dio 63.18.3 *stratiotai*), enlisted to act as a claque at the performances of the Princeps, see Perrin 2012, with sources and previous bibliography. Also see the figure of Percennius, *dux olim theatralium operarum, dein gregarius miles* by Tac. *Ann.* 1.16.3.

90 This is the prevailing interpretation among scholars; the last, Bingham 1999, with previous bibliography.

91 Cass. Dio 59.28.11. See Yavetz 1984, p. 36.

92 Tac. *Ann.* 16.5.1.

93 Scriptores Historiae Augustae *Comm.* 15.6: *Sane cum illi saepe pugnanti ut deo populus favisset, in risum se credens populum Romanum a militibus classiariis, qui vela ducebant, in amphitheatro interimi praeceperat.*

94 Herodian. 4.6.4–5. It could either be praetorians or *equites singulares Augusti* or both.

95 Sablayrolles 1996, pp. 108–109, 381, 418–419. For the relation between *vigiles* and thermal baths in Ostia, refer to the detailed illustration by Caldelli 2012, p. 294.

96 Dig. 1.15.3. 1–2 and 5, Paul *de officio praefecti Vigilum* also see Dig. 7.17.1 (*idem et in balnearibus furibus*).

97 *CIL*,VI 266 = *FIRA* 3.165. also see F.M. De Robertis, Lis fullonum (*CIL*.VI, 266): *Oggetto della lite e causa petendi* in ANRW II.14, pp. 791–815; Sablayrolles 1996, pp. 113–120; A. Magioncalda, *L'epigrafe della* lis fullonum: *una nota prosopografica*, in G. Barberis, I. Lavanda, G. Rampa, B. Soro (eds), *La politica economica tra mercati e regole: scritti in ricordo di L. Stella*, Rubbettino: Soveria Manelli, 2005, pp. 221–226.

98 *CIL*,VI 3052: *[- - - ? c]ohor(s) VII vi[gi]l(um) (thermis) Neron(ianis), / (centuria) Faustini Harius frumentari(us); c(o)h(ors) VII vig(ilum), / (centuria) Faustini termis Ner(onianis) / Harius Primus.* The graffito is on a wall near an aedicule. According to Henzen, it is the same inscription repeated twice, with minimal changes. It is difficult to interpret the indication 'frumentarius' at line 2.

99 Suet. *Nero* 12.

100 Herodian. 1.12.5–9–1.13.1–6, Cass. Dio 73.13.3–6. The episode is widely commented by Alföldy 1989 pp. 81–126 (very critical towards Herodianus' version, as Müller 1996, p. 312 recalls); and recently by Rivière 2004 (pp. 78–81). For the annotated edition of the first book of Herodianus, apart from Müller, also see now Galimberti 2014.

101 Or, more likely, the commander of the palace guard (the Latin expression is *a pugione*). In addition to the extensive bibliography cited by Alföldy in the apparatus of *CIL*,VI 41118 (funerary altar with a dedication to *Taius Sanctus*, placed by Cleander along with *Asclepiodotus*, superintendent of finances and *a memoria*, between AD 186 and 188), see *AE* 2010, 158 (F. Mitthof) and *AE* 2011, 123 (K. Krenn).

102 Herodian. 1.12.5.

103 But Herodian. 1.12.5 speaks of theatres.

104 The suburban villa where Commodus at that time resided was that of the Quintili, on the Appian Way (Cass. Dio 73.13.4).

105 Herodianus says that by the protestors' side were "*oi tes poleos pezoi*", correctly interpreted by Grosso, recalled by Galimberti, as "soldiers of the urban cohorts"; others believe they were the praetorians (see the bibliography mentioned by Galimberti 2014, p. 129).

106 Herodian. 1.13.1.

107 Grosso 1964, p. 298, n. 1.

108 *Supra*, p. 93.
109 See Part III, Chapter 7. Three years later, once again, the praetorians will be opposed to the people of Rome: when the word of the murder of Pertinax spreads, these soldiers closed the gates of their camp and placed sentinels on the towers. Persuaded by Didius Iulianus' promises and driven by greed (which never fails to be emphasized by Dio), the praetorians devised a stratagem to lead Iulianus to the Palatium from the camp where he was. So, they closed the ranks, protected also by shields and spears; within the ranks they placed Iulianus and then they went out and succeed despite the people's invectives, who did not cheer as they usually did when the imperial escort appeared (on the episode, Herodian. 2.5.9–2.6; Cass. Dio 74.11.1). See Rivière 2004, pp. 81–82.
110 If the joint dedication of the two corps continued for nearly fifty years, the last diploma that recalls them together is from 180–184. See Part III, Chapter 7.

7 The urban soldiers and the city[1]

Helmut Freis' *Die cohortes urbanae* (1967) remains fundamental for the history, organization and functions of the *cohortes urbanae* under the command of the *praefectus Urbi* and includes study of those cohorts temporarily or permanently stationed at Ostia, Puteoli, Lugdunum and Carthage.[2] If little new can be added to Freis' treatment of the cohorts' hierarchical structure, conditions of service and their veterans,[3] new documentation permits clarifying the organization and functions of the Roman and Carthaginian cohorts.[4] Increased epigraphic evidence for soldiers, junior officers and centurions is important for the cohorts in Rome and Italy as well as Carthage, although minimal for the Lyon cohort, as Table 7.1 shows.

This chapter, assessing and commenting on the new evidence, is organized as follows: (1) the urban cohorts in the first century; (2) *castra praetoria* and *castra urbana*: the barracks of the *milites urbani* in Rome; (3) public order and the military role of the urban cohorts; (4) the cohort of Lyon; (5) the cohort of Carthage. We include the urban prefect only in so far as this post pertains to the functions of cohorts at Rome.[5] An Appendix both catalogues texts published since 1967 (noting some new studies on known texts), and identifies for Rome unpublished documents and work in progress. Not least, new evidence suggests redating the final separation of the urban cohorts from the praetorians

Table 7.1 Epigraphical evidence for urban cohorts (before and after 1967)

Place	Number of inscriptions already known by Freis	Number of inscriptions published after Freis or yet unpublished
Rome	127 (+ 6 *laterculi*)	18 + 10 yet unpublished)
Italy	57 (+ 2 Sardinia and 1 Alps)	19
Africa	40	10
Lyons	16	2
Other	3 Narbonensis (*cohortes* XIII, XIV, I *Urbana*)	1 Sicily 1 Balkans 2 Orient 2 unknown origin

to the late Antonine, rather than the Severan period,[6] and considerable revision on current views of the garrisons at Lyon and Carthage.

The urban cohorts in the first century

New evidence now requires modification of Freis' view (accepted by many) that Augustus' structure of the cohorts, adjusted under Claudius and Vespasian, remained in place until the Antonine era, when significant changes occurred; the Severi initiated a new era of the *milites urbani*. Augustus created three urban cohorts,[7] originally numbered X, XI, XII as appendages to the nine praetorian cohorts[8] and placed under the *praefectus Urbi* to operate in the city and its immediate vicinity.[9] Freis' posits each cohort at 500 men, although their size fluctuated over time and recruits at 20–21 years of age. Nevertheless, the roles and responsibilities of the Urban Prefect and the urban cohorts evolved: the Augustan and Julio-Claudian phases could, in this sense, be described as 'experimental'. Rather than following a definite plan, the *Princeps* proceeded slowly in organizing the garrison of the city with repeated experiments and progressive adaptation. Just as the final organization of the *Vigiles* and the Praetorians took decades, so, too, did defining the tasks of the urban cohorts.[10]

Gradual modification of an original design can be discerned. First, relocation of cohorts, initially restricted to the capital and surrounding cities, came to include Aquileia.[11] This change may belong to Tiberius' measures for public order and a more rigid distinction between the Urban and Praetorian Cohorts.[12] Second, the epithet *urbana* first appears in a late Augustan or early Tiberian text.[13] Thus an inscription of AD 14, recording the career in ascending order of the *primus pilus, A. Virgius Gallicus* (eventually a *praefectus castrorum* and *praefectus fabrum*), still demonstrates the original connection between the urban cohorts and the praetorians. Earlier, he had been a tribune in *praeto(rio) divi Aug(usti) et Ti(berii) Caesaris Aug(usti) cohortium XI et IIII praetor(iarum)*, hence promotion from the eleventh cohort (later XI *urbana*) to the IV *praetoria*.[14] Indeed, *milites urbani* and Praetorians are recruited from the same areas (*Etruria, Umbria, Latium vetus et coloniae antiquae*: Tac. *Ann.* 4, 5); their veterans appear jointly in diplomas and *laterculi* with dedications to gods and emperors; both received *praemia* directly from the emperor and shared the same burial sites; and both corps fought on the same side in 69.[15]

Claudius changed the political role of the cohorts at Rome and increased their number.[16] Two urban cohorts were added, as now confirm four epitaphs of soldiers in a XIV *cohors urbana* (Q. Fabius Q. f. [- - -], A. Avillius Verus, T. Camurius Priscus, Sex. Munatius Marcellinus) and one of a soldier in a XV *cohors urbana* (Sex. Licinius Mansuetus), dating to the mid-first or early second centuries.[17] Suetonius' *Puteolis et Ostiae singulas cohortes ad arcendos incendiorum casus collocavit* (*Claud.* 25, 2), often cited for assignment of two new urban cohorts to Ostia and Puteoli, is questionable: Suetonius does not specify the name of the cohorts (Praetorians? or, given the reference to fires, *Vigiles*?); these are not necessarily new cohorts.

Location of the urban cohorts XIII–XVIII in the first century remains unclear. between Claudius and 68–69, the varied postings (*pace* Freis 1967), can be summarized as follows:

cohort XIII: between Rome, Lugdunum, Ostia and Carthage until Hadrian;

cohort XIV: in Ostia 42–46/47 under Claudius; in Rome under Domitian;

cohort XV: in Puteoli under Claudius;

cohort XVI: its existence deduced from *CIL* VI 395 and from the fact that we know the cohorts XVII and XVIII;

cohort XVII: perhaps in Lugdunum under Augustus, but certainly there 21–65; in Ostia 68–69; XVII and XVIII are never called *urbanae*;

cohort XVIII: associated with the events of 69 and the Vitellians.[18]

Details of these postings now require revision and will be addressed more fully in the following Sections on public order and the cohorts in Lyon and Carthage.

Although exact details of all Flavian reforms of the army are elusive, the garrison at Rome was certainly affected. The number of *cohortes urbanae* rose to four (X, XI XII, XIV); a *cohors I* (*Flavia*), stationed at Carthage in the second century, was created (see 'the cohort of Carthage'); and at Ostia and Puteoli *vexillationes* from troops at Rome (Ostia) or *frumentarii* (Puteoli) may have replaced whole cohorts. Thereafter the location and role of the urban cohorts remained stable in the early second century.[19]

Castra praetoria and *castra urbana*: the barracks of the *Milites Urbani* in Rome

Pace Freis, from AD 23 the urban cohorts shared the *castra praetoria* on the Viminal with the praetorians.[20] Coarelli, however, now argues for the construction of a *castra nova*, exclusively for the *cohortes urbanae*, in the Late Antonine period, and not, as traditionally assumed, under Aurelian:[21] a statue base with a dedication to the *Genius centuriae* by a *miles cohortis urbanae*, dated 182, suggests the creation of the camp in Regio VII as early as Commodus' reign.[22] Other literary and epigraphic evidence would favour an Antonine date for distinct barracks of the *Urbani* and praetorians – or at least the increased presence of the *milites urbani* in north-central Rome.

Ulpian's reference to the *urbana castra*[23] is generally taken as the largest barracks in the city, the praetorians' camp on the Viminal, but the same phrase could also be used for denoting the whole of the military housing in the city, the *castra praetoria* as the camps of the *Equites Singulares*, *Vigiles*, *Classiarii*, etc. A possible reference to the *castra nova* of the *cohortes urbanae*, a new camp no longer shared with Praetorians, has not been considered.

At some time in the second century, a change occurred regarding the choice of burial areas and of places for sacred dedications of the *milites urbani*. The documents relating to these soldiers are overall approximately 180. Where the place of discovery is known, four areas of concentrations emerge[24]: the first one (with a group consisting of more than forty records) is the vast area that extends around the Via Lata and, following the route of via Flaminia, joins Piazza S. Maria del Popolo and the Milvian bridge. Focusing on the dating, the highest concentration of these records occurs between the second half of the second century and the third.[25]

To give just a couple of examples, a stela from the via del Corso (now in the Barberini collection) presents the full-length portraits of three brothers, originating from *Ulpia Poetovio* and belonging to three different military corps, with different roles[26]: the deceased, *Iulius Iulianus*, is a veteran of the praetorian cohorts; his two brothers, *Iulius Glaus* and *C. Iulius* (without *cognomen*?) are respectively a soldier in the II Parthian Legion and a tribune of the XII *cohors urbana*. The inscription dates from the time of Caracalla and it is legitimate to imagine that the tribune took charge of the choice of the necropolis for his brother's monument, which was in an area near the new camp.

In a burial dedication found near Santa Maria del Popolo and kept in the Attavanti's house (via del Corso) a woman, *Statilia Helpis*, commemorates her husband, who was a soldier of the eleventh *cohors urbana*, a comrade of her husband, who served in the XII cohort, and her young daughter who died at the age of three[27]: it seems that *Statilia Helpis* has been commissioned executor and, in this capacity, she probably gained permission to provide funeral honours for the two deceased soldiers in a cemetery next to the site where the two were serving and had stayed in the second half of the second century.

After two epitaphs, it can be interesting to note the votive dedication offered by *M(arcus) Cocceius Rogatus*, a *cornicularius* of the X cohort tribune *Julius Proculus*, which comes from the 'vicolo del Fico', also dating from the second century.[28]

So, a number of inscriptions, dated to the second and third centuries and related to urban soldiers, dot the initial section of the via Flaminia. Although funerary inscriptions[29] could indicate a preference for burial along the via Flaminia,[30] or even the existence of one or more *stationes*, the number and chronological concentration of these inscriptions of the *milites urbani*, strongly suggest the possibility of a stable presence of the urban cohorts in this area. By comparison, it is notable that a *statio* on the third mile of the Appian Way (south from Rome) has yielded numerous epitaphs and dedications of soldiers (including centurions and officers) stationed in the capital and its environs (Praetorians, *Frumentarii*, *Equites Singulares*, legionaries of the *II Parthica*) but not urban cohorts as if they had a station or their camp elsewhere.[31]

Other factors, as we'll see, indicate a gradual estrangement and eventual separation of Praetorians from the *milites Urbani* in the final two decades of the second century – a trend not unrelated to the increasing military role of the urban cohorts.

Public order and the military role of the urban cohorts

Freis, following Ulpian, discerned the functions of the *cohortes urbanae* as assistance in the criminal and civil jurisdiction of the urban prefect[32] and generally protection of the rights and safety of citizens.[33] The latter duty entailed control of markets and commodity prices (especially for the supply of meat),[34] the proper conduct of exchanges (new under Hadrian), and weights and measures (from Marcus Aurelius). The urban cohorts – in conjunction with the praetorians – also policed the roads, especially on days of spectacles, places of assembly (such as *collegia* or taverns), and persons considered at risk, namely slaves and those of particular trades or engaged in conduct considered potentially dangerous. Hence, from Septimius Severus on, *beneficiarii* of the urban Prefect had the task of drawing up specific lists of such persons of interest.[35]

Yet Fries' account is not exhaustive. As repeatedly emphasized, it is clear from Suetonius (*Aug.* 49.1) and Ulpian (*Dig.* 1.12.1.12) that *milites urbani* and Praetorians had different relationships with the *Princeps* and the city.[36] The term 'city police' is now commonly used to encapsulate all tasks of the urban cohorts. Several scholars, downplaying issues of foreign policy and defense, assign to the army a primary role in maintaining public order in some Eastern provinces, where the military presence was stronger; they argue against a unitary security system created, coordinated and implemented by Rome.[37] Similarly, the tasks of the *cohortes urbanae* are debated: some believe that the transition from Republic to Principate involved a change in approach to containing violence: minimal superstructures were replaced with a more purposeful system of persecution and repression[38]; others see Augustus' motivation for creation of an urban garrison in a desire to ease tensions and to maintain social order, since the City was growing at a extraordinary rate.[39]

The long, complex history of the urban cohorts, however, does not easily fit into rigid and unambiguous positions. The urban cohorts could have been assigned a deterrent function against potential clashes between the various components of society and likewise against external pressures originating from the territory just outside the city.[40] Although the term 'police' in the modern sense may seem incorrect, the urban cohorts surely performed tasks related to maintaining public order in the city and its environs.

Employment of the urban cohorts in military operations (beginning with the Flavians) would seem to contradict Suetonius' *in urbis custodiam* (*Aug.* 49.1),[41] unless he refers just to the time of their creation. Examples of the *milites urbani* in provincial campaigns or militarily active outside the city are infrequent and not always certain.[42] Indisputably, the civil war of 69 brought both the urban cohorts and the Praetorians into action against the German legions, although Tacitus emphasized the rarity of the battlefield use of both units and clearly distinguished the praetorians from the *urbanus miles*.[43] Fifteen years later, cohort XIII was certainly involved, perhaps together with the XIV,[44] in Domitian's Dacian War (84–89) and maybe also in suppressing a rebellion in Mauretania (between 85–86?), as numerous texts show. The centurion, Q(*uintus*) *Vilanius*

Nepos, received decorations three times from Domitian for two different Dacian campaigns and his German war. *C(aius) Velius Rufus*, a *primus pilus* well known from his career inscription at Heliopolis, also received decorations perhaps as tribune of the XIII *cohors urbana* and "dux exercitus Africi et Mauretanici ad nationes quae sunt in Mauretania comprimendas". Two military diplomas of 85 and possibly the epitaph of [*T. Flav]ius Capiton* from Heraclea Lyncestis (late first century) also reveal active duty of some *milites Urbani*.[45]

Other evidence is less secure. Some see the *cohors urb(ana)* from a fragmentary dedication on a statue base in Trajan's Forum as proof of the *cohortes urbanae* in Trajan's wars,[46] but its date is a thorny problem: inscriptions honoring military corps exhibited in the arcades of Trajan's Forum date from Trajan to Marcus Aurelius.[47] The *Historia Augusta* records Marcus Aurelius' extraordinary levy in 175 during the Marcomannic Wars. Recruitment from the *cohortes urbanae* is not explicitly mentioned, but can be inferred: in the emergency the emperor also recruited gladiators, slaves, bandits and *diogmitai*.[48]

The exceptional character of the urban cohorts' participation in the civil war of 68–69 should not be combined with their role in the Dacian Wars. Their contribution to the campaigns of Domitian and Trajan may have been limited to cohort XIII (and cohort XIV?), although the exact location of cohort XIII and its intended function are unclear until Hadrian's reign (see sections on 'The urban cohort in Lugdunum' and 'The urban cohort in Carthage', *infra*). If the evidence for urban cohorts in the Marcomannic Wars is circumstantial, it nevertheless seems to belong to a new development.

The construction of new barracks for the urban cohorts should possibly be redated to the end of the second century (see *supra* the section on 'public order'). Different factors may have contributed to this decision: the greater role of the Praetorians (and perhaps *Milites Urbani*) in provincial campaigns, with an attendant increase in troop numbers; conflicts at the upper chain of command (between the prefects?) for ensuring safety in the city, which later led to the Severan reform of the urban prefect's duties; and the needs of policing, which in a couple of decades required an increase in the number of *stationes* in the city to control the main roads.

Issues of public disorder in the second half of the second century, well documented in the sources and widely discussed, may now be supplemented by new epigraphic evidence relevant to the urban cohorts.[49] A new reading of a dedication relating the equestrian career of the Beneventan *Cn(aeus) Marcius Rustius Rufinus* shows him in 180–184 as centurion in *cohors XV urbana* (not *legio* XV *Apollinaris*). *Pace* the current opinion, the *cohors* XV *urbana* did not survive Vespasian's reign, but this new reading suggests its survival or restoration in the late Antonine era. De Carlo imagines a connection with Marcus Aurelius' Danubian campaigns.[50] Further, the last diploma (on present evidence) jointly recording *cohortes praetoriae* and *urbanae* also dates 180–184[51]; thereafter only distinct diplomas for the two corps were issued.

All available evidence for the period when the military role of the urban cohorts was intensified (or systematized) and these soldiers began to have separate camps from praetorians points to a separation of their roles and responsibilities in the late Antonine rather than the Severan period. If so, this separation anticipates rather than follows the more precise definition of the prefect's powers as commander of the urban cohorts.

A well-known incident of urban unrest in 238 is apropos.[52] The attack of the crowd and gladiators of Rome on the *castra praetoria* and the soldiers and veterans stationed there is generally understood as the end of a long, slow evolution of the relationship between civilians and soldiers in the city. Herodian's language for soldiers in Rome, however, is imprecise. His *stratiotai* are usually taken as the 'new' Praetorians recruited in the German provinces and the Danubian and Balkan areas. But the 'inhabitants of Italy', called to oppose the 'soldiers' and to support the population of the city (7.12.1–2) could be the urban cohorts.

The urban cohort in Lugdunum

Tacitus' reference to a Lyon cohort in AD 21 (during the revolt of Florus) forms the basis of Freis' theory that an urban cohort was stationed there from the beginning of the Principate to protect the *officium procuratoris*.[53] His reconstruction of the Lyon garrison's chronology is as follows:[54]

42–46/47: cohort XVII, after XIII moved to Rome;

68–69: cohort XVIII; XVII transferred to Ostia;

70–96: *cohors I Flavia urbana*;[55]

Hadrian (117–138) 193: cohort XIII; the *cohors I urbana* went to Carthage.[56]

François Bérard's studies now suggest the following revisions of Freis' views, especially for *cohors XIII*[57]: a permanent garrison at Lyon probably does not belong to the Augustan period and no trace of a permanent garrison can be found before the Flavians[58]; throughout the second century (until 192), Lyon was entrusted to the Urban Cohort XIII (later a victim of the civil war of 197), until it was replaced by legionary detachments. The common view that cohort XIII, having spent its first half century in and around Italy,[59] was permanently assigned to Lyon after a brief tenure at Carthage, now invites objections.[60] Clearly this cohort, at least between the end of the first and early second centuries, was involved in frontier operations elsewhere (see section on public order above). Thereafter its soldiers possibly remained in Africa until the final suppression of the Mauretanian revolt.[61] Only later was it reposted to Lyon, perhaps at the end of Trajan's reign.

The urban cohort in Carthage

The Carthaginian cohort, absent in literary texts, is attested only by fragments of burial inscriptions and *laterculi*. In Freis' reconstruction[62] its history is linear:

Flavian era: cohort XIII *urbana* (*CIL*, VIII 1025 = *ILAfr* 379);

Hadrianic era: *cohors I urbana* after cohort XIII's transfer to Lyon.

Deployment of a military contingent to Carthage is linked to the size of the city, the number of its inhabitants (calculated at 50,000–70,000), and its importance, particularly after its refounding by Octavian and the monumental reorganization of its urban space in the second century.[63] Other elements also contributed to enhance Carthage's prestige: the nearby presence of the *proconsul*, the city's role in supplying Rome, and the existence in its territory of extensive imperial properties.[64] These developments necessitated that the African capital have a garrison.[65]

The traditional view posits two cohorts at Carthage: one sent from Rome and another from the nearby *legio III Augusta*, before its transfer to Lambaesis. Already in the 1950s excavations in the Carthaginian urban district of 'Sayda' yielded numerous fragments of lists with consular dating and an epitaph (not earlier than the Flavian era) of a soldier of *cohors I urbana*. New discoveries and subsequent studies suggest modifications to the traditional view.[66] Noël Duval, for example, prefers to think of a single cohort in Carthage (perhaps *miliaria*) at the beginning of the Principate: the *legio III Augusta* was to provide the *officium proconsulis* of perhaps a whole cohort. He shares the widespread opinion that after 75 the *cohors XIII urbana* was assigned to Carthage.[67]

The name lists of soldiers of *cohors I urbana*, dating from the late second century (published in 1998), jointly record *Urbani* and *Vigiles*.[68] Particularly interesting, at Carthage like Ostia, firefighters worked alongside the urban cohorts monitoring the grain depots and preventing fires. This would not be new: the joint action of different military corps is attested at Italian port cities (Aquileia, Ostia, Puteoli) besides some provincial towns.[69]

The current state of research does not seem to support the idea of a single set of decisions about the garrisons of Lyon and Carthage, linked by a shared (or only apparent?) rotation of the *cohors I urbana* and *cohors XIII*. The latter, probably in Italy in the first century,[70] until its deployment in the Dacian campaigns of Domitian and Trajan, finally received its definitive posting to Lyon in the early second century.

As for the first, current evidence, sporadic and dating from different periods, indicates the cohort's peregrinations or fragmentation. As for Africa, Freis observed that the cohort of Carthage gradually extended its control over territory through *stationes*[71]; new epigraphic attestations now confirm the commitment of military forces in African territory, especially for suburban tasks assigned to detachments of the *cohors I urbana* and other cohorts (*vexillationes?*) not necessarily stationed in Africa.[72] Besides Africa, before and after second century, soldiers and centurions of the *cohors I urbana* are also attested at Genava, Ariminum and Rome.[73]

If, as we have here argued, urban cohorts were not trained for combat, there was no reason to create new ones for frontier wars: if the *cohors I urbana* was created by Domitian to assist existing military troops in provincial campaigns, its duties in frontier conflicts were as 'support' units, especially logistics, as subsequently the police function and the service in the Carthaginian port could rather suggest.

Perhaps after completing its original campaign task, this cohort abandoned the epithet *Flavia* and in the adjustments of the late first and early second centuries found a permanent home in the African metropolis.

In sum, in this chapter I retraced the vicissitudes of urban cohorts, indicating the moments of continuity and those of breakage; and identified a significant change after the mid-second century AD.

The structure and organization of the *cohortes urbanae* and *cohortes praetoriae* remained fluid until the time of Claudius, when the fluidity has been gradually replaced by a process of systematization. The *officia praefecti Urbi* experienced various modifications in its composition and location, especially in the Flavian and the Severan periods.

A separation in roles, responsibilities and barracks of the praetorians and the urban cohorts should now be redated about twenty years earlier to the late Antonine period, rather than to Severan, the same era in which military tasks – as combined to urban policing – were permanently assigned to the urban cohorts. At the same time a change takes place also in awareness and in self-representation of the *milites urbani*, as we can see in their iconography: these soldiers are now always represented with arms, whose importance is further emphasized by symbolic proportions. This clue, together with the clothing of these soldiers (the *sagum*, the typical military cloak, worn instead of the *paenula*), seems to suggest a different way of viewing the position of soldiers of the urban cohorts, more enhanced in the military sense.

The cohort XIII has a unique history: it was probably in Italy in the first century, participated in the Danubian campaigns of Domitian and Trajan besides suppression of the Mauretanian revolt, and only in the early second century received its permanent post at Lyon. The *cohors I (Flavia) urbana*, possibly created by Domitian as support for cohorts in frontier combat operations, later abandoned its Flavian epithet *Flavia* after its posting to Carthage.

Appendix: epigraphic evidence after 1967

Texts are arranged in five sections, depending on their origin (A. Rome, B. *regiones Italiae*, C. Lugdunum; D. Carthago; E. Other provenances) and numbered according to each record (which might include more than one soldier, as indicated). Within each section, a basic reference (*AE* = *Année Epigraphique*, EDR = Epigraphic Database Rome) is provided for each document published after the Freis' volume. Some unpublished texts from Rome are sorted according to cohort membership. Epigraphic lemmas include the following elements: category of the soldier, his name, his cohort, his century, possibly his charge, veteran status, the type of inscription and origin, and a date. Documents mentioned in the text are preceded by an asterisk. Uncertain documents are preceded by a question mark. At the end of each section (A, B, C, D, E), new studies of the texts are not indicated systematically.

A. Roma

1) A. Ferrua, in *Riv.Arch.Crist.* 3, 1926, 77 = EDR 6591, M. Manganaro, 13-12-2006 (*miles*): *A(ulus) Helvius Severus, cohors* XII, *centuria Prisci.* Funerary inscription, maybe reused in the cemetery of Pamphilus, Via Salaria (2nd cent.)

2) *AE* 1972, 46 = C. Carletti, in *Rend.Acc.Linc.* 25, 1970, 206 no. 3 (*miles*): [- - -] *bus, cohors* X, *centuria* [- - -] *cani Principalis* = EDR 75191, F. Feraudi Gruenais, 6-2-2002. Funerary inscription, maybe reused in the cemetery of Bassilla, Via Salaria (2nd cent.)

★3) *AE* 1978, 23 = A. Ferrua, in *Rend.Acc.Linc.* 33, 1978, 37 no. 44 (*principalis*): [- - -]*mus T.f., a quaestionibus praefecti urbi, cohors* X = EDR 76884, F. Feraudi Gruenais, 12-4-2000. Funerary inscription, from the catacomb of San Sebastian, Via Appia (3rd cent.)

4) *AE* 1979, 38 = A. Ferrua in *Rend.Acc.Linc.* 34, 1979, 31 no. 15, two soldiers (*milites*): *M(arcus) Minatius Marcellus* and *Q. Salenus Pudens, cohors* XII. Funerary inscription, unknown origin (late 2nd cent.)

★5) *AE* 1983, 52 = G. Barbieri, in *Il Lapidario Zeri di Mentana*, Roma 1982, 95, no. 39 (*miles*): *Sex(tus) Licinius Pom(ptina tribu) Mansuetus* from *Dertona* (*regio* IX), *cohors* XV. Funerary inscription, unknown origin (mid 1st cent.)

6) *AE* 1983, 54 = G. Barbieri, in *Il Lapidario Zeri di Mentana*, Roma 1982, 98, no. 41 (*miles*): *C(aius) Pompenna Valens, cohors* XI. Funerary inscription, opistograph (*CIL*, VI 33965 – *ILS* 5209), from the necropolis between Via Salaria and Via Pinciana (second half of the 1st cent.)

7) *AE* 1984, 57 = S. Panciera, in *MNR* I 7.1, 1984, 159, V28A = Panciera 2006, 1392 (*miles*): *C(aius) Sertorius L.f. Cru. Iustus, Iguvio, cohors* X, *centuria Veturi.* Funerary inscription, from ponte Milvio (late 1st–early 2nd cent.)

8) *AE* 1984, 63 = S. Panciera, in *MNR* I 7.1, 1984, 159, V28O = Panciera 2006, 1398 (*miles*): *M(arcus) Sestius M.f. Serg. Clemens, Aenona, cohors* XI, *centuria Mari.* Funerary inscription, from ponte Milvio (late 1st–early 2nd cent.)

★9) *AE* 1984, 64 = S. Panciera, in *MNR* I 7.1, 1984, 159, V28P = Panciera 2006, 1398 s. (*principalis*): *L(ucius) Atilius L.f. Fal. Frequiens, Albintimili, cohors* XII, *centuria Macrini, a quaestionibus praefecti urbi.* Funerary inscription, from ponte Milvio (late 1st–early 2nd cent.)

10) *AE* 1984, 66 = S. Panciera, in *MNR* I 7.1, 1984, 159, V28R = Panciera 2006, 1400 s. (*miles*): *C(aius) Petronius C.f. Sab. Avullus, Mantua, cohors* XI, *centuria Proculi.* Funerary inscription, from ponte Milvio (late 1st–early 2nd cent.)

11) S. Panciera, in *MNR* I 7.1, 1984, 168, V28 O = Panciera 2006, 1396 s. (*miles*): *Q(uintus) Enius Q.f. Ste(llatina tribu). Moderatus, Mevaniola, cohors*

XI, *centuria Prisci*. Funerary inscription, from ponte Milvio (late 1st–early 2nd cent.)

★12) A. Ferrua, in *RPARA* 59, 1986–1987, 189 no. 29 = EDR 15656, M. Manganaro, 13-12-2006 (*veteranus*): *Iul(ius) Domitius Secundianus, cohors* X. Cf. *infra*, B.1. Funerary inscription, from the cemetery of Commodilla, Via Appia -Via Ardeatina (between 2nd and 3rd cent.)

13) G. Tagliamonte, in *Tituli* 6, 1987, 60–62 no. 17: (*miles*): *Q(uintus) Fabius Q. [f. - - -], cohors XIV*. Funerary inscription, from ancient Via Aurelia (mid 1st cent.)

14) *AE* 1995, 176b = S. Panciera, in *Studia Mihailov* 350, no. 240 (*miles*). *Grae(- - -) Primus, cohors* XI. Sacred inscription (*Silvanus*), from Via Livenza (3rd cent.)

15) *AE* 2000, 242 = G. Sacco, in *Epigraphai* 2000, 923–930 (*veteranus*): *L(ucius) Barbius L.f. Men(enia tribu) Primus, Vicetia, cohors* XI, *centuria Nepotis*. Funerary inscription (late 1st–first half 2nd cent.)

?16) *AE* 2001, 253 = P. Tassini, in *Le iscrizioni dell'*Antiquarium *comunale del Celio* (*Tituli* 8), Roma 2001, 125–127, no. 36 (*emeritus*): *Q(uintus) Iulius Martialis, (miles urbanus?) emeritus, militavit annis XXI*. Funerary inscription, unknown origin (AD 101–125)

17) *AE* 2004, 324 = S. Panciera, in *Cahiers Centre Glotz* 15, 2004, 299f. no. 16 (*miles*) = Panciera 2006, 1503f. no. 16: *M(arcus) Suellius Priscus, cohors* X, *centuria Aulli*. Funerary inscription, urban origin? (mid-1st cent.)

18) *AE* 2004, 207 = M. L. Caldelli, in *Libitina e dintorni*, Roma 2004, 178 no. 2 = EDR 74569, S. Evangelisti, 4-3-2008 (*miles*): *L(ucius) Acilius Crescens, cohors* XIV, *centuria Saturnini* (first half 3rd cent.)

?19) Unpublished marble fragment of a stele (*miles*): *L. [- - -] Fortu [- - -] , cohors* X (*urbana?*), *centuria Cand[- - -]* . Funerary inscription, now preserved in cloister of the Basilica di San Paolo, campata 23, 6. (not dated)

20) Unpublished marble slab (*miles*): *C(aius) Bellenius C.f. Fal(erna tribu) Victor, Caudium, cohors* XI (*miles*). Funerary inscription, unknown origin; now in Roman National Museum, depot, inv. 115604; neg. UniRoma1 no. 892 (Crimi 2010, pp. 329–330 no. 1) (late 1st–2nd cent.)

21) Unpublished marble slab with *infundibulum* (*miles*): *M(arcus) Iulius Proculus, cohors* XI, *centuria Similis*. Funerary inscription, unknown origin, from the Gorga collection (second half of 1st–2nd cent.)

★22) Unpublished central part of an incomplete marble slab; two peoples (*miles e veteranus?*): *C(aius) Rufican[us? - - -], cohors* XII; and *[P(ublius)?] Accius P.f. [- - -] , veteranus Augusti*. Funerary inscription, from Via Zanardelli (lungotevere Tor di Nona), now in Roman National Museum, depot inv. 108891; neg. UniRoma1 no. 2475 (late 1st–2nd cent.)

23) Unpublished fragment of a marble slab (?) (*miles*): *[- - -] , cohors* XII. Funerary inscription (?), found in 1909 at Via Flaminia (not dated)

★24) Unpublished incomplete marble slab (*miles*): *A(ulus) Avillius Pom(ptina tribu) Aretio Verus, cohors* XIV, *centuria Alfi*. Funerary inscription, from Via

Po no. 16, now in Roman National Museum, depot inv. 115522; neg. UniRoma1 no. 896 (Crimi 2010, pp. 331–332 no. 3 (mid 1st cent.)

25) Unpublished fragment of a marble urn (*miles*): [- - -] *P.f.* [- - -] , unknown urban cohort. Funerary inscription, unknown origin, now in Roman National Museum (late 2nd–early 3rd cent.)

26) Unpublished fragment of a molded marble slab (*miles?*): a *miles urbanus*, anonymous. Funerary inscription, from Via Bocca della Verità (not dated)

27) Unpublished fragment of a marble slab (*miles?*): a *miles urbanus*, anonymous. Funerary inscription, unknown origin, now in Ripartizione X depots (not dated)

28) Unpublished, partially reconstructed stele (*miles?*): a *miles urbanus*, anonymous. Funerary inscription, in the court of Basilica di Sant' Agnese (3rd cent., maybe first half)

New studies

★*CIL*, VI 2862. *AE* 1983, 24 = R. Sanquer, in *Archaeologie en Bretagne*, 37, 1983, pp. 31–40 (*veteranus*). *M(arcus) Ulpius Iulianus, cohors I urbana*. The author indicates four urban inscriptions (also *CIL*, VI 13196, 18544 and 25117) from Campana Gardens at the Lateran, bought and brought to Finistère by the founder of a religious community. Funerary inscription, unknown origin (late 2nd–early 3rd cent.)

B. Regiones Italiae

Regio I. Latium et Campania

★1) *AE* 1973, 177 = A. Giannetti, in *Rend.Acc.Linc.* 28, 1973, 473 = EDR 75479, H. Niquet, 1-7-1997, Fontanarosa di Cassino, on the wall of a private home (*veteranus*): *Iul(ius) Domitius Secundianus*. Funerary inscription, cf. *supra*, no. A.12 (2nd cent.)

★2) *AE* 1977, 182 = L. Quilici, in *Not.Sc.* ser. 8, 30, 1976, 325 (*centurio*): *Q(uintus) Avillius P.f. Cor(nelia) Camers, centurio* of the cohort III of vigiles and of the cohort XIII *urbana*. Funerary inscription, from Castel Giubileo, in the territory of ancient *Fidenae* (late 1st–early 2nd cent). Cf. *Not. Sc.* 1929, 264 no. 12.

Another *Q(uintus) Avillius* is known from a fragmentary inscription from *Fidenae*; kind report by M. G. Granino Cecere)

3) *AE* 1982, 164 = A.M. Villucci, *I monumenti di Suessa Aurunca*, Scauri 1980, 38–41 = R. Palmieri, *Misc.Gr.Rom.* 8, 1982 467 no. 21, *Suessa*. Cf. *AE* 1984, 183 = = M. Pagano – A.M. Villucci, in *Rend.Acc. Napoli* 57, 1982, 213, Cellole (*Minturnae*) (*tribunus*): *L(ucius) Magius Sex.fil. Urgulanianus*,

primus pilus, centurio speculatorum, praefectus vexillariorum legionum trium (VIIII, V, IIII) and tribune of a not specified urban cohort. Funerary inscription (early 1st cent.)

★4) *AE* 1986, 180 = G. Camodeca, in *Puteoli* 7–8, 1983–1984, 50–54. Cf. Camodeca 1999, 116 + 79 add. (*miles*), *Herculaneum*: *L(ucius) Lucretius Firmus, cohors XIII, centuria M(arci) Salvii Firmi* (August, AD 40 or 43). Cf. *AE* 1982, 192

?5) *AE* 1993, 440 adn. = M. G. Granino Cecere, in *Dives Anagnia. Archeologia nella valle del Sacco*, edited by S. Gatti, Roma 1993, 122 no. 5, Tufano (*Anagniae*): *P(ublius) Satrius P.f. Ouf(entina tribu)*, maybe a *miles urbanus* (first half of the 1st cent.)

6) M. G. Granino Cecere, in *Annali dell'Associazione nomentana di Storia e Archeologia* 2, 2001, 43–45, from Setteville di Guidonia (*tribunus*): *L(ucius) Cantinius Men(enia tribu) Maximus, Nuceria*, tribune of the IV *cohors vigilum*, of the XIV *cohors urbana* and of the IX *cohors praetoria, primus pilus bis, procurator Augusti hereditatium*. Funerary inscription (second half of the 2nd cent.)

7) Ricci 2001, 46 no. 23 = EDR 6528, M. Manganaro, 13-12-2006: *C(aius) Publicius Ianuarius, Fabia, Roma, cohors* XI, *centuria Britti* (*miles*). Funerary inscription, once in the villa Spigarelli, in Anzio, now maybe lost (2nd cent.)

Regio III. Lucania et Bruttium

8) *AE* 1969/70, 172 = V. Bracco, in *Rend.Acc.Linc.* 24, 1969, 236, *Atina* (*miles*): *Iun(ius) Felix, cohors* X. Funerary inscription, dedication to a pretorian soldier (2nd cent.?)

Regio IV. Samnium

9) *AE* 1978, 286 = C. Letta, Le imagines Caesarum di un praefectus castrorum Aegypti e l'XI coorte pretoria, in *Athenaeum* 56, 1978, 3–19, Lecce dei Marsi (AQ). Cf. *AE* 1996, 513 adn. = L. Keppie, in *Athenaeum* 84, 1996, 108–111, with a new interpretation (*tribunus*): *A(ulus) Virgius Gallicus, tribunus militum in praetorio divi Augusti et Ti. Caesaris Augusti* of a cohort XI and of a cohort IIII (Tiberian age)

Regio VI. Umbria

10) *AE* 1976, 197 = G. Prosperi Valenti, in *Epigraphica* 38, 1976, 157–162, Trevi (*miles*): *L(ucius) Caesernius L.f. Clemens, cohors* X. Sacred inscription (*Silvanus*) (1st–2nd cent.)

11) *AE* 1980, 406 adn. = A. Alessandri, *I municipi di Sarsina e Mevaniola*, Milano 1928, 48f. no. 82, Sarsina (*tribunus*). *L(ucius) Appaeus L.f. Pup(inia tribu) Pudens, primus pilus* and tribune of a cohort X *praetoria* and of a cohort XII *urbana*. Dedication to the Emperor Trajanus (AD 112–113)

Regio VII. Etruria

12) *AE* 1985, 390, *Luna* = G. Mennella, in *Ann. Museo Civico La Spezia* 2, 1979–1980, 217–220 (*milites*): fragmentary *diploma militare* for praetorians (cohorts I–X) and a *milites urbani* (cohorts X, XI, XII e XIV) (AD 182–184)

?13) *AE* 1989, 306 adn. = R. Cosentino – P. Sabbatini, *Miscellanea Ceretana*, Roma 1989, 110 (*principalis*). *L(ucius) Pantilius L.f. Duurus, signifer*, an praetorian or urban soldier? Sacred inscription (*Aquae Caeretane*) (first half of 1st cent.)

14) *AE* 1993, 646 = M. Iozzo, in *Epigraphica* 55, 1993, 173–180, *Volaterrae: diploma militare* for praetorians (cohorts I-X) and *milites urbani* (cohorts X, XI, XII, XIV?) (AD 163–164)

Regio VIII. Aemilia

15) *AE* 1981, 388 = G. Susini, in *Epigraphica* 28, 1965, 189 no. 18, Ganaceto (Modena) (*miles*): *C(aius) Samius Crescens, cohors* XII, *centuria Materni*. Funerary inscription (between the second half of 1st and the 2nd cent.)

16) *AE* 1999, 704 = D. Pupillo, *Suppl.It.* 17, 174 s., no. 7, Ferrara (*veteranus?*): *T(itus) Camurius Priscus, cohors* XIV (2nd cent.)

Regio IX. Liguria

17) *AE* 1988, 577 = S. Roda 1984, 152 no. 2, *Pollentia* (*miles*): *L(ucius) Mercleanius* [.] *f. Pal(atina tribu) Martialis*, unknown urban cohort (1st–2nd cent.?)

Regio X. Venetia et Histria

★18) *AE* 1997, 600 = M. S. Bassignano, *Suppl.It.* 15, 1997, 171f. no. 27, San Biagio (*Ateste*) (*miles*): *Sex(tus) Munatius Marcellinus, cohors* XIV. Funerary inscription? (1st–2nd cent.)

Italian origin

?19) *AE* 1988, 1137 = J.J. Aubert, J.R. Lenz, J. Roth, J.A. Sheridan, in *Zeit. Pap.Epigr.* 73, 1988, 91 no. 1 (*miles*): *C(aius) Iulius Liberalis, cohors* XI, *centuria Latini*. Funerary inscription, unknown origin, probably italic (2nd cent.?)

New studies

CIL, XIV 2956, *Praeneste* = M. G. Granino, in R. Neudecker – M. G. Granino, *Sculture e iscrizioni antiche nell'*Institutum Archaeologicum Germanicum (Palilia, 2), Wiesbaden 2001, 174 no. 96 illustrates the epitaph of *L(ucius) Pomp*[- - -] *Felix, cohors* XI (*urbana?*), *centuria* [- - -] *RAE* (late 1st–early 2nd cent.)

★*CIL*, X 1127, *Misenum* = A. De Carlo, in *Zeit.Pap.Epigr.* 170, 2009, 299–303 gives a new reading of inscription's line 13: *Cn(aeus) Marcius Rustius Rufinus* was a centurion of the *cohors XV urbana* (instead of the *XV Apollinaris*) in Antonine era (AD 180–184)

CIL, XI 4747, *Tuder* = G. Forni, in *Epigraphica* 46, 1984, 136 (= *AE* 1985, 366) gives a new reading of the inscription of *M(arcus) Pompusidius M.f. Clu(stumina) Genialis, cohors* XI, *centuria Crispini* (1st–2nd cent.)

CIL, XI 358, *Regium Lepidum* = C. Franzoni, in *Misc. St. Arch. e Ant.* I, Modena 1983, 105–117 studies the stele of a *veteranus Augusti, C(aius) Metellius C.f. Constans*, and of his son, *C(aius) Metellius C.f. Florinus, cohors* XII (second half of 2nd cent.?)

CIL, XI 1936, Ponte Pattoli (PG): M. Barbanera, *Riconsiderazione di un'iscrizione sepolcrale relativa a un* miles urbanus *da Perugia*, in *Arch. Class.* 42, 1990, 431–440 = *AE* 1991, 664 studies the stele of *P(ublius) Pacilius Alcaeus, cohors X* (late 1st–early 2nd cent.)

CIL, V 905 = *Inscr. Aq.* II no. 2853, Aquileia = Arrigoni Bertini 2003 considers the funerary inscription of a *miles* together with the epitaphs of soldiers of urban cohorts originating from Parma (first half of the 1st cent.)

CIL, V 909 = *Inscr. Aq.* II no. 2854, Aquileia = S. Blason Scarel, in *Alsa* 3, 1990, 32–43 (= *AE* 1991, 765 adn.) considers the inscription of *C(aius) Cornelius C.f. Successus, cohors* XII (late 1st–early 2nd cent.)

C. Lugdunum

1) *AE* 1976, 443 = M. Le Glay, A. Audin, *Notes d'épigraphie et d'archeologie lyonnaise*, Lyon 1976, 47–51, 3, *Lugdunum (veteranus): C(aius) Flavius M.f. Gal(eria tribu) Ianuarius, cohors* XIII and *decurio* of Roman colony. Funerary inscription (late 2nd cent.)
★2) *AE* 1993, 1194 = Bèrard 1993, 39–54 (*miles*): *C(aius) Numerius Luk(- - -), cohors* XIV (mid 1st cent.)

D.1 Carthago

1) *AE* 1984, 928 = S. Lancel, in Le Bohec, Duval, Lancel 1984, 36–46 *Carthago*. List of discharged soldiers of the *cohors* I *urbana* (about AD 230). For *C(aius) Sextilius Donatus*, see also *infra*, no. D.1, 3
2) *AE* 1992, 1810 = L. Ladjimi Sebäi, in *Bull.Travaux de l'INP, Comptes Rendues* 5, 1990, 9–16, *Carthago* (*miles*): *L(ucius) Arulenus [- f.] Fundanus Su[- - -], cohors* I *urbana, centuria Ti[- - -]*. Funerary inscription (early 3rd cent.?)

★3) *AE* 1998, 1539–1546 = Ennabli, Ben Abdallah 1998 published eight list of soldiers of the *cohors* I *urbana*: here it's the name of a *Licinius Germulus* (cf. *CIL*, VIII 24633 = Le Bohec, Duval, Lancel 1984, 72) and three epitaphs (1547–1549) of a [- - -]*enus Paternus*, soldier of the *cohors* I *urbana*; of a *C(aius) Sextilius Primus* and of a *Sextilia Donata*, probably connected with the soldier no. D.1, 1 (late 2nd–early 3rd cent.)

★4) *AE* 1999, 1833, *Carthago* = Ben Abdallah, in *CEDAC* 19, 1999, 8–10 (*miles*): *Q(uintus) Camerius Q.f.* [- - -], *cohors* XIII, *centuria Aelii Dextri*. Funerary inscription (AD 90–117)

D.2 *Africa Proconsularis, Byzacena, Mauretania Caesarensis*

★1) *AE* 1973, 576 = A. Beschaouch, in *Africa* 3–4, 1969–1970, 123, *Neferis / Henchir Bou Beker, Procos* (*principalis*): *L(ucius) Aurelius Festi f. Hirrius Festus, singularis tribuni cohortis I urb(anae), homo optimus, civis incomparabilis*. The same man erected two dedications (*ILAfr.* 333–334). Honorary inscription (AD 218–222)

★2) *AE* 1973, 595 = L. Maurin, J. Peyras, in *Cah. Tun.* 19, 1971, 61f., *Uzali Sar – Henchir Djal, Procos* (*veteranus*): *Q(uintus) Fabricius Apronianus, cohors* I *urbana*. Funerary inscription (2nd cent.)

★3) *AE* 1976, 741 = Ph. Leveau, in *Bull. Arch. Alg.* 5, 1971–1974 (1976), 180f. no. 9, *Caesarea, Mauretania Caes.* (*miles*): *Aebutius Rufus, cohors urbana* (!), *centuria Oletani* (uncertain era). Cf. *AE* 1979, 683 and Bérard 1988, 174, n. 64

★4) *AE* 1996, 1701 = M. Khanoussi, in *L'Africa romana* 11, 1996, 1346–1349 no. 2, *Capsa, Byzacena* (*miles*): *L(ucius) Messius L.f. Gal(eria tribu) Fructus, Myrtili, cohors I urbana*. He came from *Mertola, Lusitania* in the Author's opinion, the *cohors I urbana* "detached military elements from various cities to perform different tasks, including probably the guard of *stationes* of *IIII publica Africae*" (early 2nd cent.). Cf. *AE* 1996, 1702 (AD 209–211)

★5) *AE* 1997, 1621a = Le Bohec, Ben Abdallah 1997, 45–51, *Ammaedara, Numidia* (*miles*): *Geminius Orfitianus, cohors* XI. Funerary inscription (together with three epitaphs of *Geminii*) (about AD 197–230)

★6) *AE* 2000, 1738 = M. Khanoussi, in *Africa romana* 13, 2000, 1132–1135, *Vaga* -Béja, *Procos* (*miles*): *Q(uintus) Tuscilius L.f. Quintianus, cohors* XI, controlled, together with other soldiers, the *saltus Burunitanus* "à la lisière SE del *saltus*" (AD 151–230)

New studies

★*CIL*, VIII 1328 = B. Ben Abdallah, in *L'Africa romana* 8, 1991, 263 s. = *AE* 1991, 1677 adn., Chidibbia, Procos: *L(ucius) Aemilius Honoratus* maybe is a centurion of the *cohors I Urbana*, native of *Uthica* (2nd cent.?)

E. Other origins

*1) *AE* 1978, 374 = Bivona 1978, *Thermae* (*miles*): *C(aius) Popillius C.f. Mae(cia tribu) Priscus, cohors* X. The (*Ticiniana*) victory discussed in the text seems to allude to the conflict between Otho and Vitellius, a previous episode of the clash in *Bedriacum* (March/April AD 69) and confirms the participation of urban cohorts (cf. *Hist.* I 87, 1; 89, 1; II 21, 4; 17, 2). G. Manganaro, in *Epigraphica* 51, 1989, 161–209 = *AE* 1989, 345 adn. espressed another view: he proposes the reading *victoria* [*L*]*iciniana* as a reference to the Licinius's victory over Daia in *Campus Ergenus* (April AD 313).

   Difficult to justify the indication of such a complete onomastic formula (with the tribe too) in a epitaph of a soldier of urban cohorts in the fourth century AD

2) *AE* 2000, 1203 = B. Lörincz, in *Arh. Vest.* 51, 2000, 249–251, Császár, *Pannonia superior*. A military diploma with the names of soldiers of the *cohortes urbanae Antoninianae* X, XI, XII, XIII (AD 221–222)

Unknown origin

*3) *AE* 1993, 1788 = M. Roxan, W. Eck, in *Zeit.Pap.Epigr.* 96, 1993, 67–74 publish a military diploma of AD 85 (21–22 February), granted to a *miles urbanus* of the cohort XII, *Gaius Latinius G.f. Gal(eria tribu) Primus* (*Sebastopolis – Asia*). In the diploma praetorians (cohorts VI, VII, VIII, IX) appear together with *milites urbani* (cohorts X, XI, XII, XIIII)

*4) *Rom.Mil.Dipl.* IV 203 = *AE* 2002, 1770 = Pferdehirt 2002, 14 no. 5: a military diploma of may AD 85, granted to a *miles urbanus* of the cohort XIII (*Lucius Valerius L.f. Claudia Celer. Savariensis*)

*5) *Rom.Mil.Dipl.* IV 205 = *AE* 2002, 1771: the name of *C(aius) Valerius Niger, cohors* XIII, appears on the diploma of a centurion of a *classis Ravennatium* (AD 71)

New studies

AE 1903, 368 = *ILS* 9200 = *IGLS* 2796, *Heliopolis, Syria*: Kennedy 1983 and Strobel 1986 on the career of *C(aius) Velius Rufus*.

IGR I 779 *Selymbria, Thracia* = Petzl 2000 = *AE* 2003, 1561, with large references, on the iconography of the *miles urbanus M(arcus) Cincius Nigrinus* (Trajanic era?)

Notes

1 Ricci 2011a and Ricci 2014. It has been preferred to maintain the structure of the original paper and therefore also the last paragraphs (4 and 5) dedicated to the cohorts of Lyon and Carthage.

2 Note also on the *cohortes urbanae* the American dissertation of F. Mench (1968: *non vidi*); the praetorian cohorts, to which the urban cohorts were originally appended, are studied by Durry 1938 and Passerini 1939.

3 Basic for the *principales* of the *cohortes urbanae* remains the text of *C(aius) Luccius Sabinus*, discharged in 134 (*CIL*, IX 1617 = *ILS* 2117, Beneventum); for a new *a quaestionibus praefecti*, *L(ucius) Atilius Frequiens*, see Appendix A.9; a possible new *signifer*. *AE* 1989, 306 adn.

4 The significance of new evidence for revision of the history of military units is amply illustrated for the *legio II Parthica*: excavations at Syrian Apamea now require integration of those finds with the material from Albanum; for a synthesis see Ricci 2000.

5 On the functions of the *Praefectus Urbi*, largely modified in the Severan era, see Mantovani 1988 (with commentary on Dig. 1.12.1.12, Ulp. *Lib. sing. de officio praefecti urbi*); Rucinski 2009; for the third–fifth centuries see Vitucci 1956; Chastagnol 1960; F. Guizzi, *Per la storia della* praefectura urbi, Camerino 1981; Peachin 1996; and Ménard 2004; see also for other aspects: Panciera 1993 and Todisco 1999.

6 It's commonly believed that significant changes for the urban cohorts have occurred in Severan era (see, e.g., Nippel 1995, p. 95).

7 Suet. *Aug.* 49.1: (*[Augustus] ceterum numerum partim in urbis, partim in sui custodiam adlegit*), referring to 27 BC, although he doesn't specify the year of the cohorts' creation, their number, or specific functions. On the reforms of Augustus, see Eck 1995.

8 Maybe all with the epithet *praetoria*. Initial attestations of specific cohorts are: *cohors* X, Domitian (*CIL*, VI 1626 = *ILS* 1385); *cohors* XI, end of Augustus' Principate (*CIL*, X 4872 = *ILS* 2021, Venafrum); *cohors* XII, Nero (*CIL*, XI 395 = *ILS* 2648, Ariminum); Freis 1967, p. 37.

9 Dig. 1.12 specifies the functions of the urban prefect for public order: *Quies quoque popularium et disciplina spectaculorum ad praefecti urbi curam pertinere videtur; et sane etiam debet dispositos milites stationarios habere ad tuendam popularium quietem et ad referendum sibi quid in urbe agatur*. The first *praefectus Urbi*, *M. Valerius Messalla Corvinus* in 26 BC, resigned not long thereafter: Tac. *Ann.* 6.11; Hier. *s.a.* century 26, p. 164, 28 Helm; Della Corte 1980. On the uncertain date of the *praefectura Urbi*'s creation, see Vitucci 1962–1963; Keppie 1996; Daguet-Gagey 2000, p. 73; Sablayrolles 2001, p. 145; and the synthesis of Rucinski 2009.

10 Sablayrolles 2001, pp. 133 and 138. Thus Rivière's view (2004, p. 67 n. 12) of Augustus is reductive: "Il faudrait encore y réfléchir, mais l'action d'Auguste pourrait s'inscrire dans la continuité d'une logique de guerre civile, vivant à conserver des effectifs à sa disposition en cas de conjuration ou d'émergence d'un parti adverse."

11 Keppie 1996, p. 115f. *Milites urbani* are also attested elsewhere in this period. The praetorians in Aquileia were initially linked to Augustus' presence, although praetorian inscriptions there also postdate the Augustan period. Only one *miles urbanus*' inscription (*CIL*, V 905) can be securely dated to the beginning of the Principate.

12 Suet. *Tib.* 37.1. Rivière (2004, p. 64) places the construction of *castra praetoria* among the measures for public order, providing protection against armed attacks, the phenomenon of banditry and the incidental effects of sedition. See also, Part III, Chapter 6.

13 *CIL*, XI 4872 = *ILS* 2021.

14 Appendix B.9: found at Lecce dei Marsi in the territory of Marruvium; discussed by Keppie 1996 (with extensive bibliography) and Sablayrolles 2001, pp. 134 ff. See also Part II p. 102, n. 52.

15 Keppie 1996, pp. 110 ff.

16 For the reign of Caligula, Josephus speaks first of three (*BJ* 2.205), later of four (*AJ* 19.188) cohorts: Durry 1938, p. 12 n. 5; Rucinski (2009, pp. 170f with no. 1–2) now attempts to reconstruct Claudius' changes in the organisation of the *cohortes urbanae.*

17 Appendix A. 5, 13, 24, B.16, 18. The Lyon inscription of *C(aius) Numerius Luk*(- - -) of the XV urban cohort (Appendix C.2) also dates to the mid-first century: see section on the urban cohort in Lugdunum, *infra.*

18 Tac. *Hist.* 1.64.3 (*cohortem XVIII Lugduni solitis hiberni relinqui placuit*). See also section on the urban cohort in Lugdunum, *infra.*

19 No significant changes in the post of *praefectus Urbi* in the first and second centuries can be discerned; his coordination with the prefect of the *vigiles* as part of an "urban police system" continued: *Dig.* 1.15.4f.; Mantovani 1988, p. 207; Chastagnol 1997, p. 144; Rucinski 2009. For the peculiar "carattere consociativo" of republican magistracies, see Capogrossi Colognesi 2009, pp. 115f., 217f.

On the *officia* of the urban prefecture, the sources are rare and conflicting (Lyd. *Mag.* 1.34, based a lost fragment of Suetonius; Martial 2.17; *CIL*, VI 31959 = *ILS* 5523). For the discussion, see now Coarelli 1999.

On the dedication *CIL*, VI 40646 (*AE* 1965, 338; 1983, 28) and the hypothesis of a Caracalla's reorganization of the staff (and a re-location?) of the prefecture's *officia*, see Dietz 1980, 154 n. 410, Id. 1983, 397–404. *Contra*, Géza Alföldy, in the apparatus to *CIL* VI 40646.

20 Tac. *Ann.* 4.2.1; Cass. Dio 57.19.6; Suet. *Tib.* 37. 2; Aur. Vict. 2.4; Freis 1967, pp. 6f., based on the following considerations: the *milites urbani* in 31 did not oppose Sejanus; votive dedications of *milites urbani* and praetorians are in the *castra praetoria*; soldiers of the Praetorian Guard and the urban cohorts shared burial sites. The long passage of Cassius Dio (55.24.5–8, referring to AD 5), cataloging Roman armed forces, demands caution for analysis of different units in the city: *ton astikòn* (the *milites urbani*), the *somatophylakes* (praetorians?), the *tès poleos phrouroi* (vigiles?), and the *Bataouoi* (*corporis custodes*).

21 Coarelli 1993, p. 255; see also Sablayrolles 2001, p. 140. For Aurelian, cf. *Chron. 354*, p. 148, 9: Hic [*Aurelianus*] *muro urbem cinxit, templum Solis et castra in campo Agrippae dedicavit*; on the Libellus de regionibus urbis Romae, rec. A. Nordh 1949 (*templum Solis et Castra*).

22 *CIL*, VI 217 (cf. pp. 3004, 3755) = *ILS* 2106, found under the church of Trinità dei Monti. See also Sablayrolles 2001, p. 140; and Jolivet 2007, pp. 107–108. On the praetorians' dedications to Genius centuriae, see now Panciera 2012.

23 Dig. 48.5.16.3, vol. I, p. 847 Krueger (*Quod si qui praesens sit, vice tamen absentis haberetur – ut puta qui in vigilibus vel urbanis castris militat – dicendum est deferri hunc posse: neque enim laborare habet, ut se repraesentet*).

24 The viae Salaria and Pinciana (24 records); the via Appia (8); the via Nomentana (7). More sporadic finds are originating from the via Latina (2), the via Ostiensis (including San Paolo's collection, 4), the via Portuensis (1), the via Tiburtina (1), the via Aurelia (1), the via Labicana (1).

25 A different meaning, of course, should be attributed to dedications and epitaphs: the former may be interpreted as a signal of the existence of *castra*, so as the proximity of a *statio* at the northern exit of Rome. The laterculi give no help: a single fragment of *CIL*, VI 32518 (AD 154) may come from via delle Muratte (between via del Corso and Fontana di Trevi).

26 *CIL*, VI 2579; Freis 1967, pp. 97 s.

27 *CIL*, VI 2887, cf. p. 3377; Freis 1967, p. 100.

28 *CIL*, VI 2869, cf. p. 3377 and 3841 = *ILS* 2114; Freis 1967, p. 99. See the *M. Cocceius Iulianus* of the XI urban cohors (*CIL*, VI 2888, cf. p. 3377; Freis 1967, p. 100. Second century AD).

29 Unfortunately, with two exceptions, the original *locus* of most inscriptions erected by the *milites urbani* (e.g., *CIL*,VI 14, 218, 477, 481, 531) is unknown. For one, see n. 22 *supra*; a second is *CIL*,VI 2863 = 2931 (first/second century), seen *in quadam domo prope columnam Antonini in pariete, vico cui vulgo a le Convertite, via Lata*, but not surely here originally located. A provisional list of texts of the soldiers of urban cohorts from regio VII includes twenty published inscriptions, to which can now be added three unpublished from the via Zanardelli, via Flaminia, and via Bocca della Verità. See Appendix A.22, 23, 26.

30 E.g., the stelai of Ponte Milvio (Appendix A.7–11); most of the epigraphic evidence concerns praetorians.

31 Latteri 2002.

32 The court of the *praefectus Urbi* became the most important criminal venue in Rome: Mantovani 1988, part. pp. 207ff.; Daguet-Gagey 2000, p. 76.

33 On *stationes* throughout the city, see Freis 1967, p. 45, citing Cass. Dio 65.8.2 (AD 69): *stratiotai* sleeping on the Capitoline and *phylakes* (of the temple of *Iuppiter?*). Nevertheless, I think it is more probable that it is praetorians guarding the residence of the *Princeps*.

34 Dig. 1.12.1.11; Chastagnol 1997, p. 116.

35 On the *beneficiarii praefecti Urbi*, see also Rucinski 2009, *passim*.

36 See also Fraschetti 2000, p. 728.

37 For an army in the East as exclusively a force for suppression/exploitation of the natives see Isaac 1992, followed by Alston 1994 (Egypt) and Pollard 1996, 2000 (Syria and Palestine). *Contra*, sharp objections and detailed arguments in Wheeler 1993. A good bibliographical synthesis of the 'strategy debate' in Wheeler 2007, p. 263f. n. 1. For the '*pacata*' Asia Minor, see Brèlaz 2005.

38 Lintott 1968, 2008; Veyne 1976; and, more nuanced, Nippel 1995; on praetorians, see Rivière 2004, pp. 65–67 (see also n. 40 *infra*); cf. the broader discussion of Le Roux 2002.

39 Sablayrolles 2001, pp. 126 ff.

40 "L'*Urbs* semble parcourue de forces, plutot que soumise extérieurement à une force militaire organisée et dirigée contre les violences du populus": Riviere 2004, pp. 67 and 63 on the dangers of the suburbs. Against the idea of training the *cohortes urbanae* for violence and repression, Rivière 2004 (p. 71 ff.) also cites some episodes highlighting the difference between *milites urbani* and legionaries and between *milites urbani* and praetorians: the urban cohorts rarely use weapons.

41 However, "[les cohortes urbaines], de l'époque flavienne, étaient régulièrment utilisées dans le cadre d'opérations extérieures", in Rivière's view (2004, p. 75 n. 35), following Bérard 1988.

42 Outside the city, praetorians, not *milites urbani*, were involved in disorders at Pollenzo (Suet. *Tib.* 37.3) and Puteoli (Tac. *Ann.* 13.48): see Millar 1986, p. 313; Veyne 2006, p. 62 n. 204.

43 Tac. *Hist.* 1.89: *tum legiones classesque et, quod raro alias, praetorianus urbanusque miles in aciem deducti*. This is the context for the epitaph of a soldier in *cohors X urbana, optio a victoria Ticiniana* (prior to Bedriacum), studied by Bivona 1978 and Manganaro 1989 (Appendix E.1).

44 Rare documents of *XIV cohors urbana*, concentrated between the late first century and the first half of the second century, suggest its involvement in Domitian or/and the Trajan's Dacian campaigns. E.g., *C(aius) Valerius Meleager* of the *cohors XIV urbana* made a dedication to Herakles (*CIL*, III 14496,1 = *IDR*, II 642, Bucharest): he may be a soldier of Domitian as of Trajan's Wars. See I. I. Russu, in *St.Cerc.Ist. Veche* 23, 1972, pp. 121–123 (*AE* 1972, 496; 1974, 543). I intend to argue elsewhere.

45 *Q(uintus) Vilanius Nepos* (*CIL*, VIII 1026 = *ILS* 2127 = *AE* 1942/43, 33 = *AE* 1991, 1661 Carthage); *C(aius) Velius Rufus* (Appendix E, 'New studies'); diplomas (Appendix E.3–4); soldier of Heraclea Lyncestis (*CIL*, III 7318 = *ILJug* 3, 1243 = *AE* 1999, 1415). On Nepos's Dacian service, see Stefan 2005, 402 n. 24, 435. On the participation of *Urbani* in the wars of Domitian, see the extensive literature cited by Bérard 1988, 164–173, esp. n. 27; between the others, Maxfield 1981, p. 130; Strobel 1984, pp. 280–284; Id. 1986 (on Velius Rufus' career). See also Bérard 1994b, pp. 228f.

46 Pensabene, Milella 1989, p. 48, citing *CIL*, VI 2943; on this text see Packer 1997, pp. 375, 476; Bérard 1988, pp. 161–163.

47 The dedication, *CIL*, VI 3559 = 32989 = *ILS* 9081, long kept *in aedibus Coritianis in foro Traiano*, maybe originally was located in the nearby area: see Ricci 2008.

48 Script.Hist.Aug. *Marc.* 21.6–9: *Instante sane adhuc pestilentia et deorum cultum diligentis-sime restituit et servos, quem ad modum bello Punico factum fuerat, ad militiam paravit, quos voluntarios exemplo volonum appellavit. Armavit etiam gladiatores, quos obsequentes appel-lavit. Latrones etiam Dalmatiae atque Dardaniae milites fecit. Armavit et diogmitas. Emit et Germanorum auxilia contra Germanos*; puzzled Lenski 2009, p. 148, following Welwei (1988, pp. 22–27). See also Birley 1987, pp. 177, 199, 202.

49 E.g., removal of Cleander as *praefectus Urbi* in 190 (Herodian. 1.12.6–9, discussed at length by Grosso 1964, pp. 223–303; Alföldy 1989. See in Part III, Chapter 6, pp. 157 ff.) and management of the unrest following Pertinax's death (Herodian. 2.5.9; Cass. Dio 71.11.1; Rivière 2004, pp. 81f.). For commentary on Herodian, see now Müller's edition (1996) with German translation.

50 *PIR*² M 246; *CIL*, X 1127; for the new reading, see De Carlo 2009 (Appendix B, 'New studies').

51 *AE* 1985, 390 = *RMD*, II, no. 124.

52 Herodian. 7.10.5–7; 11 and 12; Script.Hist.Aug. *Max. et Balb.* 9.1–3 and 10.4–8, discussed by Grosso 1964; Mènard 2001, pp. 82–89; Ricci 2003, pp. 23–28; Rivière 2004, pp. 82–84.

53 *Ann.* 3.41.1: *Andecavos Acilius Aviola legatus, excita cohorte, quae Lugduni presidium agita-bat, coercuit*; *CIL*, XIII 1853 = *ILS* 2130, Vichy (*cohors XVII Luguduniensis ad monetam*).

54 Freis 1967, pp. 28–31, also following Durry 1938 (cited).

55 *CIL*, XIII 1853 = *ILS* 2119: *M(arcus) Curvelius M. f. Aniens(i tribu) Robustus, miles cohortis I Flaviae urbanae* (late first century?).

56 The Lyon cohort and its tribune took part to the persecution of the "Martyrs of Lyon" in 177: Fraschetti 2008, esp. pp. 106–123.

57 Bérard 1988 (military role of urban cohorts), 1991 (origin and history of the Carthaginian cohort), 1992 (veterans of the Lyon cohort), 1993 (Julio-Claudian gar-rison), 1994a (Lyon *evocati*), 1995 (hierarchy of the Lyon cohort), 2000 (governor's *officium* of *provincia Lugdunensis*), 2006 (officials between Rome and Lyons).

58 Duval 1991, pp. 24f.

59 See Appendix B.4, discussed by Bérard 1993.

60 See *infra*, Appendix D.3 and 'New studies'.

61 A military diploma of AD 85 (Appendix E.4) tells us about veterans "qui militaver-unt in cohorte XIII urbana quae est in Africa". Such an expression, referring to the beneficiaries of the imperial measures, could be translated as follows: "the veterans who fought [in the first Dacian campaign] in the ranks of the XIII *cohors urbana*, now engaged in Africa".

62 Freis 1967, pp. 31–36.

63 Lancel 2000.

64 Freis 1967, pp. 31 s.

65 To the epigraphical evidence known to Freis and that recently published may be added an epitaph of Crestus, *decurio corpor(is) c(ustodum)* (*CIL*, VIII 9459 = 21068) from Mauretanian Caesarea. Speidel 1979 proposed that he could be one of the bodyguards at disposal of African kings and princes in the Julio-Claudian era. Such units of bodyguards were organized along the lines of Augustus' urban garrison. The *decuriae* are indeed typical of *corporis custodes Germani.*

66 René Sanquer reported in *AE* 1983, 28 (Appendix A, 'New texts' *infra*) the current conservation in Finistère of an inscription previously part of the Campana collection, *CIL*, VI 2862, which, together with the epitaph of a tribune of the *cohors I urbana Antoniniana* (*CIL*, VI 2861), is one of the two attestations of soldiers of *I urbana* in Rome (see also *Bull. Com* 1940, p. 180, reported by Freis 1967, p. 61). Both documents date to the early third century.

67 Duval 1993, pp. 23–27 and *AE* 1993, 1695; see also Appendix D.1, 4; Bérard 1991 focuses on the *cohors XIII* in Carthage in the early second century.

68 Ennabli, Ben Abdallah 1998.

69 Caldelli and Petraccia, Ricci 2012.

70 See Appendix B.4, discussed with other documents by Bérard 1993.

71 Freis 1967, p. 33: *AE* 1954, 53 (Draa ben Youder, Africa Proconsularis): a *stationarius* of *cohors I urbana*, (second century); *ILAfr* 52 = *ILTun* 143 (Lepcis minor): soldiers of *cohors I urbana* (late first/early second century). Freis recalls that praetorians and urban soldiers of different cohorts were also present in Africa with police functions, both before and after the establishment of the *cohors I urbana* at Carthage. On the *stationarius ripae Uticensis* of the V praetorian cohort in the first century (*CIL*, VIII 25438 = *ILS* 9072 = *ILTun* 1198, Tachegga, Africa Proconsularis), see now *AE* 1991, 1668. The urban cohort XIII (assigned pace Freis 1967, pp. 33f.) to Carthage between the end of the first century and the beginnings of Hadrianic era is three times attested elsewhere in Africa: *CIL*, VIII 1583 (Mustis); VIII 11107 = *ILS* 2123 (Sullechtum): *stationarius*; and 23910 (Zoubia "between Thabbora and Thimisua"): centurion.

72 *Cohors I urbana*: Appendix D.2, 1, 2, 4; others: Appendix D.2, 3?; 5–6.

73 Genava: *CIL*, XII 2602 = *ILS* 2118 = *AE* 1995, 1044 (AD 90): *M(arcus) Caranilus Macrinus, centurio coh(ortis) primae urbanae*; Ariminum: *CIL*, IX 389 (last years of first to the early second century): *C(aius) Cadienus Iustus, veteranus coh(ortis) primae urbanae* with his son, whose cognomen is *Africanus*. For the third century epitaphs in Rome, see *supra*, n. 57.

Part IV

Policing and security in imperial Italy

Introduction: security in Italy and the role of the central government between Augustus and the Severans

The framework of the organization of the security plan through the military, paramilitary and civil forces in Rome (Parts II and III) excluded so far Italy, because it was normally outside these soldiers' field of action. In the cities and peninsular territories, however, during the imperial age, there were not few circumstances that required the involvement of the Praetorian Guards and other soldiers of the urban troops. Some of them were in fact deployed in port cities, where security issues were more frequent; or in other places when social discontent appeared in rather lively forms; or even in situations when disorder occurred over futile incidents. For these and other reasons, the cities and the territory of the peninsula were often overseen or manned by the soldiers of Rome.

This introduction would like to propose a brief review of some of the situations in which the urban forces were called upon to suppress or prevent disorders in Italy.[1] It continues with a more detailed discussion of two case studies: first, that of a military station at a major road junction in Southern Italy; and then of soldiers who leave the city to accompany the emperor in his travels throughout Italy. In both cases it is a different perspective, in terms of space (outside of Rome), from which to look at the dispositive of security in action.

All the circumstances that will be outlined in the general framework are subsequent to the Augustan age (the oldest being at the time of Tiberius, the most recent of AD 246) and, although limited in number, offer an insight of how the senate of Rome and the emperors, once called upon, reacted to instances of disturbance in Italy; and what strategies they adopted.

Instead, the situation changes when examining the military stations distributed specifically throughout the peninsula in those locations that were deemed appropriate; or when considering the soldiers who moved to escort the emperor during his travels. The dispositive that operates in these cases is, however, the same, with the appropriate distinctions, whether at the *stationes* outside Rome or at the palace, next to the emperor during his movements within the city.[2]

The image that scholarly research of the recent decades has suggested is, as we have seen,[3] that of a central power that does not deal with security in Italy. Instead, it leaves leeway to local governments that, in full autonomy, provide to the supervision of roads or prevention of fires, as well as to the safeguard of public order.[4] The passage or settlement of praetorian guards, *speculatores, equites singulares Augusti* or soldiers of the fleet on the peninsula, when not because of their local origin, is instead, in many cases, an indication of extraordinary interventions, of their involvement in special missions, of the installation of more or less temporary stations, and of the presence of the emperor or of one of his family members.

Literary sources provide information of such actions, but only of those in which the facts that required soldiers' involvement are considered worthy of mention; and therefore accounted because of the truly exceptional circumstances. Often it is not the military intervention itself that matters, but rather the emperor's ability in 'handling the emergency' that is being considered. Information on many other circumstances – of which, however, the details are unknown – is provided by epigraphic and archaeological evidence, as well as literary, unearthed outside the capital, often at a considerable distance.

We are aware that in Italy, at major ports, at the Alpine and Apennine passes, and at very important road junctions, there were military bases (*stationes*) manned by soldiers. Suetonius tells us that Tiberius arranged to reinforce the network that Augustus had already designed.[5] Although Suetonius does not specify the composition of the troops stationing at these military bases, we can reasonably assume that they were formed with soldiers taken from different units.

Some *stationes* consisted of *frumentarii*, soldiers from the legions, who were staying in the city for a certain length of time at the *castra peregrina*; they could be sent on missions throughout the territory of Italy or, indeed, at guard posts positioned, wherever deemed necessary, on the peninsula.[6] We are aware of several of these checkpoints, especially in central Italy: at Ostia,[7] at *Puteoli*[8] and along the road that connected the city to these two important ports.[9] We can imagine that they were far more numerous,[10] and either permanent or temporary.

It is obvious that not only simple soldiers, but also centurions and senior officers, stationed at the military bases. Some of the attestations relating to these men can receive different explanations. Exemplary from this point of view is the case of the centurion of the *castra peregrina* of Rome, Caius Iulius Proculus, commemorated on a valuable funerary altar of Severan era, in *Octavianum* (Toiano, near Pozzuoli).[11] His presence, and then his death, in this location in Campania can be an indication of his involvement in a security mission (perhaps an inspection at the *statio* of Pozzuoli?) due to his local origin. *Proculus* could have "died in Rome and his ashes buried in Campania or . . . deceased during a visit to his brother or caused by an illness".[12] The possible reasons that can be identified for the presence of *Proculus* in Toiano are largely the same as those that can be cited for other soldiers and officers of Rome. In my opinion, this example clearly suggests the difficulty encountered when interpreting sources, mostly reticent.[13]

It is not always possible to identify the composition of the Italic stations, which changed and varied due to the place where the military base was located, the period of time, and probably the reason for its establishment.

A very famous case that has been the subject of much debate is the one of the *stationarii* of (*Bovianum* and) *Saepinum*, two important centers of the *regio IV Italiae* (*Samnium*). The last used to be an important station along the transhumance route, where herders would move their livestock from the pastures of Apulia to the Apennines of Campania and Abruzzo, and vice versa.[14] The existence of the *stationarii* in this location is supported by the text engraved on the right jamb of one of the gates of Saepinum.[15] The text provides an extract of the letters exchanged between *Septimianus*, an imperial freedman, and *Cosmus*, a procurator, during the reign of Marcus Aurelius (AD 169–172). Saeptimianus wrote about an open hostility between the herders and those in charge of the entry points because the latter treated the herders like runaway slaves and herd thieves.[16] The procurator *Cosmus* decided to contact Bassaeus Rufus and Macrinius Vindex, the praetorian prefects, who informed the magistrates of *Saepinum* regarding their intention to open an investigation, threatening sanctions if such abuses did not cease.

From my point of view this text offers at least a few interesting pieces of information.[17] First, the jurisdictional powers of the praetorian prefects on the territory of Italy is attested at a time when we do not have much information in this regard. The prefects may have been induced to intervene on the grounds that the *stationarii*, instead of acting as guarantors to ensure the correct procedures, had actually proved to be conniving with the landowners.[18]

In other words, the governors of Saepinum and Bovianum and the police forces had allied against those (the *pastores*) who had threatened the landowners' interests. However, it is not clear how these interests were being threatened. It seems evident that in both cases – that of the *epistula* of Saepinum and that of another provision, roughly contemporary, recalled in the Digest[19] – the herders' position was rather weak. A position that was the last in a chain of tasks and responsibilities in the process of transhumance.

Other interesting point is to understand who exactly were the *stationarii* and what duties were entrusted to them. Their generic name, according to some scholars, does not clarify whether they were local guards appointed by the municipality or soldiers.[20] My idea is that, in analogy to other cases of stations in Italy,[21] and for the fact that the magistrates were not only from Saepinum, but also from Bovianum,[22] it is more likely that, in this case, the *stationarii* were soldiers (whether they were *frumentarii* or recruited from different urban forces it is difficult to determine) made available by the central power to guard a strategic point in the region with a dual purpose. On the one hand, they had to ensure the proper conduct of the operations associated with the transhumance; and on the other, they had to contain the actions of rustling of herds and banditry of other goods and animals.[23]

If the generic use of the term *stationarius* ('a station guard') does not allow to determine who the soldiers charged with controlling the territory of Saepinum and Bovianum were, we know for a fact that, in the third century, the praetorians and veterans were employed in order to prevent or repress brigandage.[24] Evidence of this is an inscription for the salvation of Emperor Philippus, of his son with the same name, and of his wife Otacilia, *mater castrorum* (mother of the camps) by the former praetorian Aurelius Munatianus, *agens ad latrunculum* ('active against brigandage'). He presided over a station at the Furlo Gorge[25] along with twenty soldiers of the fleet of Ravenna.

Aurelius Munatianus, was an *evocatus*, namely a former soldier of the urban troops, to whom were normally entrusted, as a rule, administrative tasks in Rome; and operational type of duties with a heightened degree of responsibility[26] outside the city. Since the dedication he made is dated *ad annum* (246), one could observe that Munatianus's departure from Rome and the provisions received are linked to the so-called 'military reforms' of Septimius Severus and to the new profile of praetorians, who were at that time mainly recruited from the legions. Other evidence, perhaps dating back to the middle or end of the second century, however, seems to predate – at least of half a century – the existence of *stationes* in the territory of Italy, composed of soldiers of Rome (praetorians and *equites singulares Augusti*) with the duties that have already been described.

A long time before the age when Munatianus was in force, there was news of famous 'bandits'. The contemporary Strabo talks about Selurus the 'son of Etna' and his unhappy ending: "He was sent up to Rome because he had put himself at the head of an army and for a long time had overrun the regions about Aetna with frequent raids."[27] We do not know how and who had arranged the capture.

Far richer in details is the story of the Italic Bulla who, with a band of 600 men (many of whom were imperial freedmen, perhaps who had fled from the surrounding properties, and slaves), raged in central and southern Italy for two years. Bulla managed to do so, despite the emperors' solicitude and the use of many soldiers, also thanks to a network of informants who collaborated with him.[28] In the account of the historian Cassius Dio, two officials are mentioned: a centurion (but the corps he belonged to is unknown) who was about to exterminate the gang of bandits, but failed in his attempt[29]; and "a tribune from Severus' force of bodyguards"[30] who finally succeeded, through deception, to capture the bandit.

Although the episode of the infamous Bulla is probably a historical invention, it is a fact that the slaves' revolts had not ceased with the beginning of the Principate. Tacitus reports of a revolt that had broken out among the peasant-slaves of the Roman Calabria (now Salento) in AD 24:

> During the same summer, the seeds of a slave war, which had begun to stir in Italy, were rendered harmless by an accident. The instigator of revolt was Titus Curtisius, a former private in a praetorian cohort. First at clandestine meeting in the neighbourhood of Brundisium and the adjacent towns, then by openly posted manifestoes, he kept summoning the fierce country slaves of the outlying ranches to strike for freedom, when almost providentially three biremes for the protection of seaborne traffic put in the port. As in addictions the quaestor Cutius Lupus, who in accordance with an old custom had been assigned the 'grazing-tracks' for his province,

happened to be in the district, he drew up a force of marines and shattered the conspiracy at the very outset. The tribune Staius, hurriedly sent by the Caesar with a strong force, dragged the leader and the bolder of his subordinates to Rome, where tremors were already felt at the size of the slave-establishments, which were assuming huge dimensions while the free-born populace dwindled day by day.[31]

In this account, the soldiers of Rome were present and active with differing roles. First of all, a former praetorian, a certain Titus Curtisius, was the instigator of the rebellion and the protestors' spokesman, first acting secretly at meetings (*coetibus clandestinis*); then, inciting followers to riot in public spaces. The first soldiers to appear in the role of repressors were the soldiers of the fleet who were in Brindisi to provide to travelers. The quaestor Cutius Lupus, who was there because he had to look over the grazing-tracks,[32] dispatched soldiers of the fleet as stormtroopers and nipped the revolt in the bud.

To definitely put an end to this experience and give a stronger signal, at a later time the emperor arranged to send a tribune with a handful of men from Rome (this time, most likely, praetorians) to block Curtisius and his followers.

The text by Tacitus is interesting for several reasons. We have corroboration of the periodic outbursts caused by intolerant slaves.[33] From the episode we can draw the image of Tiberius as obsessed with his own safety, eager to suppress the multiple signs of disturbance in Italy.[34]

Moreover, as in the case of Furlo, in Brindisi as a big port in southern Italy, distant from Rome and from the nearest *statio*, the soldiers of the fleet[35] are the ones involved in the repression of civil unrest. It is worthwhile pointing out that the Brindisi uprising happened well before the Flavian dynasty, when the role of the *classiarii* will be enhanced thanks to Vespasianus and his sons.[36]

Finally, Tacitus tells us that, a century and a half before the events in Saepinum, a quaestor had the specific duty to oversee the *calles* (namely the grazing-tracks); and it is yet another sign of insecurity in the southern areas of the peninsula, continually at risk of *seditiones*, oppression or abuse, due to riotous labourers, aggressive bandits or 'ingenious' herders (provided the latter two figures did not coincide).

This widespread unrest of the inner or southern regions, far from ceasing, became an endemic fact in the third century. A clue is also given, starting from this moment on, by the great number of *praepositi* 'in charge of controlling' more regional areas (*regiones* or *tractus*).[37] These were mostly equestrian officers who were responsible for maintaining public order and repressing riots in the areas under their control. The text on a statue basis for a *praepositus tractus Apuliae, Calabriae, Lucaniae, Bruttiorum* (mid-third century) leaves no doubts: "For his strong attachment shown towards his city and citizens and for his outstanding efforts to maintain peace in the region."[38]

Mid-third century is far from our time frame of reference: in this epoch the old dispositive, which had survived for two centuries, is being now replaced by a tighter control, a stronger presence of Rome, an initiative of intervention that was not provided for before or given for granted. Interventions are often actually solicited by the population (*postulatu populi*) of the Italic municipalities, aware of the impotence or of the difficulties of their governments to respond effectively to threats coming from several fronts.

It almost seems that, from the third century onwards, with respect to the Italic municipalities, there is no trace of Augustus's complex plan for the people's well-being and for the safety of their property; and that the *Securitas* has been replaced by the more modest aspiration for a *quies* (*regionis servanda*), which at least ensures survival.

To better clarify the changes that occurred in the forms of intervention from the third century on, it is important to step back in time and pay attention to a few episodes from the early decades of the Principate. At this time, in case of an occurring sedition, the Roman senate or the *Princeps* himself could decide to dispatch soldiers. In these cases, probably a protocol for deployment did not exist: the one that was adopted was an impromptu solution, devised on the necessity at the time of occurrence; although a few rules (except in the case of some 'overbearing' or too autonomous emperors) had to be followed and it is not difficult to deduce this fact from some recurring elements in historical accounts.

The first part of the Chapter 37 of the Tiberius's life is dedicated to how this emperor arranged his security.[39] Following a parable from Rome to Italy, first the historian recalls in little detail the concentration of the praetorians (and *milites urbani*) in the new *castra* built on the Viminal Hill. Second, Suetonius mentions the suppression of a riot in the theater by means of an exemplary measure; and, finally, the distribution of soldiers at the road junctions mentioned above.

More space is dedicated to the famous episode of *Pollentia*,[40] when the inhabitants of the Ligurian town intercepted the funeral procession for a centurion: they wanted to extort money from the heirs in order to organize the gladiatorial games. When the matter was over the emperor dispatched two cohorts into town: only one taken from the city (and probably made up of praetorians); and the other composed of legionnaires from the nearby kingdom of the Cottii. Arms in hand, the soldiers poured into the city and arrested a large part of the residents.

The prominence given to the episode narrated by Suetonius is explained by the exceptional nature of the intervention, which, as we have been told, involves Tiberius' initiative in an autonomous city, without having requested the opinion of the senate. The story of *Pollentia* occurs in a place with a large concentration of people (the Forum) and concerns the municipal *plebs* (or most of it). The reason behind the episode is apparently futile and there is no bloodshed. However, it is impossible to know what hardships sparked such a deed.

Two more episodes of social disturbances, which are also well known and much discussed, took place during the reign of Nero. Both concern residents of Campania, respectively Puteoli and Pompeii. The first taking place in AD 58:

Under the same consuls, audience was given to deputations from Puteoli, despatched separately to the senate by the decurions and the populace, the former inveighing against the violence of the mob, the latter against the rapacity of the magistrates and of the leading citizens in general. Lest the quarrels, which had reached the point of stone-throwing and threats of arson, should end by provoking bloodshed under arms, Gaius Cassius was chosen to apply the remedy. As the disputants refused to tolerate his severity, the commission at his own request was transferred to the brothers Scribonius; and this were given a praetorian cohort, the terrors of which, together with a few executions, restored the town to concord.[41]

The *plebs* and the administrators of Puteoli were unable to settle their disagreements: the reason is the supposed greed of the latter and the uncontrolled reactions of the formers.

The rationale behind the Puteolan episode seems, in substance, very different from what had caused that of *Pollentia*. The Puteolians themselves requested the intervention from the senate of Rome. The senate, however, proceeded by attempts: first by delegating Cassius (Longinus), whose intransigence (*severitas*), however, was not appreciated by the people: unable to count on the armed forces, Longinus failed. Given the failure, Longinus himself proposed to entrust the duty to the brothers Scribonianus Rufus and Scribonianus Proculus and to a praetorian cohort. The terror aroused by the arrival of the soldiers and the sacrifice of a few was enough to restore the altered order in the port town.

At least two points deserve, in my opinion, to be highlighted. The first is that Rome does not take the initiative, but is persuaded to intervene by the escalation of ongoing violence: up to that moment the conflict has been handled with rock throwing and fire threats, and risked to turn to the use of weapons. The second element is the station of the praetorian cohort under the command of the two *Sulpicii*. Tacitus refers to a *cohors data* (given); therefore, it was assigned and not necessarily dispatched from elsewhere[42]; while, for example, the praetorian cohort that had been sent to *Pollentia* is *immissa in urbem*, therefore specially dispatched from Rome to the Ligurian town.

This obviously made the difference in terms of the rapidity and effectiveness of intervention; and, at the same time, it is an indication of the differences in specific situations.

The second episode, which takes place a year after the previous one, is the famous fight between the inhabitants of Pompeii and those of Nuceria of AD 59, and found an echo in the famous fresco of Pompeii:

About the same date, a trivial accident led to a serious affray between the inhabitants of the colonies of Nuceria and Pompeii, at a gladiatorial show presented by Livineius Regulus, whose removal from senate has been noticed. During an exchange of raillery, typical of the petulance of country town, they resorted to abuse, then to stones, and finally to steel; the superiority lying with the populace of Pompeii, where the show was being exhibited. As a result, many of the Nucerians were carried maimed and wounded to the capital, while a very large number mourned the deaths of children or of parents. The trial of the affair was delegated by the emperor to the senate; by the senate to the consuls. Of the case being again laid before the members, the Pompeians as a community were debarred from holding any similar assembly for ten years, and the associations which they had formed illegally were dissolved. Livineius and the other fomenters of the outbreak were punished with exile.[43]

Even in this case, the gradual transition from the insults to the stones and arms (*lascivia in vicem incessantes probra, dein saxa, postremo ferrum*), causes the intervention of Rome, with the buck-passing of responsibilities between the emperor and the senators. The last makes the ultimate decision to cancel the games for ten years in order to dissolve the associations that had provoked violence; and to inflict an exemplary punishment for those who had fueled the turmoil.

What seem to be recurring elements – to which are, however, added local specificities – is the senate's competence in deciding to intervene along with the measures to be adopted; or in determining to refrain from taking any measures, except in cases of growing tension. The soldiers involved were those closer to the uprising (certainly in the case of Pollentia, most likely in that of Puteoli).

Those described above are just some of the incidents that occurred and involved the military troops of Rome. Millar proposed a good selection of events more than thirty years ago[44]; and Le Roux, a decade ago (and he has widened the viewpoint to cover cases that included the provinces) reviewed them, partly enriched, with interesting considerations.[45]

I was interested in giving an idea as varied as possible of the many situations involving the soldiers of Rome throughout the Italian territory, in connection with the prevention or the suppression of social disturbances when the local forces were not sufficient.[46] Even taken together, the cases known to us are few, evidence of the fact that, between the first and second centuries, the dispositive implemented by Augustus corresponded to the needs of the City and its people. In Italy, the local authorities kept the majority of the situations of turbulence under control; or mainly the turbulences did not degenerate to the point of attracting the interest of the government in Rome.

In Chapters 8, 9 and 10 of Part IV, some circumstances are taken into consideration that involved the presence of soldiers of Rome or of the Italic fleet in Italy: the first case, in continuity with what has just been presented in this Introduction, is about a military station in a strategic road junction of imperial Italy, at the site of Grumentum (Chapter 8); the second and the third cases regard the security services provided in occasion of the emperor's travels, respectively in Latium (Chapters 9 and 10) and in Campania (Chapter 10). These are papers that I published in the last ten years. Chapters 9 and 10 share a general introductory part, with unavoidable repetitions, nonetheless I have decided to present them in full, except for minor adjustments, updating and references considered essential.[47]

Notes

1 These episodes have received attention on several occasions (see Part I, Chapter 1).
2 See Part II, Chapter 4 and III, Chapter 6.
3 See Chapter 1.
4 "The central government was interested in dealing more with the major infrastructures of general political value, and for a long time it limited the direct intervention on local governments to the minimum required, except for reasons connected to public order" (Gabba 1991, p. 27).
5 Suet. *Aug.* 32.2: *grassaturas dispositis per opportuna loca stationibus inhibuit, ergastula recognovit, collegia praeter antiqua et legitima dissolvit*; and *Tib.* 37.2: [*stationes*] *frequientiores disposuit.*
6 *Stationes frumentariorum* existed also outside Rome (this has been treated in Part III, Chapter 6).
7 *CIL*, XIV 7 = *ILS* 2217 (sacred inscription); 125 = *ILS* 2223 = Thylander 1952, 324 (imperial inscription of AD 224 of a *statio numeri frumentariorum*); 4487 (sacred inscription?).
8 *CIL*, X 1771 (funerary inscription).
9 A young slave of a *miles frumentarius* of the XXX *legio Ulpia* (*CIL*, X 6095) was buried in Formia.
10 A *frumentarius* from the legio XXII Primigenia (Caius Alpinius Restitutus) and a freedman of his colleague of the legio VIII Augusta (Firminius Adventus) are commemorated at *Velitrae*, just outside Rome (*CIL*, X 6575 = VI 32873).
11 AE 1994, 424: *D(is) M(anibus)* / *C(ai) Iuli Proculi* / ((*centurionis*)) *leg(ionis)* XIIII *Gem(inae)* / *a peregrinis;* / *C(aius) Iulius Valerianus* / *fratri optimo.*
12 Panciera 1994 p. 615 f. = Panciera 2006, pp. 1457–1462, with addendum (p. 1463). Also see Faure 2001 and 2003, p. 418 no. 34; Le Roux 2002, p. 26. For a possible reconstruction of the career of *Proculus*, see Faure 2001, p. 307 f. and 2003, p. 418 no. 34.
13 For an overview of the reasons that explain the presence of soldiers and veterans of urban troops in Italy, see Ricci 1994, partially updated in Ricci 2010.
14 On the transhumance in the Italic and Roman economy, see Gabba 1994, pp. 155–176 (on the document of Saepinum, pp. 162 and 172), with a detailed bibliography.
15 *CIL*, IX 2438 = *FIRA* 1969 I no. 61. I suggest a selection of the detailed bibliography that regards it: Passerini 1939, pp. 251–259; Laffi 1965 with fig. 1; Gabba, Pasquinucci 1979, pp. 155–157; Corbier 1983, 1991, 528a and 2006, pp. 225–232; Lo Cascio 1985–1990.
16 *CIL*, IX 2438, lines 16–18: *quod in tra(n)situ iumenta et pastores, quos conductos habent, dicentes fugitivos esse et iumenta abactia habere.*
17 Many important issues are raised from the inscription of Saepinum: the main one concerns establishing who were the owners of the herds (imperial or private or both?) and thus the reason for the involvement of the *a rationibus*. For an overview of the positions, see Corbier 2006, which summarizes previous interpretations and Lo Cascio 1985–1990.
18 Passerini, 1939, p. 256 even writes: "The praetorian prefects had to intervene; and they were the ones, and no one else, that had the responsibility to maintain order in Italy". It is a problematic issue because it is not known with which soldiers the *statio* in Saepinum was composed of (were they *frumentarii*? praetorians? Both?). Zoz 2001 p. 517 n. 17 is more vague when referring to the *stationarii*. She says that at times they were under the direct dependence of the magistrates; other times they reported directly to the imperial power (?). On the territorial jurisdiction of the praetorian prefect that stretched beyond the limit of 100 miles from the praefectus

Urbi's action, see Rivière 2009, p. 248, who recalls the disorders occurred under the reign of Tiberius, and the intervention of the praetorians.

19 A letter of Marcus Aurelius and Commodus stating that it is the governors' responsibility (assisted by the *milites stationarii*) to help their masters search for *servi fugitivi*, to be punished directly if caught (Dig. 11.4.1.2). On the connection between 'nomadism' (even pastoral) and banditry, see the stimulating considerations of Shaw 1984, pp. 31 f., enriched with extensive references to the sources.

20 According to Laffi 1965 and Lo Cascio 1985–1990, the *stationarii* are police officers at the service of the municipalities. Instead, according to Corbier 1983, they were soldiers from different regiments detached with ad hoc tasks (of the same opinion is Kolb 2000, p. 160 f.). Also Zoz 2001, p. 516 shares the same idea (who, however, seems to contradict herself at p. 524 n. 41). According to Petraccia Lucernoni 2001, the *stationarii* are a distinct military force.

It is not clear why the *stationarii* appear only in the first more descriptive letter of *Septimianus* and then disappear in the one of the freedman *a rationibus* and therefore in the prefects's response: maybe because the governors of the towns of Samnium were ultimately responsible for what happened.

21 See what has just been said *supra*, pp. 194–195 about Puteoli and Ostia, and as detailed below in Chapter 8.

22 *CIL*, IX 2438, line 11: *faciatis scribere mag(istratibus) Saepin(atibus) et Bovian(ensibus)* and, further on, at lines 15–16: *iniuria(m) se accipere a stationaris et mag(istratibus) Saepino et Boviano.*

23 According to Riess (2007, p. 202), the police force had the task to hunt down runaway slaves and to confiscate stolen horses and pack animals. The prefects, in order to avoid damaging the imperial revenue, would have required the appointed soldiers to ignore the herders. On transhumance and the connection between pastoralism/banditry, see Fox 2007–2008, with extensive bibliography on p. 23 (at p. 16 the inscription of Saepinum is commented).

24 On the frequency, the unpredictability and the hopelessness of the attacks, see the two incidents mentioned by Pliny (*Epist.* 6.25.1): in the first, which occurred at Ocricoli (Umbria), the victim is the equestrian *Robustus*; in the second, which occurred in an unknown place, the victim is a compatriot of Pliny, *Metilius Crispus*. It is the case mentioned by Suet. *Aug.* 32.1 (*rapti per agros viatores*).

25 *CIL*, XI 6107 = *ILS* 509, Acqualagna, Urvinum (lines 10–15): *Aurelius Munatianus evo/catus ex cohorte VI pr(a)eto/ria [p(ia) v(indice) Philippi]ana agens at / latrunculum, cum militi/bus n(umero) XX classis {p}pr(aetoriae) R{r}aven/natis p(iae) v(indicis) Filipporum.* Recently the dedication of Munatianus was examined again by Braccesi 2007, pp. 49–55; and by Riess 2007, p. 210 no. 5.

26 For other evocati with operational assignments outside the urban context, see *infra*, Part I, Chapter 8, p. 203.

27 Strab. 6.2.6. The episode is recalled by Shaw 1984, p. 20 n. 46.

28 Cass. Dio 77.10. On the figure of *Bulla*, mostly considered fictitious and a metaphor of the weaknesses of power, see Shaw 1984, pp. 46–49; Le Roux 2002, p. 25; Grünewald 2004, pp. 111–123; Riess 2001, pp. 73 and 155. Among the rebels under Bulla there were also some of the Praetorian Guards dismissed by Septimius Severus after the victory in the civil wars (see Cass. Dio 74.2.5 and 75.1.1–2; Okamura 1988, p. 291 no. 33).

29 According to Grünewald 2004, p. 114 f. it is a sad destiny of this centurion to represent military inefficiency.

30 Who, according to Speidel 1994b, p. 63, was an officer of the *equites singulares Augusti*. The episode, however, is not historical, and any attempt of identification is purely arbitrary.

31 Tac. *Ann.* 4.27: *Eadem aestate mota per Italiam servilis belli semina fors oppressit. auctor tumultus T. Curtisius, quondam praetoriae cohortis miles, primo coetibus clandestinis apud Brundisium et circumiecta oppida, mox positis propalam libellis ad libertatem vocabat agrestia per longinquos saltus et ferocia servitia; cum velut munere deum tres biremes adpulere ad usus commeantium illo mari. Et erat isdem regionibus Cutius Lupus quaestor, cui provincia vetere ex more calles evenerant: is disposita classiariorum copia coeptantem cum maxime coniuratio-nem disiecit. Missusque a Caesare propere Staius tribunus cum valida manu ducem ipsum et proximos audacia in urbem traxit.*

32 *Cui* [i.e.: to *Lupus*] *provincia vetere ex more calles evenerant.* This episode recalls that of Saepinum: in either the dangerous tracks pastures are placed under control. On the episode, Nippel 1995, p. 91.

33 The inscription *CIL*, IX 2235 = *ILS* 961, from Allifae, recalls a task *ad servos torquen-dos*, once more during the reign of Tiberius. For this refer to: Alföldy 1969, p. 149 ff. and Camodeca 2008, p. 71 ff. = *AE* 1990, 222.

34 I have already discussed in the Introduction at p. 122 ff on Tiberius' real or presumed obsession with security. Other indicators, aside from Suet. *Tib.* 37, are the Tacitus's words in *Ann.* 1.7.3; and the inscription of the *praefectus statorum* accompanying Germanicus to Alessandria (both, the passage and the inscription, commented on p. 99).

35 Even in this case, as for the Furlo (even on the Adriatic), it is probably soldiers from the fleet at Ravenna. On the soldiers of the fleet at the Tyrrhenian ports (Centumcellae, Lorium e Alsium) see *infra*, Part IV, Chapter 9.

36 On this, see Ricci 2009, part. pp. 22 and f.: references herein to the sources (in par-ticular to the Tacitus's *Historiae*) and to previous bibliography.

37 Riess 2007.

38 *CIL*, IX 334 = *ILS* 2768 = *Epigrafi romane di Canosa* I, pp. 36–38, no. 27, cf. II, p. 29, no. 27 A (V. Morizio). *M(arco) Antonio / Vitelliano / v(iro) e(gregio), p(atrono) col(oniae) / Canus(ii), p(rae)p(osito) trac/tus Apuliae, / Calabriae, Lu/caniae, Bruttior(um), /ob insignem eius / erga patriam ac / cives /adfectionem / et singularem in/dustriam ad quietem /regionis servandam, / postulatu populi, d(ecreto) d(ecurionum), p(ublice).* On the figure, see PIR[2] A, 170–171, no. 881. On this charge, see Th. Mommsen *Römische Staatsrecht* II.3, 1887, p. 1075; Pflaum 1960, pp. 939 and 1041; Panciera 2006, p. 994; F. Grelle, A. Giardina, *Canosa romana*, L'Erma di Bretschneider: Roma 1993, p. 46 n. 28; E. Folcando, Il patronato di comunità in Apulia et Calabria, in *Epigrafia e territorio, politica e società. Temi di antichità romane* III, Edipuglia: Bari 1994, p. 122.

39 See Part III, Introduction.

40 Suet. *Tib.* 37.2: *Cum Pollentina plebs funus cuiusdam primipilaris non prius ex foro misisset quam extorta pecunia per vim heredibus ad gladiatorium munus, cohortem ab urbe et aliam a Cotti regno dissimulata itineris causa detectis repente armis concinentibusque signis per diver-sas portas in oppidum immisit ac partem maiorem plebei ac decurionum in perpetua vincula coniecit.* The painful case of Pollentia has been commented on several occasions, also to highlight the decisive shift towards a less tolerant attitude by Tiberius; an attitude at least partly motivated by the unrest and the discontent manifested after Augustus' death. The last, with bibliography, Le Roux 2002; Kelly 2007.

41 Tac. *Ann.* 13.48: *Isdem consulibus auditae Puteolanorum legationes, quas diversas ordo plebs ad senatum miserant, illi vim multitudinis, hi magistratuum et primi cuiusque avaritiam increpantes. eaque seditio ad saxa et minas ignium progressa ne c[aed]em et arma proliceret, C. Cassius adhibendo remedio delectus. quia severitatem eius non tolerabant, precante ipso ad Scribonios fratres ea cura transfertur, data cohorte praetoria, cuius terrore et paucorum supplicio rediit oppidanis concordia.*

42 On the presence of soldiers of Rome, also praetorians, in Puteoli, see Part IV, Chapter 10.

43 Tac. *Ann.* 14.17: *Sub idem tempus levi initio atrox caedes orta inter colonos Nucerinos Pompeianosque gladiatorio spectaculo, quod Livineius Regulus, quem motum senatu rettuli, edebat. quippe oppidana lascivia in vicem incessente[s] probra, dein saxa, postremo ferrum sumpsere, validiore Pompeianorum plebe, apud quos spectaculum edebatur. ergo deportati sunt in urbem multi e Nucerinis trunco per vulnera corpore, ac plerique liberorum aut parentum mortes deflebant. cuius rei iudicium princeps senatui, senatus consulibus permisit. et rursus re ad patres relata, prohibiti publice in decem annos eius modi coetu Pompeiani collegiaque, quae contra leges instituerant, dissoluta; Livineius et qui alii seditionem conciverant exilio multati sunt.* See Nippel 1995, pp. 89–90.

44 Millar 1986.

45 Le Roux 2002 in Part I, Chapter 1, pp. 12–13.

46 Another case will be presented in the following Chapter 8.

47 I decided to present the dossier of Chapter 9 as a table rather than as a cata-logue. Chapter 8 was originally longer and included, in addition to the paper I wrote, also two papers dedicated to the *Arusnates* and the security at the frontier (M.F. Petraccia); and to Ostia and the public security (M.L. Caldelli). The cited papers that, at the moment of the first edition, had not been yet published – as well as minor additions or corrections stimulated by new readings and insights – are indicated in the text between square brackets and preceded by the abbreviation NdA (author's note). Other trivial changes regard: the numbering of the notes and internal references; bibliographical abbreviations; the selection of the images.

8 *Praesidia Urbis et Italiae*

Grumentum and its territory – a case study[1]

The protection of public order and the security in the City and throughout the territory of the empire were an integral part of the soldier's duty. However, it is indisputable that it is not easy to identify such duty; and that, within this broad definition, are included a number of tasks, stable, occasional or periodic, carried out by different military corps at different times and places.

This is the reason why, in a meeting dedicated to the soldier's 'professions', it seemed appropriate to devote attention to an aspect of military life that required and created special skills. For the soldiers who had this experience, these skills represented a defining phase for their profiles, which could be used in future occasions. A 'special assignment' linked to the control of a city or the surrounding area facilitated or strengthened the bond between the soldiers and the place where they worked, the definition of roles, creating those conditions and the incentive for their integration into the fabric of urban life.[2]

When indicating a part of the duties entrusted to the soldiers of Rome (and in particular to the *cohortes urbanae*), there has been and continues to be an extensive use of the word 'police'. This term will hereinafter be used in full awareness of the slow and elaborate gestation that accompanied its birth, in completely 'different' contexts than the one of ancient societies. Since the 1960s, in fact, many have considered the improper use of this term[3]: the birth of the police constitutes the final outcome of a process of change in individual and collective attitudes towards 'social peace', [NdA.: the division of roles between the administration of justice and the policing][4]; and corresponds to the requirements of social order in European industrial societies undergoing class conflicts. Modern policing presupposes the birth of a public law and political economy, and relies on a network of information, partly already developed in imperial times, but only systematized in modernity, to gather facts in order to control individuals and their spatial and social mobility.

In societies of the nineteenth and twentieth centuries, with reference to the protection of public order, along with the activity of prevention, will be set up a real and actual dispositive of correction and repression. This phenomenon is preceded by the complex experience of the Ancient Règime, for which "the weight and the significance of the forms of imprisonment still do not appear outlined". So as few are the "[cross-sectional] studies that compare what changes and what remains in the long term".[5]

At the end of the nineteenth century, Otto Hirschfeld had repeatedly devoted attention to the issue of public order in the empire, demonstrating how diverse and reduced the commitment of the central government to ensure security was; how municipalities and cities would instead provide security with their own resources[6]; and how, especially in Egypt and in the East, cities made extensive use, in addition to military forces, even of civilians.[7]

In recent years, in his reflection on the duties performed by the soldiers to assure public order in the cities of the empire, Ramsay McMullen[8] focuses on key points: first, our limited knowledge about the public order in the first centuries of the Roman Empire and in the Western provinces; second, the difference of ways and purposes for the deployment of soldiers in the cities[9]; lastly, the 'non-stable' feature of the *stationes*, linked to specific episodes or contingencies, except for some metropolitan areas with permanent cohorts.

In Italy, aside from the exceptional case represented by Rome,[10] the choices and the means available to cities to ensure order in individual urban areas and its territories has not been studied much. Despite the undoubted crossing of Italy by different cohorts,[11] Fergus Millar considers the "almost total absence of imperial forces in Italy during the Principate [as] . . . one of the most distinctive features of the model adopted in the rest of the empire".[12]

It is not easy to determine the reasons behind the movements of individual or small groups of soldiers on the imperial territory of Italy. For the vast majority, especially in the centre and in the south of the Peninsula,[13] these soldiers belonged to the urban troops. Their presence far from the place where they lived and worked, came from and/or had gained from viritane assignation was due to various needs: permits, escort services to the emperor and his family, judicial assistance to local magistrates, control of imperial properties and so on.

The aim of this chapter is to present and discuss a concrete situation concerning the needs to control and to assure an important road junction – Grumentum and its territory – which connected the center to the south, and the Tyrrhenian to the Ionic and the Adriatic coasts.

A good part of the evidence of soldiers and veterans in Italy is linked to the places of origin[14]; some evidence, however, could be connected more to the place where the soldiers were on a particular mission or where they settled after their leave. These are the inscriptions found in port cities (Ostia, Puteoli and Brundisium)[15]; in centres known for being the home base to viritane assignments[16]; or even in places in Italy, and in particular of the *Latium Vetus*, where there were the imperial residences.

In focusing on those traces, two things seem immediately obvious: the absence of evidence of the *milites* of the *cohortes urbanae* in Regio III (*Lucania et Bruttii*)[17]; and the marginal presence of other soldiers, in the same region,[18] between the middle of the second and third centuries.

Some documents found at Grumentum (today Grumento nova, in the province of Potenza) and in the surrounding area[19] have seemed all the more eloquent and worthy of attention. These are four documents that mention six individuals (simple soldiers, graded officers or *evocati Augusti*).

The first document is a dedication to Mithras by T(itus) Flavius Saturninus, a soldier re-enlisted to carry out special duties, in a period between the end of the second and the beginning of the third centuries AD.[20] The second is the epitaph of Aelius Dignus, a staff member of the praetorian prefects (*beneficiarius*), in the third century.[21] The other two are also epitaphs and commemorate, respectively: a praetorian on duty, Aelius Marcianus, who died at the time of Gordian III and was buried by Valerius Valerianus (an *evocatus*)[22]; and an unknown soldier (most likely a praetorian), who served '*in barbarico*' and was buried by his brother, also a praetorian, Aurelius Asdula, no earlier than the third century.[23]

This concentration of attestations of soldiers in this area has peculiarities that attract the attention, not only – as said – because it was extremely rare to find soldiers in the mentioned area, but also and mostly because of the strategic position of Grumentum in Southern Italy. The town, situated by the confluence of two rivers, Agri and Sciaura, was in fact connected to the east with Potentia and with Venusia; to the south with the Ionian coast and Heraclea; and to the Tyrrhenian coast, through the Via Popilia (see Figure 8.1).[24] In this area it is certain that there were vast landholdings, including the most famous belonging to the powerful local gens of the Bruttii Praesentes[25]; the property, as a result of the marriage between Bruttia Crispina and Commodus, seems to have become listed, in the last decades of the second century, among the imperial estates.[26]

Shortly after the publication of Hirschfeld's research, mentioned in the Introduction, Michail Rostovtzeff focused his attention on the presence of *saltuarii* (*oreophylakes, agriphylakes*), appointed by individuals and municipalities, in charge of the security services and of '*richtige Polizeifunctionen*' on large estates, over the fields and forests of Italy and the provinces. Their duties did not extend, however, to the emperor's properties[27]: they were always, observes Rostovtzeff, manned by soldiers.

On the basis of his reflection, it is reasonable to wonder whether the presence of soldiers in this area can be connected to other reasons rather than due to their local origins[28]; or to the assignment of the land, at a time when this practice was no longer resorted to.[29]

The soldiers (and veterans and graded officers), stationed in Grumentum, being almost entirely absent in the rest of the *regio Lucania et Bruttium*, were perhaps appointed specific duties, such as the surveillance of the villa and the security of the emperor and of his family members when they were there.

I would like to add, to this first observation, a few more considerations, related to the nature and to the territorial extent of the duties carried out by soldiers. The presence of a *beneficiarius* and two *evocati Augusti* (or *Augustorum*),[30] suggests that the duties performed by the soldiers were not – before and during the entering of the estate of *Bruttii Praesentes* in the imperial properties – limited

Figure 8.1 The triangle between Tegianum, Atina and Grumentum

Source: drawing from P. Bottini (ed.), *Il Museo archeologico nazionale dell'alta val d'Agri*, Ministero Beni Culturali e Ambientali – Regione Basilicata: Lavello 1997, p. 152 (reworked image, courtesy C. Modesti Pauer).

to patrolling the *praedia*, but also involve checking the road hub [NdA: and preventing banditry], since epigraphic evidence is spread over an area that is located exactly at the intersection of the Via Popilia.

By joining the information provided by the inscriptions of soldiers and those of slaves and freedmen of *Bruttii Praesentes*, it is also questionable whether the strategically important area that Andermahr called "the triangle between Tegianum, Cosilinum and Grumentum" had its centre northernmost at Átina (Atena Lucana)[31] rather than at Cosilinum. In that case, the discovery in the same area of a small sepulchral altar in limestone in which a praetorian and a *miles urbanus*[32] are commemorated could be interpreted differently. The inscription, according to Bracco, reveals for the first time an '*urbanicianorum praesentia in Lucanis*' in an early time because it recalls an *Iulius* and it doesn't mention the Dei Manes.

The document of a soldier of the urban cohorts (and with him of a praetorian) is not necessarily an indication of a permanent installation on the territory of Italy, as maybe Bracco considered. Like the Grumentine attestations here considered, it can rather be related to the emperor's presence and, especially, to the important road hub. As for its dating, despite the paleography does not offer significant clues, I believe that the lack of the praenomen and the abbreviation of the gentilicium in the onomastics of both soldiers, more than an early dating, suggest a chronology between the mid of the second and the third century. The same chronology of the inscriptions here considered.

Notes

1 Caldelli, Petraccia and Ricci 2012, pp. 285–287 and 295–298.
2 It is impossible to provide appropriate literature on the issue, concerning not only Italy, but all the provinces where the soldiers had stationed. I will just point out the works by Parma 1994 and 1995 for Italy; and Alston's contribution (1999) for the provinces.
3 The discussion on the concept of 'public order', which in the ancient sources is defined as *pax/disciplina publica* and considered a value to be preserved, has been extensive and wide-ranging (Brélaz 2005). Pioneering, in this sense, can be considered the work of MacMullen 1966 and 1974, and of Lintott 1998, related, respectively, to the Late Empire, and to the Republic. Their reflections opened the way for studies on the phenomena of banditry, social exclusion and the forms of crowd control in the ancient world.

  For Rome and the Republican age, refer also to the works of Nippel 1988 and 1995; see also Lintott 2008. For the imperial period, refer to Sablayrolles 2001 and Rivière 2004. The marble slab of *Erclanius, occisus a malibus*, of unknown origin, is cited, with an extensive bibliography on the lexicon of bandits and criminals, by Panciera 2006, pp. 977–981. Some clarifications are in Rivière 2008a.
4 *Supra*, Part I, Chapter 1, p. 4 ff.
5 The quotation is by Antonielli (2006b, p. 5). The research work on the subject by the same author was intense. Refer also to Antonielli 2002, 2006a; Antonielli and Donati 2003. All the cited works offer specific literature on the above subject. See also Brunelli 1998, pp. 359–371 and the conference organized by Brunelli himself (2003). On the history of police in Europe, it useful the Campesi's synthesis [NdA: also refer to the volume of Napoli 2003, in French].
6 In regards to this, see the reflections of Le Roux 2002.
7 Hirschfeld 1891 and 1892.
8 MacMullen 2006 presents an updated list (pp. 123–130) – compared to the one provided by the same MacMullen (1988) – of Roman centres with the presence of soldiers.
9 As it appears from the research carried out in relation to the East Mediterranean, in particular in the Late Empire. Isaac 1993; Pollard 1996 and 2000 (Syria-Palestina); Alston 1994 (Egypt) think of a unified and coordinated security dispositive implemented by Rome in some of the eastern provinces of the Roman Empire. *Contra*, Wheeler 1993 and 2007 (at p. 263 n. 1 a useful review of the literature on the so-called 'strategy debate'). More generally, see the bibliographic review on the theme 'Polizei – innere Sicherheit' of Krause, Mylonopoulos and Cengia 1998, pp. 529–541.
10 On the tasks attributed to the urban cohorts, see Ricci 2011 (in this volume, Part III, Chapter 7.

11 The best-known case is that of the Praetorian Guards in Pompeii (considerable in numbers, at different times, and with different tasks): De Caro 1979 and Łoś 1995.

12 Millar 1986, p. 300. Further on (p. 315), Millar himself underlines that "the possibilities of effective recourse to senate or emperor for Italian communities were profoundly affected by the network of senatorial and equestrian landholding in their territories". On the use of armed slaves with paramilitary duties in Late Antiquity, see Lenski 2009 (very few information on the previous period).

13 Regiones Italiae I (*Latium et Campania*), II (*Apulia et Calabria*), III (*Lucania et Bruttium*), IV (*Samnium et Sabina*).

14 This is not the place for an exhaustive treatment of this topic; allow me to refer to Ricci 1994.

15 For Ostia, see M.L. Caldelli, *Ostia e la tutela della sicurezza di un grande porto*, in Caldelli, Petraccia and Ricci 2012, pp. 291–295, part. pp. 292–294.

16 For example, *Reate*, Capua, *Luceria*, the Pontinian hinterland (*Fundi, Privernum*, Cori). See Keppie 1984; and the addendum 'New evidence and further thoughts' of Keppie 2000, pp. 249–262 and Keppie 2003.

17 Except for one case, the subject of this communication (see the following pages).

18 This area of the peninsula did not represent a reservoir for the soldiers of Rome, as we know from Tac. *Ann.* 4.5.3. See Passerini 1939 (pp. 148–159 and 174–180); and Freis 1967, p. 54. Now also Crimi 2010, part. p. 332. For the Roman Lucania, see the collection by Simelon 1993, pp. 73–75.

19 Two other epitaphs, which initially seemed interesting, were subsequently discarded: Publius Titius Ampliatus, *cornicen cohortis primae* (*CIL*, X 217) is in fact most likely a *vigil*, as suggested by Mommsen, in the apparatus of *CIL* (*contra*, Sablayrolles 1996, p. 229 n. 177). In any case, the *Titia gens* is discreetly attested in *Grumentum* (Simelon 1993, pp. 156–157 no. 126). The unknown centurion of the legion XXI, maybe the *Rapax* (*CIL*, X 218) belongs to the *Pomptina tribus*, prevailing in the region: it is a strong clue to a local origin. It should be noted that both inscriptions seem of first-second centuries and are therefore earlier than those considered here.

20 *CIL*, X 204: *Soli invicto / Mythrae* (!) */ T(itus) Fl(avius) / Saturninus, / evoc(atus) Augg(ustorum) n(ostrorum)*. Simelon (1993, p. 74, n. 8), suggests a dating mid-second century; according to Weaver 1972, p. 58 the inscription is dated 161, during the coregency of Marcus Aurelius and Lucius Verus.

21 *CIL*, X 214 = CBI, no. 866: *D(is) M(anibus). / Aelio Digno, b(ene)f(iciario) / p(raefectorum) praetorio e(minentissimorum) v(irorum), / qui vixit ann(is) [- - -], / milita<v>it a[nnis - - -]. / Iulia Vera[- - -]/ia uxor, c[oniugi] / bene mer[enti fe]/cit, cu[m quo] / vix[it annos - - -]*.

22 *CIL*, X 215: *D(is) M(anibus) / Aeli Marciani, / mil(itis) coh(ortis) VI pr(aetoriae) P(iae) V(indicis) / G(ordianae) Maxim(a)e, st(i)p(endiorum) / XII. Huic pecun(ia) / eiusdem Marci/ani Valerius / Valerianus evok(atus) / faciundum / curavit.*

According to Simelon 1993 (pp. 73 and 75 n. 15), both soldiers were "most likely from *Grumentum*".

23 *CIL*, X 216: *[- - -] VIA[- - -] / Aur(elius) Asdula, mil(es) / coh(ortis) V pr(a)etori(a)e / fratri ben(e) meren(ti), / qui mecu(m) labora(v)it / an(nos) XII et Fruninone / est in barbarico.*

24 About the history and the excavations of *Grumentum* and on its epigraphic heritage, see Cifani, Fusco, Munzi 1999–2000; Mastrocinque 2007 and 2009; Buonopane 2006–2007.

25 On the properties of the Bruttii Praesentes, see Andermahr 1998, pp. 182–187; Segenni 2004, p. 136. Recently in Marsicovetere (district of Barricelle), just halfway between Grumentum and Atina, has recently been discovered a Roman imperial villa; excavations are ongoing (Di Giuseppe 2007, pp. 105–114; and Di Giuseppe 2010).

26 *L. Fulvius C.f. Pom. Rusticus C. Bruttius Praesens, consul ordinarius* in 153 and in 180 Scriptores Historiae Augustae, *Vita Marci*, 27.8; Cass. Dio 71.33; Herodian. 1.8.4), as recalled by Pflaum 1966, pp. 32–37. On the imperial properties, see *AE* 1998, 394b: [*Ex*] *pr(aediis) Augg(ustorum)*. For a hypothesis of the imperial interests in the area, dating back to Marcus Aurelius, see Di Giuseppe 2007.

27 Rostovtzev 1905.

28 Unthinkable, at least for a couple of soldiers: the two praetorians Aurelius Asdula and Aurelius Marcianus, certainly of Balkan origin.

29 Keppie 1996.

30 Even Munatianus, responsable of the *statio* in Furlo pass, is an *evocatus* (*supra*, p. 192).

31 The villa of the Barricelle district (*supra*, n. 25) was near Átina. The words in brackets are of Andermahr (cited in the same note).

32 *Inscr.It.* III.1, 132 = *AE* 1969/70, 172: *Iul(i) Amatoris, / mil(itis) c(o)h(ortis) III pr(aetoriae,) / Iun(ius) Felix, mil(es) / c(o)h(ortis) X urb(anae), opt[i]/mo commilitoni.*

9   *Praetoria* and praetorians
The emperor's travels and security (*Latium Vetus*)[1]

The emperors often moved away from Rome, to take part in military campaigns or to visit the provinces[2] and various parts of Italy. Their trips within the peninsula, which have unfortunately received scant attention from scholarship,[3] had to be planned with care and according to a precise ritual, which took into account the differing needs, appointments and events that would occur during the period of visit.[4] The imperial *comitatus* was composed of prominent personalities as well as friends and collaborators,[5] including household staff[6] and soldiers who, in the first place provided to ensure the emperor's safety during the journey and, once at destination, his security and the necessary peace of mind required to work or rest in the most profitable way, in the same conditions as close as possible to those of the capital. The presence of the soldiers was also in itself a tangible sign of imperial power and it cannot be entirely excluded that, when the emperor left, near those villas more assiduously frequented and loved by the *Princeps*, a military garrison remained, next to the staff in charge of custody.

Literary sources lack information on the preparation for the imperial journeys, especially when it came to travelling within the peninsula. We only have information when unforeseen events occurred or when an exception to the usual procedure wanted to be emphasized.

On the eve of the Principate, Cicero tells Atticus that Caesar stayed in Campania and was his guest with a retinue of slaves, freedmen and soldiers, for a total of about 2,000 men.[7]

When Tacitus gives an account of the Pisonian conspiracy, he tells how it was decided that the murder would take place at the senator's villa in Baia, "where Caesar was often attracted by its charming beauty. There he bathed and banqueted, free from bodyguards (*omitted excubiis*) and from the encumbrance of his own position."[8] The absence of bodyguards for Nero is therefore reported as an anomaly; evidently he felt very safe as Piso's guest.

On other occasions, the emperor's will to do without the security escort has been emphasized for the exceptional nature of this decision[9]: in one of the *Panegyrici Latini*[10] there is an account of Theodosius who, in Rome in AD 389, visited private homes without military bodyguards (*remota custodia militari*).

In Italy, in addition to enjoying local notables' hospitality, emperors resided at their own villas. Their journeys have been investigated by referring to the suburban residences, following their vicissitudes, from their construction, not infrequently dating back to old Republican owners, to their decline.[11]

Imperial estates have been studied at the beginning of the last century by Otto Hirschfeld[12]; in the mid-seventies by Martin Leppert, with his dissertation that has unfortunately remained unpublished[13] and by Dorothy Crawford; more recently, by Elio Lo Cascio.[14] At present, it is still difficult to give a full account of the proprieties, even only for the Lazio region.[15]

Taking as a starting point Hirschfeld's work,[16] I will limit myself to recall some of the most famous and attended residences,[17] and then focus on the sites where I have tracked, so far, urban military inscriptions.

On the outskirts of Rome, the sources mention the properties *ad Gallinas Albas*, at Prima Porta, on the Via Flaminia; the *Suburbanum Flaviorum*, at the fourth mile of the Via Appia; the *Suburbanum Luci Veri*, at the fifth mile of the Via Clodia/Cassia[18]; the two villas confiscated from the former owners Quintilii (on the Via Appia)[19] and Gordiani (on the Via Prenestina). In Southern Etruria, we know of the properties of Centumcellae, Lorium and Alsium.

In Latium Vetus we know of imperial properties at Tusculum, in the territory of Castrimoenium,[20] at Albanum, Nemus Dianae, Velitrae,[21] Lanuvium, Tibur, Praeneste; and, on the Lazio coast, at Laurentum-Tor Paterno, Antium, Circeii, Terracina, Spelunca, Formiae, Fundi and Caietae.

In order to conduct research on the organization of the security service for the emperor when visiting his villas in Italy, I started investigating some of the places that hosted the imperial residences to seek, through the inscriptions, traces of the presence of soldiers who following the lead of the *Princeps*. The research focuses on sites where the presence of soldiers cannot be otherwise explained, as it happened, for example, for ports, for border towns or for places of assignments for veterans (e.g. Ostia, Puteoli and the Campanian shore[22]; Aquileia,[23] Reate).

This research has soon provided some surprising facts: in Antium, although here there was a villa that the emperors loved very much and often visited,[24] I recorded only one entry of an active praetorian.[25] We find he same scarcity of evidence in Bovillae, Castrimoenium, Tusculum and other towns in Lazio.[26]

My dossier concerns the towns of Centumcellae (Civitavecchia), Lorium (Castel di Guido) and Alsium (Palo), in southern Etruria; and those of Albanum (Albano Laziale), Lanuvium (Lanuvio), Praeneste (Palestrina) and Tibur (Tivoli) in Latium Vetus.[27]

In the years 1864–1866, near the Roman dock of Civitavecchia, in a place called Prato del Turco, two inscriptions of *vigiles* and about twenty of fleet soldiers were found in a burial ground dating, according to numismatic evidence, from the beginning of the second century to the third century AD. The port of Civitavecchia was constructed by Traianus.[28] The soldiers of the fleet of Misenum and Ravenna served there and they were buried nearby, if they died before being discharged. However, the epitaphs of Prato del Turco are even more striking: in the same burial ground was buried other personnel, employed at the port or, more likely, at the imperial residence close to it.[29]

In addition to the *classiarii*, the inscriptions recall, as has been said, two *vigiles*[30] and four women[31] who, in my opinion, were buried there because they were staff members of the villa, besides being related to the soldiers.[32]

Among all the epitaphs of the fleet soldiers in Civitavecchia, one in particular is worthy of interest: the epitaph for Caius Caecilius Valens, a fleet soldier of Misenum,[33] posed by Caius Lucilius Valens, who defines himself as a *corporis custos*.[34] As mentioned, the burial ground of fleet soldiers at Civitavecchia can be dated between the second century and the beginning of the third century. Pietro Romanelli postulates shorter or longer stays of the emperors in the area and considers common ground that *Lucilius Valens* was part of the *corporis custodes*; at that time, however, the Julio-Claudian bodyguards had already been dismantled and replaced by the *equites singulares Augusti*.

The inscription has therefore been recalled by both Fulvio Grosso, in an article on the *ius Latinum* of the soldiers in the Flavian era[35]; and by Heinz Bellen, in his work on the *corporis custodes*.[36] They reach, however, differing conclusions: Grosso in fact, grounded strongly on a passage by Cassius Dio, which speaks of Keltoì (= *Germani* and thus, in his view, *corporis custodes*) at the time of Galba, suggests that the unit was not dismantled at least until the early second century, the period in which the inscription should be attributed to. Bellen instead, without having reviewed the gravestone, proposes a correction of *corpor(is) custos* into *armor(um) custos*, thus arguing that Lucilius Vales was a colleague of Caecilius Valens, charged with the arsenal of the fleet.

Only the first interpretation takes into account the fact that, in addition to a *statio* of the fleet, Civitavecchia was also the location of an imperial villa[37] that continued to be used from the Trajan era,[38] up to at least the second century[39] and partly beyond the next one. Thus, the funerary inscriptions for the fleet soldiers, the firefighters, the slaves or imperial slaves and, with a wider perspective, the three epitaphs of the fleet soldiers of Lorium[40] and the one of the trierarch of Alsium,[41] taken as a whole, make clear the catalyst function of the imperial villas on the Tyrrhenian coast that cannot emerge instead from a separate examination of the documents of the single locations.

By widening the focus, in order to make a comparison out of the context here considered, one can recall, the case of the necropolis of Surrentum (Sorrento) of the Julio-Claudian period, which housed slaves and freedmen assigned to an imperial property.[42] Close to it were unearthed inscriptions of a praetorian *speculator*, of two soldiers from the nearby Misenum fleet, and of a *centurio* of the second praetorian cohort, all of the second or third centuries AD. Although the evidences of soldiers of the Misenum fleet at Surrentum, taken in isolation, could easily be connected to the nearby headquarters of the fleet, the closeness of the epitaphs of these soldiers with those of staff members of the villa seems to characterize it in a more perspicuous way.[43]

The small dossier of the fleet soldiers from southern Etruria also allows another consideration: all the deceased soldiers of Centumcellae and Lorium had a great number of years of service (apart from a few isolated cases, the average amount is around 15–17 years). This detail suggests that the soldiers selected for such a delicate task had to have a fair length of professional expertise and, therefore, proven experience and reliability.[44]

At Albanum, permanent base of the *legio II Parthica*, the military presence is remarkable.[45] Some inscriptions do not refer to legionaries, but rather to soldiers belonging to other corps, and more specifically to the *equites singulares Augusti* and praetorians. Two inscriptions, commemorating respectively Titus Flavius Quintinus and Claudius Maximus[46] are different in characteristics and dating. The inscription of Quintinus is engraved on a marble slab and probably still belongs to the second century AD, therefore it is previous to the deployment of the legion *in loco* by Severus. His presence in the *ager Albanus* is connected, as suggested by Speidel, to the Domitian palace at Albanum.[47] Maximus instead is buried in a sarcophagus that, as well as coming from the same area, also has the typical features of those soldiers of the Second Parthian legion.

Two other inscriptions, both from the same ager Albanus, datable to the third century, mention praetorians: a mutilated text (found along with another inscription of a soldier, perhaps a legionary) recalls Aurelius Bithus, enlisted in the second praetorian cohort[48]; and an epitaph, also engraved on the lid of a sarcophagus, reused at the time of Alexander Severus, mentions an Aurelius Seneca, enlisted in the third praetorian cohort, who died at age of 23 and was buried by two heirs, one of whom was his mother.[49]

This evidence is too scant to enable the formulation of a reliable hypothesis regarding the reason of those soldiers' presence. With regard to the *equites singulares*, it should be emphasized that the two are commemorated by heirs and not by relatives; and the inscription mentioned first (Del) is prior to the installation of the Parthian legion. Even the second mentioned of the two inscriptions commemorating the praetorians recalls a *mater* and, perhaps, a sister (Da2). Both the presence of the two praetorians (not originated from the area, based on the epoch in which they lived and the foreign onomastics) and at least one of the *equites singulares* could be clue that the emperors also continued to employ soldiers of the urban troops for their safety.

A third group of evidences from Latium, concerning exclusively the *vigiles*, represents an anomaly since this paramilitary corps was normally employed in Rome (or in Ostia and Puteoli). In 1907, excavations conducted on Mr Seratrice's land at Borgo San Giovanni in the territory of Lanuvium have brought to light some sacred inscriptions on bases of lava stone (peperino), addressed by the *vigiles* belonging to different cohorts.[50] Alberto Galieti[51] points out that the discovery find-spot is "not far from the ruins of the temple which seems was once sacred to Hercules"; and Hercules as the recipient deity of such offerings is explicitly stated in some of these inscriptions.[52] A special feature characterized the bases, as observed by the first editor: they date back to the Roman Republic. They originally bore inscriptions addressed to Hercules, but the names of the vigiles have been added at a later epoch.[53] Unfortunately, the conditions in which the surviving bases or slabs of lava stone currently are do not allow a satisfactory reading of the texts.[54]

Particularly interesting, however, seems that *vigiles* belonging to different cohorts[55] appear as those addressing dedications to Hercules, in the early third century AD in Lanuvium. The reason for their presence, in my view, finds an explanation in connection with the villa where Antoninus Pius and Commodus were born,[56] apparently frequented also by the Severans[57]; and, with the privileged relationship that Severus had with the *vigiles*.[58] It is very difficult not to draw a parallel between the numerous dedications that the *vigiles* addressed to the emperor and the imperial family in Rome in this period and the ones in Lanuvium.

Sablayrolles asserts that the *vigiles*, under Septimius Severus and Caracalla, became soldiers in all respects; and as such, in addition to performing their specifically assigned duty of protecting the inhabitants of Rome from fires, they were also charged with the security service of the imperial villas. To confirm this point are the presence of two vigiles in the third century in Centumcellae and the absence of firefighters in Praeneste and Tibur, where the documentation is mostly prior to the reign of Severus.

A relatively rich and interesting framework is offered by the city of Praeneste, popular holiday destination for the emperors and their family. From Suetonius[59] we know that Augustus used to stay there willingly[60]; and here part of the urban garrison was temporarily established in the Augustan period.[61] In the Praenestinian territory Emperor Nerva's family had an estate,[62] a villa was built or rebuilt by Hadrian[63]; one or more imperial properties were here until Severus Alexander.[64]

Here an equestrian officer and a centurion are commemorated: both were enlisted during the reigns of Augustas and Tiberius, neither seem to be natives of the town, belonging to tribes different from *Menenia*; and both are alive at the time in which the inscription is recorded.

S'il est difficile d'imaginer une relation entre la sécurité d'Auguste et Tiberius et l'evocatus Augusti et préfet de la première cohorte des Corses[65], qui peut-être à Palestrina avait un terrain; plus probable est une connexion entre la cité du Latium vetus et le centurion Lucius Catius qui fait une dédicace à lui-même et à sa femme.[66]

It is difficult to imagine a relation between the security of Augustus and Tiberius and the *evocatus divi Augusti* (also prefect of the first cohort of the Corsi and of the *civitates* in Barbagia)[67] who perhaps had a plot of land in Palestrina.

It is instead more likely that such connection existed for the centurion Lucius Catius.[68] His name appears on a block of marble of considerable measures, found in the area of Paliano, which probably belonged to one of the drum-shaped tombs in vogue at the time. The centurion was assigned to the tribe of *Fabia* and might have originated from Rome. Maybe, as a commander of the selected guards (*speculatores*), he decided to build the tomb for himself and his wife in the place his emperor was fond of; and where, with his soldiers, he had often accompanied and protected the *Princeps*. But this is only a supposition.

The inscriptions of urban troops in Praeneste are about a dozen and cover the entire time span of the imperial era.[69] The bulk is represented mostly by praetorian guards, belonging to different cohorts.[70] There also is a *miles urbanus* and another soldier not better identified.

The concentration of praetorian guards has already been investigated by Karl Patsch,[71] who believes to have identified the reason in a passage of Tacitus[72]: in AD 64, at the gladiatorial school of Praeneste,[73] a revolt was quelled by a military garrison that stationed there. Patsch initially wondered whether the presence of the Praetorian Guard[74] in this town was not due to this circumstance. However, except for one case (Fa1), none of the inscriptions belong to the first century AD, but rather to the following one.[75] Then, realizing that none of praetorian guards were originally from Praeneste, Patsch himself[76] came to the conclusion that these soldiers, unrelated to the uprising, could belong to a *vexillum* sent periodically from Rome.

This second hypothesis – although not completely incompatible with the first one – provides a better reason for the presence of praetorian guards over a long period.[77] However, since it does not seem to be the strategic requirement for a permanent detachment of praetorian guards in Praeneste, it is equally legitimate to believe that these soldiers have accompanied and escorted the emperors during their stay at the villa. Two of the gravestones have the same provenance,[78] which could also coincide with a necropolis annexed to an imperial property.

An anonymous praetorian of the third cohort dedicated a small base to Jupiter Optimus Maximus in AD 167[79] on his return from an expedition (*reuersus de expeditione*). The soldier's passage in Praeneste corresponds to the end of the military campaign led by Verus; as in this case, the fulfillment of a vow may be the occasion for others votive inscriptions of soldiers discovered in the towns of Lazio.

At Tibur seven inscriptions recall soldiers belonging to different corps.[80]

Leaving aside a praetorian and a *miles urbanus*, perhaps hailing from territory of Tibur,[81] for a *eques singularis* of the third century[82] Speidel states that his presence is connected with villa Adriana.[83]

More interesting, in my opinion, are four epitaphs (dating at least to the second century AD) of soldiers of the two fleets of Misenum and Ravenna. At least three of them[84] were discovered at different times in the same 'proprietà Galli', immediately annexed to Villa Adriana and they may have been connected with the villa itself. The soldiers appear to be dedicants (to a female slave and a freed-woman) or recipients (by comrades) of the dedications.

Clemens Heinrich Konen,[85] who knows only one of the four epitaphs,[86] observes that, along with other inscriptions found in Lorium, Centumcellae, Alsium and Populonium, the epitaph confirms Suetonius' words[87]: some *classiarii* commuted between Rome, Ostia and Puteoli; and also from Rome to the imperial villas located in these places. However, the passage of Suetonius that he cites does not speak of the service at the villas, but rather of moving from the places where there was a garrison towards Rome.

Michèl Reddé[88] explains that this commuting is rather to be considered as a courier service that the soldiers provided on behalf of the emperor.[89] Their presence there, however, was non-sporadic and linked to that of other urban forces: it did not most likely end with the performance of such a particular service.[90]

While accepting the idea that among the duties carried out by the *classiarii* at the *praetoria* there also was the one of courier, still there is no explanation, in my opinion, for their absence at other residences where the emperor certainly was engaged in administrative activities (e.g. Albanum). It seems logical to think that the physical movement entailed for soldiers also a kind of transposition to the villa of the functions exercised in Rome: as for the soldiers of the fleet, the task of dealing with communications and transport, the *velaria*, the naval battles, the swimming pools, at both the coast residences as well as the ones inland.[91]

The emperors used the residences scattered throughout Italy not only as a holiday destination, but also for longer stays – as in the case of Tiberius in Capri, who transformed the villa into a second capital – as the home base where they could perform various kinds of duties and, in the first place, the administration of justice, through *rescripta*, *edicta* and *responsa*.[92]

In all these cases the availability of skilled personnel for all the required assistance (secretarial workers and couriers in the first place, but also military personnel) was essential.[93]

More often, although not specifically, scholars have examined the issue of security service provided at imperial villas, by integrating partially the scant information of the literary sources with epigraphic and archaeological material. In this regard, in a pioneering article written in 1905 entitled 'Die Domänenpolizei in dem römischen Kaiserreiche', Michail Rostovtzeff made a distinction between slaves in charge of safeguarding private estates and soldiers as guards of imperial properties. Marcel Durry, when referring to the Praetorian Guard, takes for granted their role to ensure security at the imperial villas[94] and cites documents from various locations to give examples of occasional accompanying services on part of these soldiers. At Villa Adriana, a building by the side of the so called Gymnasium, organized as a dormitory around a central courtyard, has, according to Salza Prina Ricotti, all the characteristics of a place designed for service personnel, perhaps military.[95]

It is certainly possible, in some cases, that at the imperial villas existed *stationes* of soldiers ad hoc. It could be nevertheless that praetorians, *milites urbani*, *vigiles*, *equites singulares Augusti* and soldiers of the fleet moved from the city towards different locations in Italy with various objectives, sometimes of public order; or to prepare for the arrival of the *Princeps* and carry out the security service.

It has been said that the *equites singulares Augusti*, between the first and second centuries AD, replaced the praetorians as accompanying troops.[96] When referring to the different places in Italy where the emperors stayed, praetorians and *equites singulares* do not seem, however, to have had the sole priority of such duty: if the literary evidence often only refer generically to *milites*, *excubiae*, *custodes*, *satellites*, *praesidia militum* (or similar) to protect the emperors, the inscriptions seem to reflect a far different reality.

Also the widespread view that the protection and the security escort of the emperor was guaranteed by the praetorians, in and outside Rome, has to be partly scaled down. The service of surveillance of the imperial villas seem to have been carried out, in the different cities of Latium here considered, by soldiers coming from different corps, even (as seen) near the same residence, with a differentiation of roles. In addition to acting as bodyguards, soldiers were required to intervene in case of fire and to assist estates with pools of water or by the river or the sea, and to watch the villa along with civilian personnel.

This common ground of a mixed composition of soldiers has been proven by documents of all the *praetoria* here analysed from different eras, although most of the material collected dates back to the second and the third centuries.[97] What stands out, always at a chronological level, is the small number of attestations for the first century and in particular for the Julio-Claudian era, of which we are relatively better informed through literature. Even so, it must be remembered that the information is certainly linked to the randomness of the findings and, in most cases, to the epoch in which the *praetoria* were built or intensively used.

In addition to having soldiers from different corps, what also stands out is that they came from different cohorts (or *centuriae*): it suggests the deployment of detachments of soldiers from Rome or, in the case of soldiers of the fleet, from the ports closest to the imperial villa. It is difficult to say what the selection criteria were: on the basis of biographical information in the epigraphic texts, for the soldiers of Centumcellae and Lorium what has been suggested here is that a preferential requirement was the length of service and, therefore, the soldiers' experience accumulated through service over the years.

If the presence of the praetorians and *equites* at the side of the emperors was predictable, less expected was the one of soldiers of the fleet and *vigiles*;

although, as we have seen, equally justifiable. In the Tyrrhenian Sea, in addition to the base of Misenum, there were detachments of *classiarii* in various locations in Campania (Baiae, Stabiae, Puteoli, Neapolis, Surrentum, Capri) and Latium (Ostia, Centumcellae, Rome),[98] which obviously, in case of need, the emperor and his officials could easily turn to. And the soldiers of the fleet, serving for longer periods of time with a lower pay than the praetorian guard,[99] had a personal attachment to the *Princeps*, which in the third century seems to have been alive and appreciated.[100]

The fleet soldiers' involvement for service by the imperial villas could be extended over time from the coast in Campania to other areas of the Tyrrhenian coast, as well as in the inner Tibur. These soldiers, probably after having been in Rome for some time, would carry out a wide range of duties such as preparing water shows, moving large and small *velaria*, loading, transporting and watching goods on inland waterway, policing imperial villas and protecting the emperor.[101] In the case of Centumcellae, aside from the duties that have already been indicated, the soldiers were also charged with the port area.

The loyalty to the imperial dynasty certainly played an important role in determining the involvement of the *vigiles* in policing some imperial villas, especially at Centumcellae and Lanuvium, between the end of the second and the beginning of the third centuries. As in the case of the fleet soldiers, one should consider that the *vigiles* followed the emperor providing an extra service, thanks to their expertise in controlling fires, maybe together with soldiers of the fleet, as in Ostia and Misenum.

Other questions raised by this material receive plausible answers, although not definite: one may wonder, for example, if the service carried out ended once the mission of accompanying the emperor was accomplished; or if it was periodically repeated. The overall picture that emerges is that of a discreet versatility required by soldiers in order to perform the duties; and of a collaboration between civilian and military personnel also reflected by their remains housed in the same necropolis (Surrentum, Centumcellae and, perhaps, Alsium).

I want to mention, for a final reflection, two texts, one by Tacitus,[102] the other taken from the Marcus Aurelius's Meditations,[103] which respectively include and exclude the bodyguards from the *aula principis*.

Tacitus points out that, after Augustus' death, Tiberius "as imperator, had given the watchword to the praetorian cohorts; he had sentinels, armed men and the court staff; soldiers escorted him into the forum, and to the curia"; he used to already act in fact as a successor.

Marcus Aurelius, instead, according to Mario Pani for moralistic reasons, did not include bodyguards among the members of the court, because "at court one can live without bodyguards, or refined garments, luxury furnishings, ostentation and splendor; one can keep a standard of living similar to that of a private citizen, without assuming for this reason an undignified attitude".

The two excerpts may reflect two different points of view, or rather, two successive stages of evolution of the historical-political thought: the Julio-Claudian dynasty and the end of a phase in Roman imperial history. In light of what emerges from the evidence coming from Rome and from Italy, both raise substantial questions about *how* the role of these soldiers was conceived; *how* that imperial power (the *maiestas*) upon which they were called to protect had changed over time; and *what* was actually guaranteed by their service.

Between the end of the Republican era and the mature stage of the empire, the role of the praetorian guard – as a strong expression of the power – changes and other soldiers start cooperating with them, as the emperors' security escort and protection outside of Rome.

These soldiers provide a more domesticated image of the power they represented. They precede, announce, escort, sometimes deal with contingent emergencies (fires, structural failure of buildings, flooding and different sorts of accidents); they mix at the villas with civilian personnel and transmit reassuring messages. In other words, those principles that the emperor-philosopher, in the late second century, elaborates under an ethical point of view, start gradually to take form.

Dossier

The documents are sorted according to the place of origin (indicated with an uppercase letter in progressive order: A = Centumcellae; B = Lorium; C = Alsium; D = Albanum; E = Lanuvium; F = Praeneste; G = Tibur) and according to the corps the soldiers belonged to (indicated with a lowercase letter in progressive order: a = *cohortes praetoriae*; b = *cohortes urbanae*; c = *cohortes vigilum*; d = *milites classiarii*; e = *corporis custodes* or *equites singulares*) and, within them, in chronological order, following roughly the order in which they are discussed in the comments. In the tables are also included inscriptions of centurions or tribunes of praetorian cohorts, worthy of consideration as possible evidence of the presence of a detachment of that cohort in the territory; inscriptions of veterans or *evocati* are not included.

In addition to the essential bibliographic reference and any adjustments, what is indicated are: a dating proposal; the soldier's name; the soldier's corps; the type of inscription when it is not a sepulchral inscription; if the mentioned soldier is the dedicant and not the recipient of the epitaph.

By no means is this repertoire exhaustive (it is possible that some documents may not have been included), but rather a practical prospect of reference for the documents examined in the text.

As for the inscriptions of the *Latium vetus*, the work has been simplified by searching the files of *CIL* XIV, relative to the examined sites, kindly provided by Professor Maria Grazia Granino Cecere, without whose

	Edition	Soldier's name	Military corps	Notes
Ac1	CIL, XI 3520	Canius Eutychus	Vigil, tesserarius, third cohort	
Ac2	CIL, XI 3521	P. Nunnienus Sabinianus	Vigil, first cohort	
Ad1	CIL, XI 3522	M. Acutius Faustinus	Classiarius, fleet at Misenum	
Ad2	CIL, XI 3523	M. Antonius Aristo	Classiarius, fleet at Misenum	
Ad3	CIL, XI 3524	P. Art[orius] Pa[stor]	Classiarius, fleet at Misenum	
Ad4	CIL, XI 3525 = 7583	T Aeatius Verus	Centurion, fleet at Misenum	The dedicator is an *optio* in the same fleet
Ad5	CIL, XI 3526	C. Caecilius Valens	Classiarius, fleet at Misenum	
Ad6	CIL, XI 3527	C. Carminius Provincialis	Classiarius, fleet at Misenum	Inscription discussed in the text
Ad7	CIL, XI 3528	Ti. Cl(audius) Se[---]	Classiarius, fleet at Ravenna	The dedicator is a soldier of the same fleet
Ad8	CIL, XI 3529	T. Clodius Naso	Classiarius, fleet at Ravenna	
Ad9	CIL, XI 3530	[Da]sumius [--]us	Classiarius, fleet at Ravenna	
Ad10	CIL, XI 3531	[C. D]omitius Regin.s	Classiarius, *suboptio*, fleet at Ravenna	
Ad11	CIL, XI 3531a	M. Helvius Maximus	Classiarius, fleet at Ravenna	
Ad12	CIL, XI 3532	C. Iulius Saturninus	Classiarius, fleet at Misenum	
Ad13	CIL, XI 3533	P. Memmius Valens	Classiarius, fleet at Misenum	The dedicator is a soldier of the same fleet
Ad14	CIL, XI 3534	M. Petronius Maximianus	Classiarius, fleet at Misenum	The dedicator is a soldier of the same fleet
Ad15	CIL, XI 3535	A. Valerius Cassianus	Classiarius, fleet at Misenum	
Ad16	CIL, XI 3536	C. Valerius Fronto	Classiarius, fleet at Misenum	The dedicator is a soldier of the same fleet
Ad17	Not.Sc. 1919, p. 222, no. 272	[- - -]prius Lon[- - -]	Classiarius, fleet at (Misenum?)	
Ad18	Not.Sc. 1919 p. 223, no. 285 = CIL, XI 7584	S(e)x. Congenius Ve=us	Classiarius, fleet at Misenum	
Ad19	Not. Sc. 1940, p. 194, no. 1	Unknown	Classiarius, fleet at Misenum	
Ad20	Not. Sc. 1940, p. 195, no. 2	Unknown	Classiarius, fleet at (Misenum?)	

help this brief catalogue could not have been written and to whom in particular goes my gratitude.

A. Centumcellae

For the inscriptions of Centumcellae, only in a couple of cases an approximate dating is given, since for most of them it was not possible to carry out an autopsy (except for Ac2; Ad1–Ad4; ad8, ad10, ad12–14; ad20, for which today the readings are diminished compared to the CIL edition). Nevertheless, as has been indicated in the text, the inscriptions of the burial ground of the soldiers of the fleet are datable to the period between the second and (first half of?) the third century AD.

The last six inscriptions of the burial ground of the *classiarii* edited in *CIL*, XI (nos. 3537–3542) refer to individuals (mainly women) who are not part of the fleet and have been mentioned in the text.

B. Lorium

In *CIL* XI Bormann observes (cf. *supra* p. 525, no. 3520 ff.):

> *ad Torre in Pietra, ubi quae inventae sunt inscriptiones sepulcrales militum classiariorum n. 3735.3736.3737 relatae a Marinio indicant stationem classiariorum fuisse (note 1 : hanc stationem consentaneum est coniunctam fuisse cum statione classium praetoriarum, quae fuit Centumcellis, quo Roma ibatur via Aurelia).*

	Edition	Soldier's name	Military corps	Date and notes
Bd1	*CIL*, XI 3735	Aufidius Livianus	Classiarius, *optio*, fleet at Ravenna	2nd cent.–first half of 3rd cent. The dedicator is a soldier of the same fleet
Bd2	*CIL*, XI 3736	[C.] Iulius Alexan[d]er	Classiarius, *suboptio*, fleet at Ravenna	2nd cent.–first half of 3rd cent.
Bd3	*CIL*, XI 3737	L. Valerius Papirio	Classiarius, fleet at Misenum	2nd cent.–first half of 3rd cent.

C. Alsium

	Edizione	Soldier's name	Military corps	Date
Cd1	*CIL*, XI 3719	Ti. Claudius Maximus	Classiarius, *trierarchus*, fleet at (Misenum?)	2nd cent.?

D. Albanum

	Edition	Soldier's name	Military corps	Date and notes
Da1	CIL, XIV 4214	Aur(elius) Bit[hus]	Praetorian	3rd cent. (beginning?)
Da2	Not. Sc., 1914, p. 194	Aurelius Seneca	Praetorian, third cohort *Pia Victrix Severiana*	AD 222–235. The inscription has been recalled several times. Among the others also by Pavan 1962, pp. 90–92; Tortorici 1975, pp. 126–128 no. 32 = AE 1975, 159; Modugno Tofini 1989, p. 58 s. no. 3
De1	CIL, VI 3255 = XIV 2287 = ILS 2211 = Speidel 1994, no. 672	T. Flavius Quintinus	*eques singularis Augusti*	2nd cent, first half
De2	CIL, VI 3246 = XIV 2286 = Speidel 1994, no. 673, cf. Ricci 1994, p. 18	Claudius Maximus	*eques singularis Augusti*	3rd cent., beginning

E. Lanuvium

	Edition	Soldier's name	Military corps	Date and notes
Ec1	Not. Sc. 1907, p. 126, no. 6 = Eph.Epigr. IX, 1910, p. 382, no. 602 = ILLRP 129a; CIL, I² p. 987. Galieti 1911, pp. 39 f.		Vigil, 6th cohort	Group dedication to Hercules, AD 202
Ec2	Bull. Com., 1907, p. 363 = Eph.Epigr. IX, 1910, p. 383, no. 605 = ILS 9246. Galieti 1911, p. 42; Epigraphica, 50, 1988, pp. 130–132, no. 14 (G. Formi)	Unknown	Vigil, 7th cohort	Fulfillment of a vow to Hercules and Iuno Sospita
Ec3	Not. Sc. 1907, p. 126, no. 5 = Eph. Epigr. IX, 1910, p. 385, no. 614. Galieti 1911, p. 40	MESES+ [- - -]	Vigil, 4th cohort	3rd cent. (beginning?)
Ec4	Not. Sc. 1907, p. 126, no. 7 = Eph.Epigr. IX, 1910, p. 335, no. 615. Galieti 1911, p. 41	Unknown	Vigil, unknown cohort	3rd cent. (beginning?)
Ec5	Eph.Epigr. IX, 1910, p. 392, no. 642. Galieti 1911, p. 4)	Unknown	Vigil, 6th cohort	3rd cent. (beginning?)
Ec6	Not. Sc. 1907, p. 126, no. 9. Galieti 1911, p. 42	Unknown	Vigil, 2nd cohort	3rd cent. (beginning?)
Ec7	Not. Sc. 1907, p. 129 = Eph.Epigr. IX, 1910, p. 388, no 624. Galieti 1911, p. 42	Unknown	Vigil, tesserarius, 7th cohort	3rd cent. (beginning?)

F. Praeneste

	Edition	Soldier's name	Military corps	Date and notes
Fa1	CIL, XIV 2948. Muzzioli 1970, pp. 57 f.	L. Aufidius C.f. Celer	Praetorian, eques, 4th cohort	1st cent. AD, first half
Fa2	CIL, XIV 2951. Muzzioli 1970, pp. 57 f.	[- - -] Firmus	Praetorian, 3rd cohort	1st cent. AD, first half
Fa3	CIL, XIV 2958 = CIL, XI 4995 and p. 1380, da Ferentillo = Eph.Epigr. IX 1910, p. 432 f.; Ricci 1994, p. 42 f. no. 13	L. Valerius L.f. Primus	Praetorian, 8th cohort	1st cent. AD
Fa4	CIL, XIV 2953	L. Geganius Victorinus	Praetorian, 9th cohort	1st cent.–first half of the 2nd cent.
Fa5	CIL, XIV 2905 = Eph. Epigr. IX, 1910, p. 432	Unknown	Praetorian, 3rd cohort	AD 167. Fulfilment of a vow to Iuppiter, *reversus de expeditione*
Fa6	Unpublished (CIL, XIV archive)	Unknown	Praetorian (maybe 2nd cohort)	1st–2nd cent.
Fa7	Unpublished (CIL, XIV archive)	M. Antonius Processus	Praetorian, 3rd cohort	1st–2nd cent.
Fa8?	Unpublished (CIL, XIV archive)	[- - -] Tauru[s - - -]	Praetorian, unknown cohort	1st–2nd cent.
Fa9	CIL, XIV 2952 = Speidel 1994a, no. 746	T. Fl(avius) Paternus	Praetorian, 10th cohort	2nd cent., end–3rd cent., first half. Dedicators: Flavius Seuerus, of the XXVI cohors voluntaria (!); L. Aelius Candidus, eq(ues) sing(ularis) Aug(usti)
Fa10	CIL, XIV 2955. Andermahr 1998, p. 329, no. 324	L. Mantennius Sabinus	Tribune, 3rd praetorian cohort	3rd cent., first half. Funerary inscription of L. Mantennius Severus, son of Sabinus
Fb1	CIL, XIV 2956. Freis 1967, pp. 53 and 131	L. Pompe[ius] Felicissim[us]	*Miles urbanus*, 11th cohort	2nd cent. AD?

G. Tibur

	Edition	Soldier's name	Military corps	Date and notes
Ga1	*Inscr.It.* IV.1, no. 154	M. Caeli[us – – -]	Praetorian, 3rd cohort	2nd cent. AD (?)
Gb1	*CIL*, XIV 3633 = *Inscr.It.*, IV.1, no. 169; Freis 1967, pp. 54 and 132; Ricci 1994, p. 13	Val[erius] [S] uperus	*Miles urbanus* and *optio*	2nd cent. AD (?)
Gd1	*AE* 1996, 517	Q. (H)oratius Severus	Classiarius, fleet at Misenum	2nd cent. AD. Horatius is a dedicator Coming from 'Fondo Galli', at Villa Adriana, as Gd3 and Gd4
Gd2	*CIL*, XIV 3627 = *Inscr.It.* IV.1, no. 160	C. Numidius Quadratus	Classiarius, fleet at Misenum	2nd cent. AD (?). The dedicator is perhaps also a comrade
Gd3	*CIL*, XIV 3630 = *Inscr.It.* IV.1, no. 167	Sex. [– – -]nius Seneca	Classiarius, fleet at Ravenna	2nd cent. AD (?). The dedicator is perhaps also a comrade. Coming from 'Fondo Galli', at Villa Adriana, as Gd1 e Gd4
Gd4	Unpublished (*CIL*, XIV archive)	[- Iu]llius Aq(u) i(l)a	Classiarius, fleet at Misenum	2nd cent., second half – 3rd cent., first half. Coming from 'Fondo Galli', at Villa Adriana, as Gd1 and Gd3
Ge1	*CIL*, XIV 3623 = *Inscr.It.* IV.1, no. 153 = Speidel 1994a, no. 674	Aur(elius) Disza	*Eques singularis Augustustorum nostrorum*	3rd cent.

Notes

1 Ricci 2004.
2 Halfmann 1986, who, on pp. 110, 127, 160, 206 and 217, gives examples of Praetorian Guards (soldiers and officers) who followed the emperors. On imperial travel, especially in Histria and Dalmatia, I shall just mention the contribution of A. Stamc, *Diadora*, 20, 2001, pp. 3–119, unfortunately in Croatian, with English summary.
3 As Barnes 1989 rightly observes (review of Halfmann's work).
4 Suet. *Cla.* 38.2 reveals what is expected during an imperial visit, when reporting: *Ostiensibus, quia sibi subeunti Tiberim scaphas obviam non miserint, grauiter correptis eaque cum invidia, ut in ordinem se coactum conscriberet.*
5 Pani 2003.
6 [NdA: on the staff responsible for the imperial family's first aid during such travels, see now Ricci 2015, part. p. 362 n. 3, with bibliography].
7 *Ad Att.* 13.52 (45 BC): *cum secundis Saturnalibus ad Philippum vesperi venisset, villa ita completa a militibus est ut uix triclinium ubi cenaturus ipse Caesar esset vacaret; quippe hominum duo milia..*
8 *Ann.* 15.52.1 (AD 64): *Coniuratis tamen metu proditionis permotis placitum maturare caedem apud Baias in villa Pisonis, cuius amoenitate captus Caesar crebro uentitabat, balneasque et epulas inibat omissis excubiis et fortunae suae mole.*
9 For the sources, see Halfmann 1986, pp. 110 s. (though regarding to journeys outside Italy).
10 *Pan. Lat.* 2 [12].47.3 : *ut crebro civilique progressus non publica tantum opera lustraveris, sed privatas quoque aedes divinis vestigiis consecraris, remota custodia militari publici amoris excubiis.*
11 D'Arms 2003 (new edition revised of D'Arms 1970). On Tiberius' villa in Capri, see P. Gros, in JRA, 17, 2004, pp. 592–598 (discussion on the book by Krause 2003).
12 Hirschfeld 1902, reprinted with some integrations in *Kleine Schriften*, Berlin, 1913, pp. 533–544.
13 Leppert 1974. For his analysis, Leppert has selected twenty-three among the most famous imperial residences.
14 Crawford 1976; Lo Cascio 2000a, in part. pp. 103–106.
15 On the topic of imperial villas, also refer to the proceedings of the first research seminar of the Museum of the city of Monte Porzio Catone [N.d.A. now edited by Valenti 2008].
16 Who has as a privileged point of view offered by the literary sources, supplemented at times by the epigraphic and archaeological ones; and has been widely resumed and partially integrated by Leppert 1974. [N.d.A. Now see also Maiuro 2012, who enriches the panorama of the imperial villas provided by Hirschfeld at the beginning of the last century].
17 For the bibliography, refer to the above-mentioned authors. In the following notes, I will give some supplementary bibliography.
18 Also see Mastrodonato 1999–2000, referenced by Panciera 2003 (p. 49 f.), with new literature on the same topographical and archaeological context.
19 On the most significant residential sites on the Via Appia (at the III and V mile), which passed to imperial domain during the middle of the second century, see Spera 2002.
20 See Granino Cecere 1995.
21 Ghini 2001.
22 D'Arms 2003. Numerous are the inscriptions of praetorians in Pompeii, recently investigated by Łoś 1995.

For Sperlonga, it's only Tacitus (*Ann.* 4.59.2) who informs us of the presence of a security service at the villa built by Tiberius: while the emperor was eating, suddenly rocks at the entrance collapsed killing some servants. Sejanus rushed to defend the emperor and *opposuit sere incidentibus, atque habitu tali repertus est a militibus qui subsidio venerant.*

23 See literature mentioned by Durry (1938, p. 45 with n. 2; p. 60; and p. 243, with n. 2). According to him, the many inscriptions of Praetorian Guards found here are to be referred to Augustus' campaign in Illyria (Suet. *Aug.* 20), as well as to the Civil War in AD 68–69.

24 In Antium emperors dealt with an extensive amount of correspondence related to administrative paperwork. Millar (2000, p. 373) cites two letters written by Traianus in the imperial villa, in November AD 99, and in a date between autumn 99 and March 101, or between 107 and 113.

25 Caius Roscius Saturninus, a praetorian of the third cohort (*CIL*, X 6673); and the veteran of the Praetorian Guards Lucius Veratius Afer, *decurio* and *quaestor* in Antium, who received a dedication by some centurions (*CIL*, X 6674). In Rome, many Praetorian Guards indicate this latial town as their place of origin. For Antium refer to Lugli 1940; Morricone Matini, Scrinari 1975; but also Solin 2000, in particular p. 639.

26 In the territory between Bovillae and Castrimoenium and at Tusculum I traced just four funerary inscriptions (Freis 1967, pp. 53 and 131: a *miles urbanus* of the twelfth cohort; *CIL*, XIV 2430: a praetorian (?) of an unknown cohort; *CIL*, XIV 2523: a centurion of the third praetorian cohort; and *CIL*, XIV 2619: a soldier of the praetorian cavalry). They are too few to advance founded hypotheses regarding the reasons of their burial in that area. It should not be forgotten, however, that Vitellius sent cohorts to Bovillae (Tac. *Hist.* 4.2).

27 Other case studies of military escort or presidium in Italy in the imperial era can be found now in Caldelli, Petraccia and Ricci 2012; and Granino and Ricci 2015.

28 According to what is attested in Plin. *Epist.* 6.31.15–17 and Rut.Nam. 1.237–250 (which also recalls the nearby 'Thermae Taurinae'). Extensive bibliography in this regard is selected by Zevi 2000, p. 509 with n. l.

29 On the Palace of Traianus, see in particular Correnti 1990. This author believed on the existence of rooms "for the quartering of the Praetorian Guards"; he imagines that the villa has been attended up to AD 170, when Marcus Aurelius prefers the residence of Lorium.

30 Ac1 and Ac2.

31 In addition to the uncle who makes a dedication to his nephew (*CIL*, XI 3537). The women are: Iulia Salo (*CIL*, XI 3538), Lacinia Italica, or Italicensis (*CIL*, XI 3539) and Nonia Colonica (*CIL*, XI 3540). The first two pose the burial dedication to their husbands (in the latter case said *Caesaris servus*); the third is buried by her spouse. In a fourth inscription (*CIL* XI 3541), a young female slave, *Polychronia*, is commemorated by her *dominus*, a Marcus Valerius Maximus, by the compatriot Titus Flavius Capito and by her friend Titus Eraulius Valens. On the frequency of the *gentilicium* Valerius among the soldiers of the fleet, see Mócsy 1968, pp. 308–311.

32 Romanelli 1961, pp. 19–23 considers so the three devoting to *Polychronia*, recalled in the previous note.

33 Ad5.

34 *CIL*, XI 3526, The text of the inscription is the following: *D(is) M(anibus). / C(aio) Caecilio V/alenti, mil(iti) cl(assis) pr(aetoriae) / Mis(enatium) ((triere)) Salami/na, milit(avit) ann(is) VIII, / vix(it) ann(is) XXXI. / C(aius) Lucilius V/[a]/ens, / corpor(is) custos, / fecit b(ene) ma(erenti).*

35 Grosso 1965, part. pp. 545–550.

36 Bellen 1981, pp. 69–71.

37 Spurza 2002 tries, on the basis of the choice of the place of construction, a distinction between real 'palaces' and 'residence-headquarters', "which usually were in an elevated position overlooking the harbor; or otherwise were placed on a promontory overlooking the harbor side" (p. 130). To the same category, according to him, belonged the complex here studied of Ostia, Nero's villa in Antium, the Traianus' palace in Centumcellae, the Tiberius' villa in Misenum and the *villa Iovis* in Capri. The reference is evidently to *villae maritimae*, on which in particular see Lafon 2001.

38 Plin. *Epist.* 6.31: *Evocatus in consilium a Caesare nostro ad Centum Cellas (hoc loco nomen).*

39 Fronto *Epist. ad M. Caesarem* 3.20 (AD 144–145): *Lectulo meo teneo. Si possim, ubi at Centumcellas ibitis, itineris idoneus esse, VII idus vos Lorii videbo deis faventibus; Epist.* 5.59 (AD 140–143): *Quis enim tibi alius dolorem genus, quem scribis nocte proxima auctum, quis alius eum suscitavit, nisi* Centumcellae, *ne me dicam?;* Scriptores Historiae Augustae, *Vita Commodi* 1.9 (AD 173): *Auspicium crudelitatis apud Centumcellas dedit anno aetatis duodecimo;* following a *balneator* and a *paedagogus* who work at the villa will be mentioned.

40 Two *optiones*, one from Misenas and the other from Ravennas, and a simple soldier of the fleet, also from Miseno (Bd1, Bd2, Bd3). On Lorium and the imperial visits, see the synthesis of Morizio 1995 who recalls the education received here by Antoninus Pius who then built a *palatium*, later frequented by Marcus Aurelius and Commodus.

41 It is not excluded that the trierarch of Alsium (Cd1), was charged with the *villa Alsiensis* mentioned in *CIL*, Xl 3720 (as a *procurator Augusti* also mentioned).

42 D'Arms 2003, pp. 76 and 114.

43 The inscriptions of the *speculator* and centurion are respectively *CIL*, X 684 = 2119 and *CIL*, X 686 = 2120. To be added to the two fleet soldiers indicated above (*CIL*, X 685 = 2122 and *CIL*, X 687 = 2121): the veteran of the fleet *CIL*, X 719 (bilingual inscription, Greek/Latin) and a twenty-two-year-old soldier from Misenum of an unpublished inscription. Both are reported by Magalhaes 2003, pp. 93 and 160–162, no. 19. She repeatedly assumes the existence of a *statio* of praetorians and *classiarii* in Sorrentum between the Augustan and Tiberian Age; so as the task of accompanying and surveillance of the emperor's personal by the centurion.

44 Already Bastianelli 1954, pp. 25–27 noted that all the fleet soldiers of the Civitavecchia's necropolis had served for multiple years and assumed that a barrack was built for military quarters, "located where probably now is the Michelangelo's fort" (p. 27).

45 On the presence of soldiers next to Domitianus at the Albanum, see *infra*, Part IV, Chapter 10.

46 Respectively De1 and De2.

47 For the *equites singulares Augusti* mentioned in inscriptions of other parts of Italy, Speidel suggests, amongst other hypothesis, the one of a military escort service during the emperor stays (Speidel 1994a p. 363, no. 676; p. 365 f., no. 680 and 681). On the Albanum, see Darwall-Smith 1994.

48 Da1.

49 Da2.

50 Ec1–Ec7.

51 Galieti 1911, p. 40 ff.

52 With Iuno Sospita, in Ec2.

53 Only one, Ec1, explicitly dated to AD 202.

54 I could see directly the inscription Ec3: the first two lines are not comparable, either paleographically (letter forms and engraving technique) or for the content, with the ones that follow. However, between the two parts of the text there does not seem to be a chronological distance of more than two centuries.

55 The fourth, the sixth, the seventh (and maybe the second) cohorts are recalled twice.

56 Scriptores Historiae Augustae *Vita Ant. Pii* 1.8: *Ipse Antoninus Pius natus est XIII k(a)l(endis) Oct(obribus) Fl(avio) Domitiano et Cornelio Dolabella cons(ulibus) in villa Lanuvina. Educatus Lorii in Aurelia, ubi postea palatium extruxit, cuius hodie reliquiae manent; Vita Comm.,* 1.2: *Ipse autem natus est aprili Lanuvium cum fratre Antonino gemino;* see also 8.5 (where the activity of the *venator* in the Circus of Lanuvium in the age of Commodus is mentioned) and 16.5 (prodigies occur in the villa in Lanuvium).

57 On the site of the residential complex, see Cassieri and Ghini 1990, with previous bibliography. "The villa, originally private, became imperial property, probably after the death of Commodus, as would seem to be proved by the presence of a barrack for vigiles in Lanuvium" (p. 172). The idea was already advanced by Galieti (1935, p. 141), who imagined a *vexillatio* would have safeguarded public policy, as at other imperial villas. What, however, confirms the frequentation of the *villa Aureliorum,* still in the early third century, is the discovery of numerous images of members of the Antonine family and the portrait of Severus Alexander. The dedication of the imperial freedman and *procurator Euphrates* can be traced back to the time of Antoninus (*CIL*, XIV 2087 =VI 246 cf. p. 3756 = *ILS* 3652, from the same area).

58 Sablayrolles 1996, pp. 51–55.

59 *Aug.* 72.5 (*ex secessibus praecipue frequentavit maritima insulasque Campaniae aut proxima urbi oppida, Lanuvium, Praeneste, Tibur, ubi etiam in porticibus Herculis templi persaepe ius dixit. Ampla et operosa praetoria gravabatur*) and 82.1 (*Itinera lectica et noctibus fere eaque lenta ac minuta faciebat, ut Praeneste vel Tibur biduo procederat*).

60 Cf. also *CIL*, XIV 2910 (dedication to Lucius Caesar) and 2911 (inscription that recalls Tiberius); the *fasti Praenestini* of Verrius Flaccus (*Inscr. It.,*ì XIII 2.17); and the prenestine altars with dedications respectively to the *divus Augustus* (Jacopi 1973, p. 17, no. 77, fig. 35 bis), to the *Securitas Augusta* (*CIL*, XIV 2899 = *ILS* 3788) and the *Pax Augusta* (*CIL*, XIV 2898 = *ILS* 3787), on which Zevi 1976 (see Part I, Chapter 3, pp. 43–44).

61 Suet. *Aug.* 49: *Neque tamen umquam plures quam tres cohortes in urbe esse passus est easque sine castris, reliquas in hiberna et aestiva circa finitima oppida dimittere assueverat* (author emphasis).

62 According to Hirschfeld 1902, pp. 69 s., who recalls the brick-stamp *CIL*, XV 2314.

63 Marcus Aurelius spent his time in *secessu Praenestino*: Scriptores Historiae Augustae, *Vita Marci* 21. 3: *Sub ipsis profectionis diebus in secessu Praenestino agens filium, nomine Verum Caesarem, execto sub aure tubere septennemn amisit.*

64 Two *fistulae* come from Lugnano, near Praeneste: one bearing the name of *Iulia Mamea, mater Aug(usti)* (*CIL*, XIV 3037 = XV 7880); the other with the names of Geta and Antoninus (*CIL*, XIV 3036 = XV 7879).

65 *CIL*, XIV 2954 = *ILS* 2684: *Sex. Iulius Sp.f. Pol(lia) Rufus, evocatus divi Augusti, praefectus I cohortis Corsorum et civitatum Barbariae.*

66 Ricci 2017 (in print). See also infra p. 240 n. 1.

67 The epitaph of *Rufus* has often been cited in relation to the administrative organisation of the Barbagia during the Roman period and to the duties of the *praefectus*.
On the career of Rufus, see in particular: Meloni 1958, part. pp. 18 s., 78, 82, 88
and 272 and Meloni 1988, part. p. 239; Saddington 1987, part. pp. 268, 270; Zucca
1987, part. p. 349; Spanu 1998, p. 65 A. 275. *Sulla carriera del personaggio*, PME I 114,
Suppl. I 1987.

68 *Eph.Ep.* IX, 891 = *AE* 1895, 124. Vd. Arnaldi 1995 = *AE*, 1995, 259. Cf. *SupplIt
Imagines* – Latium I, no. 806, now in EDR 122224 (G. Di Giacomo, 26/7/2012):
*L. Catius L.f., M.n., Fab(ia), / centurio speculator(um) / Aug(usti) sibi et Caleiae P.f.
Buculae / uxori.*

69 Fa1–Fa3 (first century AD), Fa4–Fa8; Fb1 (first or second century), Fa9–Fa10 (from
the beginning of the third century). I have excluded the inscriptions of two *evocati*
(*CIL* XIV 2959 and 2954, both of the late first or second centuries), which may be
explained by other reasons.

70 The third cohort (Fa2, Fa5, Fa7, maybe also Fa6, and the tribune Fa10); the fourth
cohort (Fa1); the eighth cohort (Fa3), the ninth cohort (Fa4); the tenth cohort
(Fa9).

71 Patsch 1893.

72 *Per idem tempus gladiatores apud oppidum Praeneste temptata eruptione praesidio militis,
qui custos adesset, coerciti sunt.*

73 On the so-called 'Tondo' of Zagarolo, an amphitheater or an arena adjoining the
imperial *ludus*, see Fora 1996, pp. 10 and 90, no. 49, with previous literature.

74 Mentioned in some of the inscriptions collected in the dossier (Fa1, Fa2, Fa3, Fa4),
aside from the one of the *miles urbanus* Fb1.

75 The inscription Fa5 is clearly dated AD 167.

76 With whom Durry 1938, pp. 60 and 279 agrees. From Praeneste is perhaps *Lucius
Geganius Victorinus* mentioned in the text Fa4.

77 To the inscriptions known by Patsch from *CIL*, XIV, are to be added other three
unpublished of the first and second centuries (Fa6–Fa8).

78 Fa1 and Fa2: Muzzioli 1970, p. 57 f. speaks of them as coming from the "Casale
Rodi, on the hill east of the road that branches off from the main Via towards the
north-west . . . [where] some archaeological pieces of which I have not been able
to determine the origin are kept. It is very likely, however, that they belonged to an
existing villa in this location."

79 It is the soldier Fa5. On the cult of Iuppiter Optimus Maximus at Praeneste, see
Granino Cecere 1989, p. 150 f. and no. 16.

80 Not related to the service at the villa was most likely Lucius Graccius Constans,
tribune of the first praetorian cohort who, at the beginning of the third century, buried his wife Numitoria Moschis (*CIL*, XIV 3628 = *Inscr.It.* IV, no. 158).
Numitoria had first married Lucius Cominius Maximus, *primipilus bis*, then *procurator* during the reign of Antoninus Pius, prefect of the Legio II Traiana, tribune of
the urban cohorts (*CIL*, XIV 157 = *ILS* 2742 = *Inscr.It.* IV no. 157; Pflaum 1960,
no. 189; Dobson 1978 p. 265 s., no. 149). Cominius Maximus was buried in a villa
he owned north of Tibur (Mari 1991, p. 84, no. 36); in the same place is buried his
wife. Cf. also Granino Cecere and Ricci 2006.

81 Respectively, Ga1 and Gb1. The dedication to the first from an *amicus*, a local priest,
seems to confirm the hypothesis.

82 Ge1.

83 Speidel 1994a, p. 363, no. 674, supposes that the dedicant is a *secundus heres* because the *primus* remained in Rome.

84 Gd1, Gd2 and Gd4.

85 Konen 2001, no. 127.

86 Gd3.

87 After having discharged most of the troops of Vitellius *classiarios uero, qui ab Ostia et Puteolis Romam pedibus per uices commeant, petentes constitui aliquid sibi calciarii nomine, quasi parum esset sine responso abegisse, iussit posthac excalciatos cursitare; et ex eo íta cursitant (Vesp. 8.5).*

88 Reddé 1986, pp. 148, 204, 447. Already Romanelli 1961, p. 22 and Starr 1941, p. 18 and 177 had mentioned the warships carrying the emperor's orders in the entire empire.

89 Crogiez 2002 is not convinced of the use of warships and soldiers of the fleet for the transmission of official information. Of the same opinion, in the same volume, is Bérenger-Badel 2002, p. 227 f.

90 Romanelli (1961, p. 22) mentions the service of couriers probably carried out by soldiers of the fleet from Centumcellae (and Alsium) to Rome; when referring to the attestations of Lorium, he suggests that they had broader functions. Eck (1996a, pp. 343 f.), rather than considering the *classiarii* as the 'postmen by sea', suggests that the *frumentarii* and, maybe, the *speculatores*, were responsible for such task.

91 On the 'water architecture' at Villa Adriana, see Manderscheid 2000 and 2002.

92 Leppert 1974, pp. 33–55 dedicates a section of his investigation to the villas as places for the exercise of the law by the emperor, citing several famous cases, such as the porticoes of the Temple of Hercules at Tibur (Augustus, in Suet. *Aug.* 72.2; and Claudius, in Sen. *Apoc.* 6.4 and Suet. *Cla.* 34); the *praetorium* of Baiae, where the edict of Claudius on the right of citizenship to the Anauni AD 46 is prepared (*CIL* V 5050); and the Albanum, where Domitianus decides a dispute between *Firmani* and *Falerienses* (Suet. *Dom.* 8; *CIL*, IX 5420, *Falerio*).

93 Eck 1996a.

94 Durry 1938, pp. 59, 276, 278 f. (the literary attestations mainly refer to the Julio-Claudian age).

95 Salza Prina Ricotti 1978–1980, pp. 287–294; 1982, pp. 37 f.; 1992–1993; 2001, pp. 157–161.

96 Halfmann 1986, p. 110. Helmut Halfmann's view must be further clarified and corrected in part: it is perfectly possible that the praetorians (and, in particular, their cavalry) have constituted the bulk of the urban contingent in military campaigns: see E.g. Tac. *Ann.* 1.24.1–2 (*equites* following Drusus in the military expedition to Pannonia). Also see, in my catalogue, Fa5, from Praeneste.

97 Albanum: praetorians and *equites singulares Augusti*, in the second and third centuries; Tibur: soldiers of the italic fleets, praetorians, *milites urbani* and *equites singulares Augusti*, in the second and third centuries; Lanuvium: *vigiles*, in the third century; Praeneste: praetorians and *milites urbani*, between the first and third centuries; Centumcellae and Lorium: *vigiles* and *classiarii*, between the second and third centuries.

98 Sirago 1983–1984, p. 94 ff.

99 Economic factors appear to have been meaningful in the decision to acquire or not a *comitatus*. See Suet. *Tib.* 46: *Pecuniae parcus ac tenax, comites peregrinationum expeditionumque numquam salario, cibariis tantum sustentavit; una modo liberalitate ex indulgentia vitrici prosecutus, cum tribus classibus factis pro dignitate cuiusque, primae sescenta sestertia, secundae quadringenta distribuit, ducenta tertiae, quam non amicorum sed Graecorum appellabat.*

100 Always Sirago 1983–1984, pp. 108–110. On the privileges granted to the *classiarii* in Antonine age, see Eck 2001. Officers and graduated soldiers of the fleet are co-conspirators in an attempt to sink Agrippina's ship at Bauli, by order of Nero (sources and account in D'Arms 2003, p. 95 f.).

101 For the involvement of the fleet soldiers in the repression of riots and prevention against banditry, refer to what has been said *supra* (Part IV Introduction, p. 192 ff.).

102 Tac. *Ann.* 1.7.3: *defuncto Augustus signum praetoriis cohortibus ut imperator dederat; excubiae, arma, cetera aulae; miles in forum, miles in curiam comitabatur.*

103 M. Aur. *Ta eis heautón* 1.17 (*infra*, in the text, translated). Both texts are mentioned in Pani 2003.

10 Emperors on the move

Security in the Campanian cities and in the *Albanum Domitiani* (first century AD)[1]

The emperors would leave Rome for various reasons. Some had greater mobility and travelling was a way to do propaganda and to reveal their ecumenical program: their journeys, besides being frequent, were often directed towards the provinces, and some in particular. The most well known case is that of Hadrianus and his inclination for the eastern area of the empire as for Egypt.

When moving away from Rome, regardless of the chosen destination, a similar device and a scrupulous ceremonial was set in motion. Whether emperors moved towards the most distant provinces, or their destination was one of Italy's cities,[2] the '*Reisebegleitung*' of the emperor was prepared, by the *Princeps* himself together with his collaborators, with the utmost care and prudence: with a retinue variously composed of members of the imperial family, *comites*, officials and service personnel, designed and adapted to the specific needs of the moment in which the journey was undertaken and of the place of destination. Some variations could intervene for what François Chausson[3] defines as "a lighter infrastructure", with short stop overs along the journey and a less obsessive planning.

The presence of an escort service was fundamental both during the process of moving and once the destination was reached, to ensure the best conditions for the emperor and his entourage to work and rest. It is also not excluded, that sometimes, in the most popular and beloved villas, even when the emperor was not present, next to civilian personnel,[4] a military garrison remained there.

The organization of security during imperial travels has received scant attention by scholars. Among those authors who have studied the urban troops, Marcel Durry[5] is the one who most frequently recalls episodes in which the praetorians are alongside the emperor outside the city walls. In the volume of 1977 devoted to the emperor's role, and his collaborators and interlocutors, Fergus Millar deals with the emperor's travels in Italy in two different chapters[6]: in the first, there is no references to security services; while in the second there is an excursus on the different forces that protected the emperor in Rome and during his journeys. In Helmut Halfmann's work, a few pages and some examples are devoted to the issue.[7]

The services of imperial security receive more space in an article written by Alexandra W. Busch in 2012, which investigates the particular aspect of the relation between the imperial architecture and 'prophylactic' measures. Following on the current research on imperial security, in particular during the second century,[8] Busch reviews the palaces of Rome and the major suburban villas, trying to prove that, in the city as in Italy, unlike what happens in the rest of the empire, the emperor's security is provided by men rather than by the structure of the buildings. In the first three centuries of the imperial age, in her view, the soldiers were entrusted with the delicate task of protecting the *Princeps*; in some cases, freedmen were invested with this function.[9]

The reason for the lack of attention to this topic is certainly due to the few explicit attestations, but also to the *communis opinio* about *who* was charged with this task during the emperor's travels: if one is asked whose was the responsibility to oversee and ensure the emperor's safety, the instinctive response would be that, for this purpose, Augustus had created in AD 6 the Praetorian Guard. If this response is well founded,[10] it is, however, also true that, expressed in such a way, it risks being reductive and not taking into consideration, for Rome itself and especially for travels outside the city, a couple of issues that should not be neglected.

First, the birth of the praetorian cohorts is part of a larger plan designed and partially implemented by Augustus himself for the security and the control of Rome, which envisaged more participants and more intervention levels. Next to the praetorian cohorts, other military and paramilitary corps co-existed and had an active role in the city, with diverse duties, partly intertwined, in the implementation of the dispositives for the safety of the imperial family.

Just to give a few examples in relation to the men involved, in occasion of events that the emperor was attending and for which a massive participation of the urban population was expected,[11] in addition to the praetorians, also *milites urbani* and *vigiles* came into action in order to prevent fires and ensure the protection of public order in the streets and near buildings. If the emperor went to a place where water could be dangerous, it was appropriate to ensure the presence of men who would know how to deal with this danger; and were able to swim and to row, such as the *classiarii* of Misenum and Ravenna, already present in the city with other duties.

The *milites cohortium urbanarum*, in the first century, do not seem to leave Rome[12]; and Gaius, when building the mighty bridge of boats between Baiae and Puteoli, is followed by praetorians.[13] When Claudius decided for a naval battle on Fucino, the praetorians were once again responsible for the security service around the lake area, while the *classiarii* were involved in the battle.[14]

In Rome, Agrippina, first wife and then mother of an emperor, had the right to be escorted by the Praetorian Guard and by the *Germani*.[15] Probably after being deprived of the right to have a personal escort, the daughter of Germanicus was accompanied by soldiers of the fleet, travelling on a trireme from Antium to Bauli, where her son was waiting for her with criminal intent.[16]

So we can legitimately imagine that, in Rome as elsewhere, the Praetorian Guard and their officers, in order to perform the task assigned to them in the best possible conditions, acted in close and functional collaboration with other military or paramilitary corps. All of them had been created and empowered to act for what might be called 'the security plan' for the citizens and the emperor.

A second aspect that should never be disregarded and that in our case is of particular importance is the diachrony: the Augustan plan for maintaining public order and security in the city, outlined here in its essential aspects, was not implemented in its entirety by Augustus, but rather developed and perfected by his successors. This is certainly not the time and place to deal extensively with this issue. I can simply recall a detail of no small significance: the praetorians, as a distinct body from the urban cohorts, their camp, the buildings and all the facilities that were part of it, and most likely even the first structure of the command hierarchy, is an accomplishment of the Tiberian age and an idea of the praetorian prefect Sejanus.

Moreover, if the Praetorian Guard, the urban soldiers and the *vigiles* recur all along the imperial history of Rome until the deep reforms of Constantine, other figures, which have to deal with the *Princeps* and his security, have shorter life or represent 'adjustments' to the original plan.

This long introduction, before moving on to some examples, illustrates the ambiguity of the information we have and the difficulties of interpretation that it raises. In Italy, in the period here considered, documentation is more straightforward: we will find mostly praetorians (and *speculatores*), with a convergence of data among different sources. It will be only in the second century that the emperors, in order to guarantee their own safety, adopt different solutions related to non-constant variables; not least among them was the reputation for reliability that some corps had gained at the expense of others.

It is widely known that some of the imperial residences in central Italy have been identified: in most cases, historians' accounts have been reflected in the archaeological remains and in the richness of their furnishings. Inscriptions are an equally valuable indicator: in particular, those on *fistulae* with the names of the recipients of the concessions; and the epitaphs of freedmen and slaves employed, with the most varied duties, in the management of properties.[17]

In order to find evidence of the presence of a security escort, I believe that the most appropriate path to follow is the one that integrates the different reliable information: so, in addition to the historical accounts, the archaeological remains, the 'military' portraits and reliefs, as well as epitaphs and inscriptions of soldiers who were in the places where the emperors spent their time.[18]

The latter are tools that must be considered very carefully, and undergo rigorous verification: the soldiers of Rome, the same who accompanied the emperor during his travels and that belonged to special corps, were recruited in Italy and mostly in the central and northern regions. An isolated epitaph, especially if the imperial visit was sporadic, is not in itself sufficient evidence of the prolonged presence of the soldier in a specific place, which instead could correspond to where his family came from or to a property. Deserving of attention are therefore those documents found in non-episodic form and that find a more reasonable explanation in connection with the emperor's presence.

In some famous villas, such as those of Tiberius at Tusculum[19] and at Speluncae,[20] the Antianum of Gaius and Nero[21] and the Sublaqueum of Nero,[22] according to the criteria that I have just outlined, there is no documentation – or actually unsubstantial – regarding security services. The residential complex of Hadrian in Tibur, of great interest even because preceded in the first century by a *praetorium* – for chronological reasons falls outside the framework here considered and it will be treated elsewhere.[23] Proceeding roughly in chronological order, on this occasion I will focus on the travels or on the visits of the Julio-Claudians, and their soldiers, in some towns along the Campanian coast[24]; and on those of Domitianus in his beloved residence in Albanum.

Augustus and the Julio-Claudians between Capri and the Campanian coast

Many residences, renowned for their dimension and magnificence, were scattered along the Campanian coast, so rich in archaeological remains. Not a few locations on the Bay of Naples enjoyed considerable importance in being greatly appreciated by many of the most influential senators and the emperors themselves, who went there to enjoy the hospitality of generous hosts.[25] Only the imperial villa at Baia – a special case in the multifaceted landscape of the Campanian coast – was attended from the Julio-Claudians up to the Severi.[26]

Despite frequent imperial travels in this area, the absence of sources[27] and the conditions of the archaeological findings are not of great help in deducing how the security service was organized. I will confine myself to a few, but very different, examples; as we shall see, what is obvious in Capri and Surrentum is the contrast between certainty, frequency and duration of the emperor's visits and a documentary deficiency in relation to the security measures adopted; while both in Pompeii that Puteoli, different types of attestations offer a more detailed picture of the military presence.

It is known that the emperor's attachment to Capri dates back to Augustus, who considered it among his favourite destinations in Italy.[28] Since 29 BC, the island seduced the emperor who wanted to acquire it from the Neapolitans, to whom he gave Ischia as a compensation. Since then, Augustus considered it his personal property: he planned how to better highlight its beauty, even through imposing or eccentric items of furnishing[29]; there, he was surrounded by young Moors and Syrians, one of which, Masgabas, his favourite, was buried in Capri. Before dying, Augustus even spent four days of rest and enjoyment on the island.[30]

Tiberius stayed in Capri for a long time,[31] and he carefully planned the conditions of his permanence. Tacitus supposed that among the reasons for choosing Capri there was the fact that no one could land there without being seen by the guards (*nisi gnaro custode*).[32]

It is certain, thanks to several clues, that the emperor was accompanied by the praetorians: Sejanus often came here, and a passage in Suetonius refers to a praetorian punished for stealing a peacock in the gardens.[33] Examining the archaeological remains, Clemens Krause, relying on evidence not totally decisive, speculates that, inside the so-called *Pharus*, the space between the corridors and the external wall could serve "for the staff at the lighthouse and perhaps as a camp for the Praetorian Guard".[34]

The soldiers in Capri accompanied the emperor even during his periodic departures from the island: during one of these journeys, Suetonius again reports of an incident: when his litter snagged on brambles, Tiberius clubbed to death a centurion of the first cohort, guilty for having failed in his duty to explore the terrain.[35]

The inscriptions found on the island do not lack information regarding the staff at the service of the emperor: Mommsen himself was impressed by the large number of epitaphs that recalled the names and often the duties of imperial slaves and freedmen.[36] If these inscriptions do not represent but a relative percentage of the number of slaves and freedmen who were actually charged with the Villa of Augustus and Tiberius, one is nevertheless struck by the almost complete absence in the rich dossier of soldiers' epitaphs.

A lacunous epitaph merely gives the number of years of life and service of a soldier of an unknown corps.[37] Far from being certain, whether he was a soldier of the fleet or not: the proximity of the Misenum *statio* certainly does not allow us to exclude it. It is also difficult to specify the date of the epitaph: supposing that the villa, after the successful exploits of the Julio-Claudian age, experienced a period of relative eclipse, it is nevertheless also true that it continued to be popular until the end of the second century, when Lucilla, Marcus Aurelius' daughter, was exiled to the island.[38]

The existence of an imperial villa in Surrentum, beyond the possible identification with the remains of the complex overlooking the sea in the city center, is likely to be made both by Suetonius' account that Augustus forced Agrippa Posthumus to reside there between 7 and 5 BC[39]; and, as in Capri, by the large number of inscriptions of Augustan *liberti*, all of them not prior to the time of Claudius.[40]

The inscriptions of the staff employed in the villa tell us of the existence of a civil guard service.[41] None of the six epitaphs of soldiers discovered in the area refers instead to a security service; or to the first century. Despite these difficulties, Martins Magalhaes does not exclude the existence of what she calls a military *statio* in Surrentum because, in her own words, "an imperial property justified a surveillance".[42]

The relatively meagre dossier that has just been illustrated exemplifies, in my opinion, the difficulties raised by the attestations of soldiers in places like *Capreae* or *Surrentum*, where even the presence of the emperor and his family was not episodic. Scant and questionable is the actual evidence of the soldiers' passage; and, for *Surrentum*, there is no literary evidence, whereas the archaeological remains are not very eloquent.

More evocative are the signals that reach us from two other well-known centres of Campania, Pompeii and Puteoli, where the presence of the soldiers can be, even symbolically, related to the emperor's safety.

In Pompeii approximately fifteen between funerary inscriptions (partly collected in the necropolis of Nola) and wall graffiti of praetorians or *speculatores* of the Julio-Claudian period were unearthed.[43] De Caro wonders what may be the reason for their presence in Pompeii: the heterogeneous nature of the documentation is an indication of the following passages of soldiers over a period of time of more than fifty years; and of soldiers selected ad hoc from different cohorts.[44]

Andrzej Łoś' hypothesis to connect at least some of these soldiers to the well-known events of AD 59 – when the famous riot between Pompeians and Nucerians broke out in the Campanian town – deserves attention.[45] To restore order, Łoś assumes – but Tacitus does not tell us – that the senate sent soldiers of the urban garrison. They would have stayed here and, a few years later, in AD 62, would have helped the population to compensate for damage caused by the earthquake. Since it is certain that the intervention of the praetorians in Pompeii was not limited to a single episode, I don't see why the epitaphs of soldiers cannot be connected to other interventions and on other occasions, such as imperial journeys or special missions in a period of time that extends to include the first century as a whole.

The year before (AD 58) the Pompeian riot, in Puteoli strong tensions between *plebs* and local *ordo* had broken out, followed by clashes. To allow the delegates of the senate of Rome to act and to quell the unrest, Nero sent the Praetorian Guard.[46] In this era,[47] Puteoli was still the main port of the Tyrrhenian: the need for an intervention of the central power to restore order is quite clear, even if limited to the episode.

The clues of the praetorians's presence in Puteoli is not limited to the Julio-Claudian era. There are in fact the impressive remains of an honorary arch of the Trajan era, built to celebrate the construction of the road connecting Puteoli with Neapolis. The arch frieze was made with blocks of marble pushed together and carved as a niche, where praetorians in civilian clothes were represented in high relief; a bas-relief, only partially preserved, also portrayed the Praetorian Guards, this time wearing their full dress uniform, escorting the princeps. One of the high-relief portraits in a niche was made reusing the back of a base in honour of Domitian, erected by the plebs and by the senate of Puteoli as a sign of gratitude for the building of the via Domitia.[48]

The splendid remains of the Trajan arch, now partly preserved in the Pergamon Museum in Berlin and partly in Philadelphia Museum,[49] are of no small interest to our subject: they also provide incontrovertible evidence of the role that, in the aftermath of the elimination of Domitian, the Praetorian Guard continued to play in the representation of imperial power. As the guarantors of the security of the Princeps, the praetorians are beside him on public monuments that were intended to have an impressive impact.

We can conclude that the ancient texts – written texts or texts of images – do not always remain silent on the presence of soldiers, at least in Campania,[50] where they are present, with or without the emperor, in each case for his precise desire and as an expression of his authority. The episode of the bridge of boats built by the Praetorian Guard to connect Baia and Pozzuoli has been mentioned. The soldiers themselves, for the people of the towns in Campania, are from time to time protagonists of entertainment, agents of repression or, as in the reliefs of Berlin and Philadelphia, guarantors of order in its positive aspect.

Domitian and his *Albanum*

The last point of our excursus regards the *Albanum Domitiani*. Cassius Dio[51] mentions Vespasianus' predilection for his birthplace of *Aquae Cutiliae*, where he used to stay during the summer; where he would receive the ambassadors[52]; and where he returned, when suffering from a disease that caused his death in AD 79.[53]

His younger son Domitian showed a completely different attitude by storing away his childhood memories, and by repressing, without too much difficulty, the nostalgia for his Sabine origins, he gave clear signs of the will to leave a powerful imprint of his regime even through residential constructions. With this in mind, as well as proceeding to the recovery and the new layout of the preexisting complex of the Julio-Claudian dynasty,[54] Domitian took possession of the residence of Albanum that had already been of Pompeius first and then of Caesar.

Of the three houses he frequented,[55] *Albanum* can be considered his favourite and the one that saw him committed almost like the Domus on the Palatine.[56] Here the presence of military forces engaged in the emperor's protection is quite clear[57]: where the epigraphic signs show no evidence of the composition and the organization of the security service,[58] literature and, above all, archeology aid us, with buildings and images.

The villa of Albanum has never been subjected to systematic excavations and only two areas are reconstructed: the cryptoporticus and the theatre. While certainly a security service protected the emperor's safety when attending the *Ludi Minervae*,[59] no trace of this service comes from the theatre.

A very suggestive clue comes from the great cryptoporticus that bordered the main terrace of the complex and constituted a sort of vestibule to the palace. In this passage area connecting the outside with the inside, according to Henner von Hesberg's reconstruction, was placed the security personnel, in two large niches at the sides of the vestibule.[60] On the walls of this sort of waiting room, where the people waited to be received while enjoying the view of the garden, was painted a series of standing figures, most likely staff at the service of the emperor, although it is unclear whether they were civil or military personalities (see Figure 10.1). The artist's intention was to present a kind of tableau vivant to visitors, creating the effect of a crowd following Domitianus.

The suggested message (and largely also the adopted language) is clear and very similar to that which not many years later will be resumed in the relief panels of the arch of Traianus in Puteoli. It's a recurring message in public

Figure 10.1 Standing figure (maybe a soldier) on the wall of the Cryptoporticus
　　　　of the Albanum Domitiani

Source: photos von Hesberg for the DAIRom, S5593-06 and S5593-10.

iconography of the soldiers of Rome. Just to mention a couple of examples:
the Claudian reliefs of the Louvre and the Flavian panels of the Chancellery
building.[61] Even in the cryptoporticus in Albanum, the procession of people,
all at the same height of the visitor, looks like leading him to the emperor and
gives the effect of a real presence to those who passed by as a sort of enhancement of the service of the guards in the entrance niches.

Another portait of a soldier on the wall of a private room of the Albanum,
completes the suggestive picture of the bodyguards around the youngest of
the Flavians (see Figure 10.2): it is an historical relief, now fragmented, representing Domitianus as a *sacerdos pius* in the act of sacrificing; next to him,
with an eye towards the viewer, an armed soldier, perhaps an officer.[62] It is
impossible to tell whether the intention of the artist was to allude to bodyguards; or rather give a symbolic message; or even a mythical reference. It is

Figure 10.2 Historical relief recovered from the Albanum Domitiani: the prince
is represented as sacrificing; next to him a soldier, or rather a
mythological figure

Source: reconstructive drawing by H. von Hesberg, *E cornu taurum. Zu fragment von Staatsrelief aus dem Albanum Domitians*, in *Rome et ses provinces. Hommages Balty*, Le Livre Timperman: Bruxelles, 2001, p. 248.

certain, however – even if the part of the relief that interests us most has been lost – that in the soldier's picture we cannot distinguish elements connotating a specific military corps.

A research extended to other parts of Italy, where the emperors had their residences, could enrich the framework, outlined so far, of the duties performed by bodyguards and escort soldiers, and their coordination with other members of the emperors' retinue; or of the service at the residences in the first century of the imperial age.

If it is not the case to draw general conclusions from such heterogeneous materials and disparate sources, it is, however, appropriate to argue about what can be acquired from the excursus here conducted.

First, under the profile of the troops in charge of security. In the first century, also due to the absence of other soldiers, *speculatores*, the Praetorian Guard and, in special cases, the *milites classiarii* and the *statores* and their officers, were responsible for the emperor's protection. Different sources, with their peculiar language, broadly agree in outlining the function, which they constantly maintained despite the turbulent events in which they were involved at the end of the Julio-Claudian era. As Harriett Flower rightly observes, the Pozzuoli reliefs represent a non-verbal response to the query on the role played by the Praetorian Guard in the aftermath of Domitianus: instead of seeing their duties and public appearances reduced, they continued to carry out their role of loyal attendants of the new emperor.

The epigraphic, archaeological, historical sources are not significant enough to allow us to determine the amount of military forces and/or the duration of the service, in case of prolonged imperial stays at the residences in Italy. If we combine the almost absolute silence of the attestations of Capri; and the fact that the soldiers in charge of the security service (for example, at Praeneste) died young, we may suppose '*vexillationes*' of soldiers drawn from different cohorts (Puteoli, Praeneste); and a short duration of the detached service, with frequent alternations (Capri and, perhaps, Surrentum).

The isolated epigraphic attestation of a soldier in Capri cannot be considered eloquent. We would have expected more, given the duration of the retreat of Tiberius and the due importance that it has in the historical accounts. However, the very nature of the island, beloved by the emperor for its *solitudo* and the scarcity of landings; the presence of a large group of slaves and freedmen, including certainly some in charge of guarding and security; along with the frequent visits of Seianus and his men did not make essential, most likely, the need of a substantial or permanent military garrison.

Much different is the picture depicted by the *secessus Domitiani* and, I would say more generally, by all the attestations on the security escort services provided to the last of Flavians. The design itself of the architectonic structures, the reliefs, the decoration of the interiors, the historians' accounts, all variously provide clear evidence of the presence of the Praetorian Guard at the side of the emperor in the splendid residence that Domitian transformed in a "little Rome".[63]

The frequency with which the images of soldiers recur can be considered both representative of the bond between the soldiers and the emperor; and of the will of Domitian to emphasize, among the people, the idea of his own safety in the city as in the beloved *Suburbanum*. With similar and greater force than with literature, the reliefs and the frescoes at Albanum depict praetorians as diligent guardians of the most important residences[64] and faithful companions of the *Princeps* at official events in the city, and at his arrival or departure for a successful campaign.

In AD 84, Domitian prepared to leave with the Praetorian Guard to quell the rebellion on the Rhine led by Saturninus[65]; and then, subsequently, for the Danubian campaigns. That the emperor's rapport with the praetorians was privileged[66] is also confirmed by other evidence: the addition of a tenth praetorian cohort to the nine restored by his father[67]; the prudence with which Domitianus chose his praetorian prefects among the men considered to be the best and most faithful, differing in this way from both his father's and his brother's conduct. The young emperor entrusted the prefects with duties of great prestige and strategic importance, as in the case of Cornelius Fuscus, placed in command of the troops during the Dacian Wars.[68]

Our sources seem in fact to confirm that the Praetorian Guard acted as escort and security personnel of the Julio-Claudians in the imperial residences in Italy as in Rome. Even when the Princeps was in Rome, they moved episodically to carry out duties of restoration of an altered order in the towns of imperial Italy. With Domitian this function reached what we might call a representative overexposure.

It is certain that things, a little later, were bound to change: in the second century, the emperors had very different attitudes and took away from the Praetorian Guard the exclusiveness they had in protecting the Princeps. The tumultuous events and the ambiguous role during the conspiracies that had seen the deposition of some of the Julio-Claudians, which Vespasian first and then Domitianus tried to make the people forget, carried too much weight in the public's memory, and in that of the successors.

In their travels in Italy and to their villas, where they concentrated many aspects of their government, the second century Roman emperors opt for a security service made up of mixed composition, at least in part permanently stationed, as is made evident by the apparatus set up at Centumcellae,[69] at Tibur,[70] at Lanuvium.[71] Not a few of the inscriptions of *equites singulares Augusti*, scattered around Italy,[72] prove the hypothesis: in addition to policing missions, these soldiers gradually depose the Praetorian Guard – who rather intensified their operational role in the main theaters of conflict – from their exclusiveness of protecting the emperor.

Notes

1 [N.d.A.] The contribution here presented, originally written in French (Ricci 2017, in print) has been significantly modified: most of the introduction, which contains the framework of military and paramilitary forces at the disposal of the emperors (see here Part II), has been expunged; so has the whole section dedicated to Praeneste, in order not to create an overlap with Part IV, Chapter 9.

2 In research there is a clear imbalance (in favour of the first) between the attention given by scholars to imperial travels in the provinces, and to those in Italy; as underlined for example by Spurza 2002, p. 124, on the basis of Barnes 1989 (review of Halfmann's book).

3 Chausson 2012, p. 24.

4 Civilian personnel were used as guards, such as the slave *circitor* who, in the early decades of the first century commemorates his son in an epitaph from *Surrentum* (see, *infra*, n. 41).

5 Durry 1938, p. 276 f., 387 n. 2 and *passim*.

6 Millar 1977, part. chap. II.3 (The Emperor in Italy); and III.2 (Military escorts and bodyguards).

7 Halfmann 1986, pp. 103–104.

8 See, e.g., the recent work by Boatwright 2010.

9 Busch 2012; resumed, without any substantial novelty, in Busch 2013.

10 Supported by the words of Suet. *Aug.* 49 (*partim in sui custodia adlegit*), referring to the tasks assigned to the soldiers in the capital.

11 Suetonius, speaking about the frequency and the magnificence of the spectacles curated by Augustus, perhaps due to the heterogeneity of the military personnel involved, makes general reference to 'guards': *Quibus diebus custodes in urbe disposuit, ne raritate remanentium grassatoribus obnoxia esset* (Suet. *Aug.* 43). The soldiers were assigned separate seats at the theatre, maybe due to their role to protect the emperor (Suet. *Aug.* 44: [The Princeps]*militem secrevit a populo*).

12 On the possible stationing of an urban cohort at Ostia, starting from the reign of Claudius (Suet. *Cla.* 25.2), see Bérard 1988, part. pp. 176 nn. 77 and 181; Bérard 1991, part. p. 17; lastly, Caldelli in Caldelli, Petraccia and Ricci 2012, part. pp. 291–295.

13 Suet. *Cal.* 19: *comitante praetorianorum agmine*; and Cass. Dio 59.17.6. Vd. Kleijwegt 1994.

14 Tac. *Ann.* 12.56: *Claudius triremes et quadriremes et undeviginti hominum milia armavit, cincto ratibus ambitu, ne vaga effugia forent . . . in ratibus praetoriarum cohortium manipuli turmaeque adstiterant.*

15 Tac. *Ann.* 13.18.3: *Cognitum id Neroni, excubiasque militares, quae ut coniugi imperatoris oli[m], tum ut matri servabantur, et Germanos nuper eundem [in] honorem custodes additos degredi iubet.*

16 Tac. *Ann.* 14.4.3–6: *Quippe [Agrippina] sueverat trireme et classiariorum remigio vehi.* The role of *Anicetus* in the assassination of Agrippina (Suet. *Nero* 35.2; and, before, 20.2–3; Tac. *Ann.* 14.62.2) "illustrates the power that the soldiers of the fleet were gaining at the expense of the praetorians", according to D'Arms 2003, p. 99.

17 For the *familia Caesaris* as "indizio indiretto più forte della presenza di una villa imperiale"; and, more generally, on the variety of sources that attest to the existence of imperial properties, see Maiuro 2012, p. 164.

18 Some centres of Latium have been examined: Ricci 2004 (in this volume Part IV, Chapter 9); Granino Cecere and Ricci (2004 and 2015). Many of the attestations, however, belong to a later period.

19 Valenti 2008, with literary, archaeological and epigraphic documentation.

20 Where, during a dinner in the famous grotto, an accident occurred (Suet. *Tib.* 39 and Tac. *Ann.* 4.59.3). It would have caused the emperor's death if Sejanus did not make a shield with his own body to protect him, as the *milites* who promptly rushed to the scene could see. On the facilities of the villa, see Cassieri 2008, with bibliography and photographic evidence; on the extension of the *praedium*, refer to Quilici 2009, part. pp. 201–322.

21 On the structures of the villas in Antium, see Jaja 2008. For the sources related to the imperial stays and the *familia Caesaris*, see also Maiuro 2012, p. 266 f., with other literature. Millar (2000, p. 373) quotes two letters written by Traianus at the imperial villa of Antium, in November 99 and between 99 and 101 (or between 107 and 113). The inscription *CIL*, X 6667 recalls a Titus Flavius Aug(usti) lib(ertus) Evangelus, who was *tabularius praetorii Antiatini.* Yet from Antium there is an attestation of a twenty-one-year-old praetorian still in service, in a family tomb (*CIL*, X 6673). More interesting is the attestation of an *evocatus Augusti* who, still in the first or at the beginning of the Second century, makes a dedication to a *frater* (*CIL*, X 8294). For the rest, in agreement with what has been reported from literature, there are epitaphs of veterans (*CIL*, X 6669, 6671, 6672, 6674).

22 Mari 2008. In general, on the imperial properties along the Tyrrhenian coast, see Maiuro 2012.

23 Ricci 2004 (in this volume Part III, Chapter 9, pp. 213–215), part. pp. 329–331. Of great interest is the analysis of the graffiti of Villa Adriana (some of which certainly drawn by soldiers) conducted by Molle 2012.

24 Before starting, I would first like to spare a thought on the excellent work that John D'Arms has done for many years with meticulous care, insight and passion to rebuild the frequentation of the Roman villas in Campania. I have strongly felt the lack of such a work and such an insight for the villas of *Latium vetus*.

25 An importance that declined with the advent of the Flavian dynasty, and then rose again as the *otium Albani lacus Baianique torpor* of Domitianus. All this, as known, has been the subject of careful research by John D'Arms: I will just mention, besides the aforementioned D'Arms 2003, the previous D'Arms 1975 and 1977.

26 Millar 1977, pp. 26–27. In particular refer to Borriello and D'Ambrosio 1979 and Maniscalco 1997. From Baia are Not. Sc. 1887, p. 83 (*EphEp* VIII 385, Baiae = EDR 123282 (A. Parma, 15-11-2012): [D(is)] M(anibus). / [Iu?]liae [S]atur/[n]inae, quae vi/ [xi]t ann(is) p(lus) m(inus) XVIII, / mens(ibus) VIIII, d(iebus) IIII, Iu/lius Atenororus, / miles, coiugi be/nemerenti fecit.

The villa of *Vedius Pollio* in Posillipo remains an imperial property at least until the death of Hadrianus, administered by an imperial freedman of procuratorial rank (De Caro and Vecchio 1994).

27 Except the passage of Tacitus (*Ann.* 15.52.1, AD 64): Nero goes to Piso's villa, undisturbed, *omissis excubiis*.

28 Suetonius (*Aug.* 72.2–3) talks about his preference for locations by the sea; and in particular for the islands of Campania. Cass. Dio 52.43.2 (Capri) and 72.12.6; Suet. *Tib.* 40.

29 Strabo 5.4.9 speaks of Capri conquered by the people of Neapolis; and of Augustus who established his residence there; Suet. *Aug.* 72.3: *sua vero quamvis modica non tam statuarum tabularumque pictarum ornatu quam xystis et nemoribus excoluit rebusque vetustate ac raritate notabilibus, qualia sunt Capreis immanium beluarum ferarumque membra praegrandia, quae dicuntur gigantium ossa et arma heroum.*

30 Suet. *Aug.* 98.

31 Neri 1990 reflects on the reasons for Tiberius's choice (on the basis of the famous passage of Tac. *Ann.* 4.57.1 and of Suet. *Tib.* 40 and ff.): in his view, the choice did not result in a crisis of the concept of Rome as *domicilium imperii*.

32 Tac. *Ann.* 4.67.

33 Suet. *Tib.* 60: *Militem praetorianum ob subreptum e viridario pavonem capite puniit.* On the praetorians in Capri, see the reflections by Bingham 1997, pp. 51–53.

34 Krause 2003, p. 258.

35 Suet. *Tib.* 60: *In quodam itinere lectica, qua vehebatur, vepribus impedita, exploratorem viae, primarum cohortium centurionem, stratum humi paene ad necem verberavit.*

36 *CIL,* X 691–713.

37 *EphEp* VIII 670; Federico, Miranda 1998, p. 355, no. E38 (E. Miranda). Cf. EDR 123903 (M. Stefanile, 4-11-2012), where the inscription is dated between the second century and first half of the third century: [- - - - - - -] / [- - -]ORI / [.] F(- - -), / milit(avit) / ann(is) XXII, / vixit an(nis) / XL.

38 Where her brother Commodus had her killed (Scriptores Historiae Augustae *Comm.* 5.7).

39 Suet. *Aug.* 65.1. This would be the Agrippa's villa, which later became a Tiberius' property.

40 Including that of the *verna Capretanus Euplutus* (Magalhaes 2003, pp. 169–171, no. 26 (Claudian era). Only three of the many inscriptions of imperial slaves and freedmen seem attributable to the Flavian dynasty, as noted by Cristilli 2011, p. 183. This point now in Camodeca 2007, part. p. 157.

41 *CIL*, X 711 = *ILS* 1712: *Lalemus Aug(usti servus) circitor, / natione Lycao, Donato / filio et sibi et suis fecit.*

42 Fleet soldiers are commemorated in three epitaphs (*CIL*, X 685 = 2122; *AE* 2005, 327 = EDR 101526, G. Camodeca, 12-8-2009; *CIL*, X 687) and they could have been buried for other reasons and in later times in this area, where they were "both for maritime communications . . . that for the safety of the shoreline" (Parma 2002).

The presence of a centurion of the Praetorian Guard (*CIL* X 686 = ILS 9191 = Magalhaes 2003, p. 160 f., no. 19 = EDR 102197, G. Corazza, 18-10-2009) can instead be connected to a local origin. The only inscription that refers to a *speculator*, twenty-six years old, who died on duty (*CIL* X 684 = 2119, *Surrentum*, from second or third century, according to Magalhaes 2003, p. 93) is certainly after the first century AD. A dedication from a wife to her husband, a veteran of the fleet, belongs to the third century (*CIL*, X 719).

43 *CIL*, IV 1266: a *sacerdos* (scil. *centuria*) *Martialis*; 1711 (the same centurion appears perhaps in 1717 and 1733), 1994, 2145, 2157, 4310, 4311, 4688, 8405, the unidentified *commilitones* of 4618 and *Not. Sc.* 1897, p. 275. See De Caro 1979, part. pp. 85–95, who published other four; and Łoś 1995. Other two tombs of praetorians have recently been discovered in excavations outside the Sarno Gate, 'Porta Sarno' (cf. C. Avvisati, in *Il Mattino* of 23/05/1999 p. 25, referred to by F. Senatore).

44 Maiuro (2012, p. 281) is sure about the existence of an imperial residence of the Julio-Claudian era.

45 Tac. *Ann.* 14.17, on which Galsterer 1980.

46 Tac. *Ann.* 13.48. On the episode see D'Arms 1975; Leppert 1974, p. 279; Łoś 1995, pp. 168 s. In this volume, Part IV, Introduction.

47 Helmut Galsterer rightly emphasizes that (Galsterer 1980).

48 The bibliography on this issue is extensive: it is remembered and discussed in the fine article by Flower 2001. See also Zevi 2009, p. 102.

49 A model in mould of the monument was made for the Roman Exhibition on the Augustan Age of 1938, and is now preserved at the Museum of Roman Civilization (MCR 871). See Cagiano de Azevedo 1939.

50 It must not be forgotten that *Puteoli*, cosmopolitan city and main port of Rome until the expansion of the one in Ostia, was divided by Augustus, on the urban model, in *regiones* and *vici*: to them probably, as to Rome, was given the function to control the territory and its facilities.

51 Cass. Dio 66.17.1.

52 Suet. *Vesp.* 20.21.24.

53 There are no inscriptions of soldiers from here. Significant instead is the number of soldiers on duty and attestations of veterans at *Reate*, centre of viritane assignments.

54 Von Hesberg 2009.

55 Mart. *Epigr.* 5.1. On Albanum and on the villa of Circeii, also *Epigr.* 11.7.3–4. On Circeii, see also Fink 2012; and Maiuro 2012, p. 269 f.

56 Numerous are the sources that mention it: among the others, Plin. *Epist.* 4.11.6 (end of AD 89); Tac. *Agr.* 45.1 (AD 93).

57 For the later period, the presence of the Second Parthian Legion certainly also had the function of protecting the emperor and his family. Busch 2012, p. 121 f. in particular refers to Scriptores Historiae Augustae *Car.* 6. 7: when Caracalla died he was accompanied by two prefects of the *legio II Parthica* and two tribunes of praetorians.

58 *CIL*, XIV 2270, 2271, 2284 (*veteranus Augusti*); 2286 and 2287, *equites singulares* in service (at least the second century); 4214 and *Not. Sc.* 1914, p. 194 (G. Mancini) = Pavan 1962, pp. 90–92 = *AE* 1975, 159: two praetorians of the third century. See in this volume Part IV *supra*, p. 211 ff.

59 Cass. Dio 67.1–2, Suet. *Dom.* 4.4 e Juv. 4.99. On the necroprolis of Albanum within the imperial grounds, see Crea 2006.

60 Von Hesberg 2009, part. p. 329.

61 On the reliefs of the Chancellery building: Magi 1945, part. pp. 81–82. More divulgative is Rankov 1994, pp. 46–48.

62 Liverani 1989, p. 18 nos. 2 and 3, with images; von Hesberg 2001, part. pp. 245–253.

63 Suet. *Dom.* 19. Von Hesberg 2009, p. 333. On the Albanum, aside from the works of Lugli 1917 and 1918, see the most recent ones of Darwall-Smith 1994 and Liverani 1989 and 2008. On Domitian and the court in the Villa Albana, see Jones 1992, pp. 27–28.

64 On security of *Domus Flavia*, see lastly Wulf-Rheidt 2012, with previous bibliography.

65 Cass. Dio, 67.1; Jones 1992, p. 144.

66 Vd. Durry 1938, p. 377 f.; Jones 1992, pp. 59–61 and *passim*; Flower 2001, p. 643.

67 Durry 1938, pp. 80 f.

68 Suet. *Dom.* 6.1 and Cass. Dio, 67.6.6. Jones 1992, pp. 59–61.

69 *Classiarii, vigiles* and, perhaps, a *corporis custos*: Ricci 2004, pp. 321–324 and 336–338 (in this volume Part IV, Chapter 9, p. 209 ff.

70 *Equites singulares Augusti* and *classiarii*: Ricci 2004, pp. 329–331 and 341 (in this volume Part IV, Chapter 9, p. 213 ff). Of the four epitaphs of *classiarii*, at least three are from the 'Fondo Galli', located in the immediate vicinity of the villa.

71 *Vigiles*, between the end of the second and third centuries: Ricci 2004, pp. 325–327 and 339 (in this volume Part IV, Chapter 9, pp. 211 ff).

72 See in particular: Speidel 1994a, no. 676 (Luceria), no. 680 (Ruvo di Puglia), no. 681 (between Ruvo and Luceria). *CIL*, XI 3891 = Speidel 1994a, no. 678, from Capena, recalls *Titus Flavius Victorinus*, second century: the deceased is a *decurio singularium Augustorum*. In these cases it is often difficult to distinguish whether they accompanied the emperor or held policing duties.

Epilogue

Securitati Caesaris totiusque Urbis

Securitas: history of a concept

My aim, in writing this book, was to follow the evolution of a discourse, that of security in the imperial period, starting from the words that express it, by the protagonists that are doing it, and from the places where it was adopted.

The first chapter of this book was dedicated to a review of the main literary sources (archaeological, epigraphic and numismatic) related to *Securitas* and to synonymous expressions that define it. The review has highlighted how, over a century (from the middle of the first century BC to the middle of the first century AD), the word *Securitas*, in the literary sources, has gradually enriched its meaning, shifting from a philosophical to a more properly political scope. The use of the term with political connotations reaches its peak over the fifty years between AD 50–100, until, in the second century, this word, while being spread through the epigraphic and numismatic propaganda, is used again mainly in its original meaning and in a private setting.

In the climate of literary and philosophical circles of the late Republic and Early Principate terms like *Tranquillitas*, *Concordia* and *Pax* and, with a more timid presence, *Securitas*, become keywords of the cultural debate. These concepts are fed by Hellenistic philosophies, with reference to the state of mind produced by a condition of uncertainty – so as was determined by the civil wars – and the consequent need to gain a new balance, both in private and in public life.

The *Securitas* gradually acquires a full value in communicative language and, in the artistic, monetary and epigraphic representations, contributing to, along with the imperial *virtus* par excellence, the *Pax*, the transmission of the message of the end of the emergence and of the restoration of the normal course of life, after long decades of conflicts.[1]

The insistence on the issue of the threat to the peace of mind of the individual, of the insecurity of the existence leaves room, at an early stage of Augustus' reign, for radically new actions and messages: the threat has ceased, civil wars do not return, and peace was re-established, to last and to enable prosperity. Peace, tranquillity, and with them security have been re-established thanks to *Princeps* who is personally responsible for their maintenance and spreading.

That is the reason why I chose to focus on the imperial 'security policy' rather than on the interventions on public order. What at first may seem the use of different words to express the same concept instead proposes a change of perspective.

What I hope has emerged in the first chapter is the process started in the late Republican era which did not end with the advent of the Principate, but rather a few decades later, during the age of Nero: it is at this time that Seneca pulled the strings of an old discourse and made the *Securitas* one of the most characteristic and characterizing *virtutes imperatoriae*. The idea of a *Securitas Augusti*, a meaningful formula that sends a clear message, is due to Seneca's reflection, conducted mainly in the *De Clementia*: security must be a prerogative of the *Princeps* and, at the same time, a guarantee of protection of the people of Rome.[2]

It is no wonder that the parable of the imperial *Securitas* reaches its highest and most organic point with Nero. The discourse that had been prepared by Cicero and became action, implemented for the first time with Augustus, takes a long time to come to a full definition; the same time is required by the Julio-Claudians, and then by Vespasian, to define and revise, if needed, the means of action of the military and paramilitary in Rome.

Despite the message of *Securitas* being persistently transmitted in philosophical works and in the propaganda (especially on the coins), Nero did not interpret the proposal of his preceptor. His successors, the emperors of the years of the crisis, and the Flavians, recover the full value of Seneca's message, instilling the certainty of stability, despite the recent conflicts and the shifts in the summit of power. At the dawn of the new century, the concept of Securitas – along with the *Tutela*, *Pudor* and *Libertas* – is recovered and synthesized by Nerva and his son in the new formula of *Securitas Temporum*.[3] This expression has a richer and more extended value than before, combining the political content with the economic one, as a guarantee of materialistic well-being and the recovery of a paternalistic function of the *Princeps* who acts on behalf and with the support of the Senate.

An integrated plan

I am convinced that Augustus' work on the security plan was the result of a conscious even if not organic project, not isolated from the rest of the measures that the *Princeps* introduced, for example, for the good conditions of buildings and roads. This does not mean that the program did not undergo adjustments over time. So it is that the determination of his decisions combined with the progressivity of the action. That is Augustus' characteristic approach, but also a measure of caution that he takes especially in the early decades of his reign, when the memory of the traumatic civil wars was still close and the presence of military or paramilitary forces in the city could evoke terror not yet dormant.

The lack of an organic plan for security explains the absence of an organic account by historians: what can be reconstructed is taken from episodes or interventions, seemingly isolated, and the overall presentation of the work of individual emperors. Also if an organic plan had existed, perhaps it would not have been made explicit: if, for instance, the intention of a comprehensive plan had been manifested in relation to the armed men in Rome, it would have been rejected vigorously by the people of Rome. The project therefore developed gradually, and responded to the needs mainly concerning the life of the inhabitants of the city and of the *Princeps*; and was destined to undergo adjustments beyond the reign of Augustus. This does not mean, however, as claimed by Werner Eck,[4] that the *Princeps* proceeded in relation to the emergencies; and therefore the idea of what to do and why (if not completely how and when to do it) is not very present in the mind of Augustus.

Historical research has focused on what motivated Augustus to act, on his will of control and domination. What has emerged, depending on the points of view, is the figure of a benevolent monarch or that of an authoritarian centralizer; more often, that of a fine political and a careful evaluator, able, through some devices of great subtlety and effectiveness, to get the best results and longer-lasting effects through his deeds.

In this book, rather than the reasons of the propaganda and the means adopted to spread it, attention has focused on the articulation of a plan in order to make the reasons of action concrete. The interpretative key was not so much these reasons (well known and investigated) for which he acted, but rather the ways that made them appropriate and effective to the eyes of the Romans. Ways were certainly studied in order not to create trauma or to awake the ghosts of the past. In other words, rather than focusing the attention on intervention procedures in the case of disorder or urban violence, so often investigated by scholars, I was interested in highlighting the thoughtful construction of a preventive plan against civil unrest.

Before engaging soldiers, this plan mobilized the inhabitants of the districts of Rome and their magistrates, who lived in a particular territory and thus had every interest that civil unrest and violence would not erupt; and felt induced, through personal involvement, to ensure prevention so that such incidents would not occur.

In the second chapter of this book, I have therefore tried to draw Augustus' lines of action in reorganizing the territorial structures of Rome and coordinating the governors, at varying degrees, to ensure the security of the city and the safety of its *Princeps*. Thus Augustus or the 'art of governing', in the most traditional and ancient meaning of the expression: the art of applying to the State the good rules of economy, extending a form of surveillance and control over individuals' conduct and the community that was not less attentive to that practised for their own private sphere.

The law is the first tool that the *Princeps* uses in his security policy. Sure of his control over the state and in order to be perceived as a guarantor of the community's well-being, Augustus proceeds to better define one of the oldest notions of political and legal Roman culture, the *maiestas* and the *crimen* connected to it. Having experienced what had happened with Sulla and then with Caesar, the *maiestas* is now inserted in a political and religious sphere: from this point on, it will not include only the Roman state in the persons of its representatives (*maiestas populi Romani*), but also the emperor and his family (*maiestas Principis*).[5]

More specifically, the assurance of Augustus' safety is intended, as it already was for the magistrates, as an extension of the *maiestas populi Romani*, adjusted and enriched through the precise definition of the categories of possible attacks to the constitutional order. Violation of *maiestas*, from Augustus onwards, can be expressed not only through actions, but also in words and in writing, including the form of insults or ridicule to the *Princeps* even when absent or to his portrait.

The *Princeps'* legislative action or that of his representatives (especially through the *lex maiestatis*, the *lex de vi* and the *lex de collegiis*) appears in harmony with the institutional action and in particular with the security plan based on prevention, put into effect on two separate but converging levels: the organization of military forces to ensure the *Princeps'* safety; and the intervention on the urban fabric for people's safety in public places.

Augustus, however, links other tactics to the law; and the security plan appears as the desired outcome of a complex series of interventions that had the legislative activity as an instrument, the city fabric as a theatre, and at the top of the bureaucratic pyramid, the *Princeps* himself and the new prefects as protagonists; and, at the base, the military, paramilitary and civilians representing the operational network.

The organization of the military and paramilitary forces was complex and required a very long time to be accomplished. In this book I tried to highlight how the cohorts created or reformed by Augustus had received and then kept a number of skills that had as a key point the security of the *Princeps* and of the city; and the implementation of prevention and intervention strategies that involved the cooperation of the civil forces.[6]

In this organization, the first substantial innovation in terms of Augustan security measures has been identified. With the unscrupulousness that characterizes much of his work, in the formal respect of the ancient tradition, Augustus created the basis for infringing the principle that forbade soldiers to be allocated in Rome. For his personal safety, acting in substantial continuity with the commanders of the late Republic, Augustus recruited private bodyguards, the *corporis custodes*. But he wanted to link the functions of the Praetorian Guard and the *speculatores* to the representation of imperial power, wherever manifested, in the imperial residence and in public spaces, in Rome and outside of it, during his travels in Italy and away from it.

To avoid resistance and opposition to the numerical increase and to the growing importance of the Praetorian Guard and the *speculatores* and, more generally, to the presence of armed men in the city, Augustus has nevertheless adopted some extremely cautious measures: these soldiers did not reside in the capital,[7] they did not have their own camp nor a fully defined hierarchical framework. The daytime presence was probably limited to a few cohorts;

their involvement in arrests, and in encouraging opponents and suspects to suicide or murder was, in the Augustan age, rare, if not entirely absent. What was just perceptible was their intervention in cases of breach of public order in Italy, in accordance with the role and responsibilities conceived for the imperial power.[8]

A second innovation, as disruptive as the first, is the way in which Augustus arranged for the security in the city, both outdoors and in enclosed places[9]: in the streets, the *vigiles* and the *milites urbani* exert preventive and emergency services, against theft, violence and various disorders; or when fires are likely to cause disasters and alter the public peace. The same men are coordinated with the praetorians on occasion of spectacles, when riots can also occur for trivial reasons, as a reaction to the victory of the enemy faction or due to a too warm support to the favourite: outside the buildings, *vigiles* and soldiers of the urban cohorts are mobilized; inside them, the praetorians guarding the *Princeps* are called to intervene.

Augustus' plan, as said, is not limited to the intervention of the military: it had as a theatre the fabric of the city, and as protagonists the magistrates who work there and new figures who flanked them. At the top of the administrative pyramid we find the prefects and the *magistri vici*, re-functionalized governors of the neighbourhoods ('enriched with new or enhanced features'); at the base, there are the military, paramilitary and civilians representing the operational network.[10]

The *lex de vi* is put in place and reinforced through preventive action of military and paramilitary forces that act synergistically with the *vici* and their directors/administrators; and ensure, through the control of gathering places, against the violation of the association ban to prevent the risks of seditious meetings. While the safety of the individual is not – as it had never been – the State's duty: how Sablayrolles states "etait affaire privée".[11] It becomes the responsibility of the Senate or of the emperor but only when a private disturbance puts at risk the peace and the security of citizens.

Like the *milites urbani*, the *cura Urbis*, could count on the cooperation of the inhabitants of the *vici* to exercise their supervisory functions at markets and meeting places; and on the *vigiles*, for night patrolling; in the same way the *vigiles* had to rely on the *magistri vicorum* and their collaborators and on the urban soldiers to exercise their function of guard on citizens and against the consequences of fires. It seems clear that the security of the city is not entrusted to the action of a single magistrate or paramilitary corps; nor only to soldiers and their officers.[12] Furthermore, the republican figures didn't disappear from the scene abruptly, but they continued to do their part: if in the past in order to hold off those risky situations intervened the *aediles* and their collaborators, the same function is now distributed among several figures. The underlying principle is to coordinate all the different forces acting metaphorically under the eyes of the *Lares* and *Genius Augusti*, present at the corners of the *vici*.

The Augustan plan will be accomplished within a few decades and, with no essential changes, it will maintain the original *ratio* for at least two centuries.

The conditions that allowed the people of Rome and the Senate to accept the plan were the gradualness of the process and what has been called here 'an integrated system',[13] where military and paramilitary forces, officials and their subordinates are all called to participate and contribute, according to their skills and abilities.

What could have been the reason for this multiplication of skills and partial overlapping of tasks? A reason that can be considered sufficient in itself can be identified in the will of not allowing a military body nor a district association to have the exclusiveness of action. What has sometimes been considered an unnecessary multiplication of figures and a duplication of tasks is a reflection of the typical Roman way of conceiving and constructing the magistrates' powers; it is possible to make out in it the features par excellence of Augustus's modus operandi: continuity with the past and respect for tradition regenerated with a powerful innovative force.[14]

For all the above-mentioned reasons, the organization of the neighbourhoods on the one hand (with its civil and religious implications) and the day and night patrolling of the streets on the other (with its political and military implications) can be seen as tiles of the same mosaic, which remain unchanged and only adapted in the following two centuries. It takes the same time frame to allow a better definition of the profiles of the 'big prefects' and to provide a comprehensive definition of their powers and of the territorial area of reference.[15]

In the chronological vagueness in which mostly are the single initiatives of the *Princeps*, the year AD 6 assumes a key value. We are at the end of the reign, and clouds appear on the horizon of the world of peace and security that he had created: a severe famine afflicts the City and, to prevent the risk of riots, Augustus sent off slaves for sale and *lanistae*, all foreigners except for doctors, teachers, gladiators and part of the household slaves.[16] In the same year there is the attempted conspiracy of Publius Plautius Rufus and the many issues on the foreign front, particularly in Dalmatia and Isauria.[17]

In that year, in closing a long interlude of incubation in search of solutions to the problem of fires and criminal offenses connected to these, Augustus organises the *vigiles* as cohorts subject to a prefect. Composed of freedmen, the cohorts of the *vigiles* will be occasionally used to quell riots.

From that same year onwards, the main cohorts of Rome (the praetorian and the urban cohorts) begin to be conceived as distinct entities; they will continue to have common accommodation for nearly two centuries. This distinction can be considered along with the creation of a new (or reformed) prefect, the *praefectus Urbi*.

In AD 6, again, Augustus established a special military treasury, the *aerarium militare* – initially fuelled by Augustus himself and later with the tax revenues – that will put a point to one of the major causes of insecurity in the recent (and still scorching) past: the veterans' claims for the *praemia militiae*.

If it seems risky to give AD 6 the meaning of the turning point of the Augustan regime, it is difficult not to view it as a key moment of a period – the last decade of the reign of Augustus – when, gone are the days that required an emergency policy, there is a clear need to create the premises for action concerning the security of the successors.[18] It is at this stage that the *crimen maiestatis* and the treason trials become the centre of radiation of a series of initiatives that the sources have improperly attributed to Tiberius; although it seems certain that the new *Princeps*, with old and new instruments, is going to stabilize the numerous initiatives and the results of decades of trial and testing.

Experimentation, maturation and crisis of a plan

In the fourth chapter of the first book of the Annals, Tacitus opposes the feeling of security that dominated in Rome when Augustus was young to the feeling of danger that began to spread in the city when the *Princeps* advanced in age:

> It was thus an altered world, and of the old unspoilt Roman character not a trace lingered. Equality was an outworn creed, and all eyes looked to the mandate of sovereign – with no immediate misgivings, so long as Augustus in the full vigour of his prime upheld himself, his house, and peace. But when the wearing effects of bodily sickness added themselves to advancing years, and the end was coming and new hopes dawning, a few voices began idly to discuss the blessings of freedom; more were apprehensive of war, others desired it, the great majority merely exchanged gossip derogatory to their future masters.[19]

The excerpt, according to Olivier Devillers, suggests a distinction between a private and a collective pole[20] of the security, which has been discussed at length; and at the same time it anticipates in a veiled form what is going to happen with Tiberius and the reopening of hope for their lost *libertas* (*pauci bona libertatis in cassum disserere*).

The provisional aspects of Augustus' plan in terms of safety will be defined and specified by Tiberius: the cooperation between the district magistrates and the *vigiles* or urban cohorts, although already functional, now becomes ensured; the distinction between the Praetorian Guard and the *milites urbani*, which has just been mentioned, is effective; the construction of the large barracks on the Viminal Hill is planned and implemented; even the quarters for the *vigiles* are most likely arranged at this time. The number of trials against those who violate the *maiestas populi Romani et Principis* has increased considerably.

However – if we exclude the construction of the large military camp on the Viminal Hill, which Augustus had not dared to think of – Tiberius does not implement anything that has not instead already been provided for by the founder of the Principate.[21]

The same continuity characterizes the phase from Tiberius to Nero, apart from various adjustments, mainly concerning the *vigiles*. With Tiberius they had gained the possibility to obtain the status of citizens; instead under Claudius they were given the first barracks in Ostia. Therefore this successful (Augustan) experiment is exported to the port of Rome. Again under the reign of Claudius there is the official establishment of the hierarchical organization of the *vigiles*,[22] which can be interpreted as an eloquent sign of the strong development of the long phase of experimentation.

The turning point that Seneca had hoped for during the Principate of Nero, in terms of security, was not achieved. It will therefore be during the biennium 68–69 and the Principate of Vespasianus that significant changes will take place, with the necessary development and overall redefinition of the role and areas of competence of the urban troops, after the parenthesis of the civil wars.[23] The direction that the emperors took was, first of all, to contain the grandstanding and the repressive aberrations that some military and paramilitary forces (praetorians and *corporis custodes*) had assumed, for their tight relationship with the *Princeps* and for their role in the succession of those at the higher ranks of power; and, in parallel, to reward those troops that had remained neutral or had openly supported the winner of the civil conflict.

During the Flavian age, the number of praetorian cohorts is resized, but their functions are not reduced, if anything, they are enhanced in terms of their role as fighters; and the *speculatores* are now permanently integrated with them. The effort and the determination to contain the 'exuberance' of the praetorians (and of their prefects) are clearly perceived and pursued through various means, not all operated directly on the same cohorts, but still competing for the same purpose: to balance the roles of the corps charged with the safety of Rome and of the *Princeps*, all, in the words of Mucianus, "soldiers of the Emperor".[24]

For similar reasons, as we have seen, the *corporis custodes* were dissolved and probably from the age of Domitian, a new corps (the *equites singulares*) was being prepared with the aim to take over what was once the responsibility of the Germans, namely their role in charge of safety, now left uncovered, and in close relationship with the emperor. These soldiers on horseback were therefore, more or less consciously, designed to reduce the role of the Praetorian Guard.[25]

The initiatives promoted by the Flavians also affect other urban military or paramilitary forces: the *vigiles* are given stable barracks in Ostia; and the *milites classiarii* are rewarded for their behaviour during the war of succession.[26]

Essentially, it seems that we can say that, in the decade between 70 and 80 (and, with continuity, during the following two decades), the guidelines on security marked by the Augustan policy have led to full consequences. There is no trace of substantial novelty, since the interventions that regarded the praetorian cohorts and the development of the *equites singulares Augusti* are nothing more than a corrective (and preventative measure) to degeneration that were not foreseeable at the time of the *Princeps*; and that the privileged relationship with the soldiers of the fleet is configured as linked to the circumstances and had no significant repercussions in the security policy in Rome and Italy.[27]

Substantial changes occur later, with the reign of Hadrianus and those of his successors. They regard the organization of *vici* and of the control system, and possibly also the functions of the soldiers, of their non-commissioned officials and paramilitary forces of Rome.[28] It is in this period of time, in my view, that the foundations are laid for what later under the Severan dynasty will become fully operational, commonly known as the 'Severan military reforms'.

In the urban fabric, the action begins with Hadrianus, continues with Antoninus and Marcus Aurelius, and is completed by Commodus. The figures involved in the administration of the districts were gradually organized in a hierarchy consisting of three levels: *curatores* and *praefecti*, at the upper level; *magistri* and the *tribuni* of the urban cohorts and of the *vigiles*, at the intermediate level; *ministri*, *vigiles* and *urbaniciani*, at the lower level. These distinct levels, where each acts within their competence, in the second century see a leading position of *vigiles*/soldiers over *magistri*/freedmen or freemen.[29]

A clear signal of the will to intervene in an authoritarian way in the number and in the autonomy of the many protagonists of the safety in the neighbourhoods has been transmitted to us by the texts of the aedicules at the *compita*: the procedures of authorization for the restoration of the *aedicula* first connected to the *magistri vici* and the *aediles*, as of AD 136, require that the applicant is authorized in advance by the magistrates and therefore by the emperor, through the prefect of the *vigiles*.[30]

It is in that same period that the cohorts of the *vigiles* start to acquire a fully military character; and that their prefect and the urban prefect begin to expand their expertise in the field of criminal jurisdiction, as well as later reflected in the sources of the Severan epoch.

Always in the age of the Antonines (with a beginning perhaps delayed of a couple of decades), there are signs of change in the functions of the soldiers of Rome, intended to have, even if indirectly, considerable repercussions for the citizens' security system. For the *equites singulares Augusti* and the praetorians, the operational function in war zones is strengthened; and the former, initially side by side of the latter, now prevail on the praetorians in the role of escorting the *Princeps* and in the value of a symbol of imperial power.[31] Both corps, free from the task of *tutela urbis*, merge the role of fighters to the one of security escorts and/or bodyguards of the emperor.[32] The praetorian prefect is the last reference point for both.

Even the soldiers of the urban cohorts, the 'police' of Rome, again around mid-second century, gradually see their profile modified, through: the more systematic participation in the imperial military campaigns; an increase in monitoring public places in the city, where the signals of unrest begin to become more numerous;[33] their accommodation in a new military camp on the via Lata, far from the old one on the Viminal hill.[34] In all cases, except the first, they are symptomatic indications of a definitive separation from the praetorian cohorts, confirmed by the role played during episodes of violence that took place in the city.

These measures should be read, I believe, in close connection with the changes introduced in the government of the city regions, where, as we have seen, the synergy between different figures is replaced, since the age of Hadrianus, by a greater centralization of powers in the hands of emperor and his direct collaborators. The construction of the *castra urbana* must also be inserted into a wider picture, that of the topography of the military camps of Rome. What prevailed in the first century was a 'centrifugal logic'[35] that aimed to keep the barracks away from the city centre and, at the same time, to make space for the soldiers in the *castra* and in large surrounding areas so that they could perform manoeuvres and drills.[36] Besides this centrifugal logic that moves praetorians, *milites urbani* and *speculatores* away from the heart of the city, a functional logic was introduced that integrated into the urban fabric the *vigiles*, whose stations and guard posts are distributed in all regions, for the obvious need for ready intervention, and the soldiers of the fleets of Misenum and Ravenna, called upon to carry out specific tasks in the city especially related to performances.

During the second century the two logics still seem to coexist when the new barracks on the Caelian hill were built, the one for the *equites singulares* in Via Tasso and the *castra peregrinorum*[37]; and, as already mentioned, the *castra urbana* along the via Lata, for the cohorts of the senatorial prefect. The latter with a strategic position in relation to the old and new tasks in the city and in the first part of the suburban Flaminia, which until then was not controlled enough unlike the roads of the south-eastern area of the city.

Even the scenes of urban warfare that I referred to earlier, more and more numerous (or at least registered as such by our sources) between the late second and early decades of the third century,[38] recall – for an affinity to the political climate, although quite diverse for the contexts and the characters – those of similar riots that occurred, in the same period, outside the *Urbs*, in cities of Italy or of the provinces. In reading their description one gains a strong feeling that historians' insistence on the emperors' *mansuetudo* and *clementia* in these circumstances implies, conversely, the clear perception of a change in register and a restriction in terms of the policy of control.[39]

And so in Rome if, in the course of the first century, with small adjustments or improvements, the Augustan security plan was the one adopted, from the second century on a substantial modification phase slowly starts to take place. Probably it is not a structured plan, but more individual interventions gradually shared and improved that, in terms of control of the social dynamics, probably reached a critical point only in the last decades of the second century.

The change in register and the intensification of the repressive nature of the imperial action seem to combine perfectly with the picture – which has recently been emerging from studies on the history of the Roman army – of the continuity between the armies of the last Antonines and those of Diocletian.[40] The third century marked, in this sense, a watershed, and represents a laboratory of interventions and long-term studies that actually start with the last of the Antonines, and has its most systematic moment with Septimius Severus, its 'revolution' between Diocletian and Constantine.

In relation to the troops in the capital, it is in fact possible to see a change of perspective in the interventions of the emperors during the second century, against which Septimius Severus' action, rather than being disruptive and shocking, is to be seen as the natural outcome of previous actions[41]: the reforms of the early decades of the third century, along with Severus Alexander's following actions in the field of the *cura Urbis*, appear to be the settling point of a slow process, and the moment to pull the strings of actions and preconditions placed, rather than the result of a completely new plan for security.

The novelty in Severus' action for the City certainly does not consist in the dissolution of the old praetorian cohorts, so irreparably compromised; nor in the new character of the *vigiles*, that had been defined over two centuries with a slow and unrelenting development; nor in the placement of *stationes militum* outside Rome and Ostia. None of these were an original initiative; and with regard to the military stations it is clear that there is political continuity that ties the positions of Augustus and Tiberius, even if temporary, against banditry,[42] and the military bases that during the second century had multiplied throughout the territory of Italy, close to road junctions, crossings, passing herds, etc.[43] and involving praetorians, *equites singulares* and fleet soldiers as required and depending on the context or strategic calculations.

The real novelty – for which the predecessors had, however, established a strong foundation – consists in the creation of new praetorian cohorts, composed of men with diverse experience and of totally different ethnic origin; and in the quartering of the Legion created for Parthian expeditions[44] in Albanum. From this moment on, the praetorians, the *equites singulares* and the legionnaires will be recruited from the same (peripheral) regions of Europe and they will build a special bond, consisting of real or acquired kinship, of cults, traditions, common participation in military campaigns; and, conversely, this obvious diversity (of culture, mentality, language, functions) will alienate them from the urban cohorts and from the *vigiles*.

An Italy traveled by troops under the pretext of creating a security net against thefts and brigandage and, especially in the case of Ostia, to ensure better communication with the central administrative offices; a capital where soldiers who do not know Latin circulate, whose sole reference point is the emperor who has chosen them and allowed them to arrive, soldiers who are now definitely far from urban police forces, ancient heritage of the original security plan; the territorial frameworks widely subtracted from the administration of their rulers and subjected to the emperor's direct control. These are all images that inevitably evoke the idea of a more 'provincial' Italy,[45] of a sophisticated and delicate balance made up of legislation, jurisdiction, military policy and administration, which now heavily falters.

Notes

1 *Supra*, Part I, Chapter 3, pp. 40–45.
2 *Supra*, Part I, Chapter 3, pp. 47–50.
3 *Supra*, Part I, Chapter 3, p. 53 ff.

4 *Supra*, p. 83 n. 8.

5 *Supra*, Part II Introduction.

6 *Supra*, Part II, Chapter 4–5.

7 He most likely had a careful attitude even with the *cohortes urbanae* and *praetoriae* and the *cohortes vigilum*. "Se i pretoriani dovettero attendere il tempo di Seiano per avere un loro campo ... non vi è ragione di credere che per i vigiles ci si sia originari-amente comportati in modo diverso. È probabile insomma che ancora nel 12/13 d.C. la forza dei vigili non fosse concentrata in caserme, ma piuttosto distribuita nei singoli quartieri della città, presso abitazioni private, come era avvenuto per i preto-riani" (Panciera 1978, p. 320).

8 One can agree with Le Roux 2002, p. 39 (who recalls Reddé 1986, p. 451) when noting that the functions of the soldiers of Misenum and Ravenna were intended "to protect citizens from sun and weather, or to provide naumachias ... it was the comfort of the citizens summoned by the emperor for amusement".

9 *Supra*, Part II, Chapter 5. It certainly cannot be said that the topic of the control on spaces to hold the potential turbulence of the population has never been addressed: masterful in that sense is the reflection conducted by Augusto Fraschetti, who closely related the control on time and that on urban spaces, both for the Augustan Rome (Fraschetti 1990) and for the city of Constantine (*La conversione. Da Roma pagana a Roma cristiana*, Bari-Roma: Laterza 2004). More recently, the practice of control of areas in Rome, primarily through the construction of large complexes of represen-tation where the private is more than ever the public, has been analysed by David Fredrick for the Flavian city.

10 In Italy and in the provinces, there are some models for the 'Augustan creations' like other Mediterranean metropolis such as Alexandria. To be considered, e.g., the νυκτερινὸς στρατηγὸς, the model for the *praefectus vigilum* (Strab. 17.12). See Fraschetti 2000, p. 727 and Sablayrolles 1996, pp. 26 f. with n. 53.

11 Sablayrolles 2001, p. 144.

12 "Facts show that it [the monarchy] was aware of it and could not abuse of it" (Le Roux 2002, p. 51).

13 *Supra*, Part I, Chapter 2, pp. 24 ff. and Part II, Chapter 5, p. 111. The same expression in Wallace-Hadrill 2003, p. 198.

14 Le Roux states that, in the context of policing as in others, "the Augustan mon-archy has refined techniques originated under the Republic by giving them an effectiveness through the use of new media" (Le Roux 2002, p. 36 and note 73). It is therefore incorrect to reduce the problem to a question of fear or arbitrary use of force: it "involved a political culture in the strong sense of the term". Le Roux, it must be said, also takes into consideration the provinces.

15 Chillet 2012 considers, appropriately in my view, the dates of the creation of the prefects, as another eloquent clue of Augustus's progressivity of action: AD 2 (*praefec-tus praetorio*); AD 6 (*praefectus vigilum*); AD 8 (*praefectus annonae*); AD 13 (*praefectus Urbi* as a permanent charge).

16 Suet. *Aug.* 42: *Magno vero quondam sterilitate ac difficili remedio cum venalicias et lanis-tarum familias peregrinosque omnes exceptis medicis et praeceptoribus partimque servitiorum urbe expulisset*; cf. Cass. Dio 55.26.1.

17 Cass. Dio 55.27.2–3 (for the foreign policy, see paras 28 and 29) and Suet. *Aug.* 19. As for Plautius (or Plotius) Rufus, see Wardle 2014, p. 164. A different view in Swan 2004, p. 184.

18 *Supra*, Part II Introduction.

19 Tac. *Ann.* 1.4.1–2: *Igitur verso civitatis statu nihil usquam prisci et integri moris: omnes exuta aequalitate iussa principis aspectare, nulla in praesens formidine, dum Augustus aetate*

validus seque et domum in pacem sustentavit. postquam provecta iam senectus aegro et corpore fatigabatur, aderatque finis et spes novae, pauci bona libertatis in cassum disserere, plures bellum pavescere, alii cupere. pars multo maxima inminentis dominos variis rumoribus differebant.

20 Devillers 2009, p. 317.

21 *Supra*, Part III Introduction, pp. 122–124.

22 Freis 1967, p. 36 ff. Le Roux (2002, p. 27 and n. 38) does not identify cases of serious friction between the commanders of the various military corps; nor does he identify precise rules that once and for all would acknowledge the responsibilities. Only over time, "the judicial powers of the commanders have been better defined and thus those that I would call the 'action domains' have been better outlined".

23 *Supra*, Part III Introduction, pp. 128–129.

24 *Supra*, Part III Introduction, p. 128.

25 *Supra*, Part III Introduction, p. 129.

26 For Vespasian's interventions towards the military and particularly in relation to the soldiers of the fleet, see Ricci 2008, pp. 21–24.

27 *Supra*, Part III Introduction, pp. 128–129.

28 About the changes of the centurion's role, between the first and the second centuries, see Le Roux (1972, p. 116 with n. 2): "un certain nombre de faits bien connues montrent, à notre sens, que la réorganisation d'Hadrien, relatée dans l'Histoire Auguste, avait eté l'aboutissiment d'un travail commencée par des prédécesseurs", referring to the Flavians.

29 *Supra*, Part III Introduction, p. 130 ff.

30 *Supra*, Part III Introduction, pp. 130–131.

31 *Supra*, Part III Introduction pp. 129–130 and Chapters 9 and 10.

32 *Supra*, Part IV, Chapters 8 and 9.

33 *Supra*, Part III, Chapter 6, p. 155 ff.

34 *Supra*, Part III Introduction and Chapter 7.

35 For the *castra praetoria* Tacitus (*Ann.* 4.2) must be cited: *et severius acturos si vallum statuatur procul urbis inlecebris ut perfecta sunt castra.*

36 Tac. *Ann.* 12.36; Cass. Dio 74.1; Speidel 1994b, p. 114.

37 I do not agree completely with Rivière 2004, pp. 68–71 about a functional logic predominating already during the second century: with the construction of *castra* legionaries in Albanum by Severus (to be considered together with the *castra nova* intended for the *equites* on the territory of the Laterans), a massive intervention takes place on both the topography of the immediate suburban territory and on the logic of control.

38 *Supra*, Part III Introduction. True only perhaps for the first half of the second century AD, what Levi says about the reduction of military intervention against demonstrating crowds: the "re-establishment of the Empire", according to Levi, also involves the elimination of the use of too brutal repressive measures in the manner of the Julio-Claudi, or at least, judging from what we have received, at their skillful dissimulation (M.A. Levi, *Adriano. Un ventennio di cambiamento*, Bompiani: Milano 2000).

39 See, by way of example, the passage from the Epitome de Caesaribus 15.9 (author emphasis): [Antoninus Pius] *usque eo autem mitis fuit ut, cum ob inopiae frumentariae suspicionem lapidibus a plebe Romana perstringeretur, maluerit ratione exposita placare quam ulcisci seditionem*; or even the insistence on the mildness that was typical of Hadrianus in Alexandria (Cassius Dio 69.8.1a). This and other incidents are mentioned by Kelly 2007, pp. 163–164 n. 62. Some repressive measures against those who disturbed public order in Gaul dates back to the Principate of Antoninus Pius and the epoch of Marcus Aurelius (Robinson 1992, p. 202)

40 Smith 1972, esp. pp. 482–485; 491–499; Carriè 1993; Carriè and Rousselle 1999; Rocco 2012 with further bibliography.
41 The same continuity between the action of the Antonines (Marcus Aurelius and especially Commodus) and that of the Severi is evident, according to Peter Eich, in the way in which the role and functions of the praetorian prefects were conceived. (Eich 2013, pp. 89–92).
42 *Supra*, Part IV, Introduction, *passim*.
43 Puteoli, Velitrae, Formiae and Grumentum of the mid-second century, Aveia, Bovianum and Saepinum, until the later of the Furlo pass, mid-third century. See Part IV Introduction and Chapters 8–10.
44 Lastly, Busch 2013, part. pp. 107–110.
45 The signals are different, not only the placement of the new legion in *Albanum*. *Pace* Smith 1972, p. 487.

Bibliography

Bibliographical abbreviations

ANRW: *Aufstieg und Niedergang der römischen Welt*. Series eds H. Temporini (Parts 1 and 2) and W. Haase (Part 2). Walter de Gruyter: Berlin, New York 1972 onwards.

BMCRE: H. Mattingly, *Coins of the Roman Empire in the British Museum*, I–V, British Museum: London 1930–1965.

CBI: E. Schallmayer, K. Eibl, J. Ott, G. Preuss, E. Wittkopf, *Der römische Weihebezirk von Osterburken I: Corpus der griechischen und lateinischen Beneficiarier-Inschriften des Römischen Reiches*, K. Theiss: Stuttgart 1990.

FIRA: S. Riccobono, J. Baviera, C. Ferrini, J. Furlani, V. Arangio-Ruiz, *Fontes iuris romani antejustiniani: in usum scholarum ediderunt*, Barbèra: Florentiae 1940 (new edn 1968, enlarged and amended).

IlJug: A.J. Sasel (ed.) *Inscriptiones Latinae quae in Jugoslavia repertae et editae sunt*, I–III, Delo: Ljubljana 1963–1986.

LTUR: E.M. Steinby (ed.), *Lexicon Topographicum Urbis Romae*, I–VI, Edizioni Quasar: Roma 1993–2000.

LTURS: V. Fiocchi Nicolai, A M.G. Granino Cecere, A. La Regina, Z. Mari (eds), *Lexicon Topographicum Urbis Romae – Suburbium*, I–V, Edizioni Quasar: Roma 2001–2008.

Not.Sc.: *Notizie degli scavi di antichità comunicate alla Reale Accademia del Lincei per ordine di S. E. il ministro della Pubblica Istruzione*, Tipografia Salviucci: Roma, from 1876 onwards.

RIC I²: C.H.V. Sutherland and R.A.G. Carson, *The Roman Imperial Coinage*, I (from 31 BC to AD 69), 2nd edn. Spink & Son: London 1984.

SupplIt Imagines – Latium 1 2005: M.G. Granino Cecere, Supplementa Italica Imagines. Latium vetus 1 (CIL, XIV; Eph. Epigr. VII e IX). *Latium vetus praeter Ostiam*, Quasar: Roma 2005.

Frequently cited corpora or repertories

Agamben 2006: G. Agamben, *Che cos'è un dispositivo?* Nottetempo: Roma 2006.

Alföldy 1969: G. Alföldy, *Fasti Hispanienses. Senatorische Reichsbeamte und Offiziere in den spanischen Provinzen des römischen Reiches von Augustus bis Diokletian*, F. Steiner Verlag: Wiesbaden 1969.

Alföldy 1989: G. Alföldy, Cleander Sturz und die antike Überlieferung, in *Die Krise des Römischen Reiches. Geschichte, Geschichtsschreibung und Geschichtsbetrachtung. Ausgewählte Beiträge*, F. Steiner Verlag, Stuttgart 1989, pp. 81–126.

Alston 1994: R. Alston, Violence and Social Control in Roman Egypt, in A. Bülow-Jacobsen (ed.), *Proceedings of the 20th International Congress of Papyrologists*, Copenhagen, 23–29 August, 1992, Museum Tusculanum, University of Copenhagen, 1994, pp. 517–521.

Alston 1999: R. Alston, The Ties that Bind: Soldiers and Societies, in A. Goldsworthy, I. Haynes (eds), *The Roman Army as a Community* (*JRA*, suppl. 34), Portsmouth, pp. 175–195.

Amoroso 2007: A. Amoroso, Il tempio della Tellus e il quartiere della praefectura urbana, in *Workshop di archeologia classica. Paesaggi, costruzioni, reperti* 4, 2007, pp. 53–84.

Andermahr 1998: A. M. Andermahr, *'Totus in praediis'. Senatorischer Grundbesitz in Italien in der Frühen und Hohen Kaiserzeit*, R. Habelt Verlag: Bonn 1998.

André 1967: J.M. André, *Mécène: essai de biographie spirituelle*, Belles Lettres: Paris 1967 (trad. it. *Mecenate: un tentativo di biografia spirituale*, Firenze 1991).

Angeli Bertinelli 1974: M.G. Angeli Bertinelli, Gli effettivi delle legioni e delle coorti pretorie e i laterculi dei soldati missi honesta missione, in *Rendiconti dell'Istituto Lombardo* 108, 1974, pp. 3–12.

Antonielli 2002 (ed.): L. Antonielli (ed.), *La polizia in Italia in età moderna*, Rubbettino: Soveria Mannelli 2002.

Antonielli 2006a: L. Antonielli, *Carceri, carcerieri, carcerati. Dall'antico regime all'Ottocento*, Rubbettino: Soveria Mannelli 2006.

Antonielli 2006b: L. Antonielli, *La polizia in Italia e in Europa: punto sugli studi e prospettive di ricerca*, Rubbettino: Soveria Mannelli 2006.

Antonielli, Donati 2003 (eds): L. Antonielli, C. Donati (eds), *Corpi armati e ordine pubblico in Italia (XVI-XIX secolo)*, Rubbettino: Soveria Mannelli 2003.

Arangio-Ruiz 1938: V. Arangio-Ruiz, La legislazione, in *Augustus. Studi in occasione del bimillenario augusteo*, Accademia Nazionale dei Lincei: Roma 1938, pp. 101–146.

Arcaria 1992: F. Arcaria, *Senatus censuit. Attività giudiziaria ed attività normativa del senato in età imperiale*, Giuffré: Milano 1992.

Arcaria 2006: F. Arcaria, In tema d'origine della giurisdizione penale senatoria, in M.P. Baccari, C. Cascione (eds), *Tradizione romanistica e Costituzione*, vol. 2, Edizioni Scientifiche Italiane: Napoli, pp. 1055–1095.

Arcaria 2007: F. Arcaria, Sul dies a quo della giurisdizione criminale senatoria, in *Fides, humanitas, ius. Studi in onore di Luigi Labruna*, vol. 1, Editoriale Scientifica: Napoli 2007, pp. 183–214.

Arcaria 2009: F. Arcaria, *Diritto e processo penale in età augustea. Le origini della cognitio criminale senatoria*, Giappichelli: Torino 2009.

Arcaria 2015: F. Arcaria, Gli aspetti processuali della vicenda di Cornelio Gallo, in F. Rohr Vio, E.M. Ciampini (eds), *La lupa sul Nilo. Gaio Cornelio Gallo tra Roma e l'Egitto*, Edizioni Ca' Foscari – Digital Publishing: Venezia 2015, pp. 107–162.

Arends 2008: J.F.M. Arends, From Homer to Hobbes and Beyond: Aspects of Security in the european Tradition, in H.G. Brauch, Ú O. Spring, C. Mesjaszet, J. Grin, P. Dunay, N.C. Behera, B. Chourou, P. Kameri-Mbote, P.H. Liotta (eds), *Globalization and Environmental Challenges: Reconceptualizing Security in the 21st Century* (Hexagon Series on Human and Environmental Security and Peace, 3), Springer: New York 2008, pp. 263–277.

Arrighetti 2000: G. Arrighetti, Filodemo fra poesia, mito e storia, in M. Erler, R. Bees (eds), *Epikureismus in der späten Republik und der Kaiserzeit* (Akten der 2. Tagung der Karl-und-Gertrud-Abel-Stiftung vom 30 September–3 Oktober 1998 in Würzburg), F. Steiner Verlag: Stuttgart 2000, pp. 13–31.

Arrigoni Bertini 2003: M.G. Arrigoni Bertini, L'uso epigrafico dei militari parmensi. Carriera, mobilità sociale, evergetismo, in M.G. Angeli Bertinelli, A. Donati (eds), *Usi e abusi epigrafici, Atti del Colloquio Internazionale di Epigrafia Latina* (Genova 20–22 September 2001) (*Serta antiqua*, 6), G. Bretschneider: Roma 2003, pp. 93–111.

Arnaldi 1995: A. Arnaldi, Un centurio speculatorum Augusti a Paliano (Frosinone), in *Scritti di Antichità in memoria di Benita Sciarra Bardaro*, Schena: Fasano 1995, pp. 235–239.

Atkins 2008: E.M. Atkins, *Cicero* (Cambridge Histories Online), Cambridge University Press: Cambridge 2008.

Austin, Rankov 1995: N.J.E. Austin, N.B. Rankov, *Exploratio: Military and Political Intelligence in the Roman World from the Second Punic War to the Battle of Adrianople*, Routledge: London, New York 1995.

Avallone 1962: R. Avallone, *Mecenate*, Libreria scientifica editrice: Napoli 1962.

Bablitz 2007: L. Bablitz, *Actors and Audience in the Roman Courtroom*, Routledge, London, New York 2007.

Bailie Reynolds: P. K. Baillie Reynolds, *The Vigiles of Imperial Rome*, Oxford University Press: Oxford 1926.

Barnes 1989: T.D. Barnes, Emperors on the Move, in *Journal of the Roman Archaeology*, 2, 1989, p. 254 (review of Halfmann 1989).

Bastianelli 1954: S. Bastianelli, *Italia romana: municipi e colonie*, ser. I, vol. 14. Centumcellae (*Civitavecchia*), Castrum Novum (*Torre Chiaruccia*), Istituto di Studi Romani, Roma 1954.

Bauman 1967: R.A. Bauman, *The Crimen Maiestatis in the Roman Republic and Augustan Principate*, Witwatersrand University Press: Johannesburg 1967.

Bellen 1981: H. Bellen, *Die germanische Leibwache der römischen Kaiser des julisch-claudischen Hauses*, Akademie der Wissenschaften und der Literatur: Bonn 1981.

Bellomo 2009: B. Bellomo, Potere e marginalità. Il defensor tra potentes e latrones, in *Mediterraneo antico* 12, 2009, pp. 257–268.

Beltrami 2008: L. Beltrami, Il De clementia di Seneca: un contributo per l'analisi antropologica del valore della clementia, in G. Picone (ed.), *Clementia Caesaris. Modelli etici, parenesi e retorica dell'esilio*, Palumbo: Palermo 2008, pp. 11–38.

Benferhat 2005: L.Y. Benferhat, Ciues Epicurei. *Les épicuriens et l'idée de monarchie à Rome et en Italie de Sylla à Octave*, Latomus: Bruxelles 2005.

Bérard 1988: F. Bérard, Le rôle militaire de les cohortes urbaines lyonnaise, in *Mélanges de l'École française de Rome: Antiquité*, 100, 1988, pp. 159–182.

Bérard 1991: F. Bérard, Aux origines de la cohorte urbaine de Carthage, in *Antiquités Africaines*, 1991, pp. 39–51.

Bérard 1992: F. Bérard, Vie, mort et culture des veterans d'après les inscriptions de Lyon, in *Revue des Etudes Latines*, 70, 1992, pp. 166–192.

Bérard 1993: F. Bérard, La garnison de Lyon à l'époque julio-claudienne, in Y. Le Bohec (ed.), *Militaires romains en Gaule civile* (Actes de la Table Ronde, Lyon 1991), Université de Lyon III: Lyon 1993, pp. 9–22.

Bérard 1994a: F. Bérard, Les evocati de la cohorte urbaine lyonnaise, in Y. Le Bohec, J.L. Voisin (eds), *L'Afrique, la Gaule, la Religion à l'époque romaine. Mélanges à la memoire de Marcel Le Glay*, Latomus: Bruxelles 1994, pp. 390–400.

Bérard 1994b: F. Bérard, Brétagne, Germanie, Danube: mouvements de troupes et priorités stratégiques sous le règne de Domitien, in J.M. Pailler, R. Sablayrolles (eds), *Les Années Domitien*, Presse Univertaire du Mirail: Toulouse 1994, pp. 221–240.

Bérard 1995: F. Bérard, La cohorte urbaine de Lyon: une unité à part dans la Rangordnung? in Y. Le Bohec (ed.), *La Hiérarchie (Rangordnung) de l'armée romaine sous le haut-empire*. Actes du congrès de Lyon (15–18 September 1994), De Boccard: Paris 1995, pp. 373–382.

Bérard 2000: F. Bérard, La garnison de Lyoin et l'officium du gouverneur de Lyonnaise, in G. Alföldy, W. Eck, B. Dobson (eds), *Kaiser, Heer und Gesellschaft* (Habes, 31), F. Steiner Verlag: Stuttgart 2000, pp. 279–305.

Bérard 2006: F. Bérard, Quelques officiales entre Lyon et Rome, in Les unitès de la garnison urbaine et leur épigraphie à Rome et hors de Rome, in *Cahiers du Centre Glotz* 14, 2004 [2006], pp. 357–369.

Bérenger-Badel 2002: A. Bérenger-Badel, La circulation de l'information d'après la correspondance de Pline le Jeune avec Trajan, in J. Andreau, C. Virlouvet (eds), *L'information et la mer dans le monde antique*, École française de Rome: Roma 2002, pp. 219–231.

Bingham 1997: S.J. Bingham, *The Praetorian Guard in the Political and Social Life of Julio-Claudian Rome*, Doctoral Thesis, University of British Columbia 1997.

Bingham 1999: S.J. Bingham, Security at the Games in the Early Imperial Period, in *Echoes du monde antique – Classical Views*, 18, 1999, pp. 369–379.

Bingham 2013: S.J. Bingham, *The Praetorian Guard. A History of Rome's Elite Special Forces*, The Baylor University Press: Waco 2013.

Binder 2001: C. Binder: *Securitas*, in *Der Neue Pauly*, Band 11, Stuttgart 2001, p. 317.

Birley 1987: A.R. Birley, *Marcus Aurelius: A Biography*, Batsford: London 1987 (rev. edn).

Bivona 1978: L. Bivona, Un 'urbaniciano' a Thermae, in *Kokalos*, 24, 1978, pp. 112–127.

Bleicken 1990: J. Bleicken, *Zwischen Republik und Prinzipat. Zum Charakter des Zweiten Triumvirats*, Vandenhoeck & Ruprecht: Göttingen 1990.

Boatwright 1987: M.T. Boatwright, *Hadrian and the City of Rome*, Princeton University Press: Princeton 1987.

Boatwright 2010: M.T. Boatwright, Antonine Rome: Security in the Homeland, in C. Ewald, C.F. Noreña (eds), *The Emperor and Rome: Space, Representation and Ritual*, Cambridge University Press: Cambridge 2010, pp. 169–197.

Boatwright 2015: M.T. Boatwright, Visualizing Empire in Imperial Rome, in L. Brice, D. Slootjes (eds), *Ancient World Views: Institutions and Geography from the Greco-Roman World*, Brill: Leiden, Boston pp. 235–259.

Borriello, D'Ambrosio 1979: M.R. Borriello, A. D'Ambrosio (eds), *Baiae-Misenum (Forma Italiae* I 14), Olsckhi: Firenze 1979.

Braccesi 2007: L. Braccesi, *Terra di confine. Archeologia e storia tra Marche, Romagna e San Marino*, L'Erma di Bretschneider: Roma 2007.

Braund 1998 (ed.): D. Braund (ed.), *The Administration of the Roman Empire (241 BC–AD 193)*, University of Exeter Press: Exeter 1998.

Breeze, Maxfield, Davies (eds) 1989: D. Breeze, V.A. Maxfield, R.W. Davies, *Service in the Roman Army*, Columbia University Press: New York 1989.

Brélaz 2005: C. Brèlaz, *La sécurité publique en Asie mineure sous le Principat (Ier–IIIème s. ap. J.-C.). Institutions municipales et institutions impériales dans l'Orient romain*, Schwabe: Basel 2005.

Brélaz 2007: C. Brèlaz, Lutter contre la violence à Rome: attributions étatiques et tâches privées, in C. Wolff (ed.), *Les exclus dans l'Antiquité*, Université de Lyon III: Lyon, 2007, pp. 219–239.

Brélaz, Ducrey 2008 (ed.): C. Brélaz, P. Ducrey (eds), *Sécurité collective et ordre public dans les sociétés anciennes* (Fondation Hardt, *Entretiens sur l'Antiquité classique* 54), Fondation Hardt: Vandoeuvres-Genève 2008.

Brunelli 1998: G. Brunelli, Presenza dei militari a Roma tra Cinque e Seicento, in F. Sonnino (ed.), *Popolazione e società a Roma dal Medioevo all'età contemporanea*, Il Calamo: Roma 1998, pp. 359–371.

Brunelli 2003: G. Brunelli, Istituzioni militari e 'controllo del territorio' nello Stato della Chiesa dell'età moderna, in *Le armi dello Stato. Militari e sicurezza interna in Italia dall'Ancien régime alla Repubblica* (Convegno, Roma – Biblioteca Alessandrina, 15–16 May 2003), unpublished contribution.

Buonopane 2006–2007: A. Buonopane, Le iscrizioni romane di Grumentum: rivisitazioni e novità da studi recenti, in *Rendiconti della Pontificia Accademia Romana di Archeologia*, 79, 2006–2007, pp. 315–342.

Busch 2005: A.W. Busch, Kamaraden bis in den Tod? Zur militärischen Sepulkraltopographie im Kaiserzeitlichen Rom, in W. Kockel, R. Neudecker, P. Zanker (eds), *Lebenswelten. Bilder und Räume in der römischen Stadt der Kaiserzeit (Palilia, 16)*, L. Reichert: Wiesbaden 2005, pp. 101–111.

Busch 2007: A. W. Busch, 'Militia in urbe': The Military Presence in Rome, in De Blois (ed.) 2007, pp. 315–341.

Busch 2011: A.W. Busch, *Soldaten in Rom. Militärische und paramilitärische Einheiten im kaiserzeitlichen Stadtbild* (Palilia 20), L. Reichert: Wiesbaden 2011.

Busch 2012: A. W. Busch, Schutz und Verteidigung kaiserlicher Residenzen und Villen im Spiegel der archäologischen und literarischen Quellen, in F. Arnold, A.W. Busch, R. Haensch, U. Wulf-Rheidt (eds), *Orte der Herrschaft. Charakteristika von antiken Machtzentren*, VML Verlag Marie Leidorf: Rahden/Westfalien 2012, pp. 113–123.

Busch 2013a: A.W. Busch, Kaiserzeitliche Wehrarchitektur im Zentrum des Römischen Reiches, in Ch. Flügel, J. Obmann (eds), *Römische Wehrbauten: Befund und Rekonstruktion*, Egon Johannes Greipl: München, 2013, pp. 113–131.

Busch 2013b: A.W. Busch, Militär im Severischen Rom: Bärtige Barbaren? in N. Sojc, A. Winterling, U. Wulf-Rheidt (eds), *Palast und Stadt im Severischen Rom*, F. Steiner Verlag: Stuttgart 2013, pp. 105–121.

Caballos, Eck, Fernandez 1996: A. Caballos, W. Eck, F. Fernández, *El senadoconsulto de Gneo Pisón padre*, Sevilla: Universidad de Sevilla 1996.

Cagiano de Azevedo 1939: M. Cagiano de Azevedo, Una dedica abrasa e i rilievi puteolani dei musei di Filadelfia e Berlino, in *Bullettino del Museo Imperiale* 10, 1939, pp. 45–56.

Cagnat 1912: R. Cagnat, *L'armée romaine d'Afrique et l'occupation militaire de l'Afrique sous les empereurs*, Imprimerie Nationale: Paris 1912.

Caldelli 2014: M.L. Caldelli, Il funzionamento delle infrastrutture portuali ostiensi nella documentazione epigrafica, in C. Zaccaria (ed.) *L'Epigrafia dei porti (Actes de la XVIIe Rencontre franco-italienne sur l'épigraphie du monde romaine*, Aquileia 14–16 October 2010), Editreg: Trieste 2014, pp. 65–80.

Caldelli, Ricci 1999: M. L. Caldelli, C. Ricci, *Monumentum familiae Statiliorum. Un riesame* (Libitina 1), Quasar: Roma 1999.

Caldelli, Petraccia, Ricci 2012: M.L. Caldelli, M.F. Petraccia, C. Ricci, Praesidia Urbis et Italiae. I mestieri della tutela e della sicurezza, in C. Wolff (ed.), *Le métier de soldat dans le monde romaine* (Actes du Ve congrès de Lyon, 23–25 September 2010), De Boccard: Paris, 2012, pp. 341–364.

Camodeca 1999: G. Camodeca, *Tabulae Pompeianae Sulpiciorum. Edizione critica dell'*
Archivio puteolano dei Sulpicii, Quasar: Roma 1999.
Camodeca 2007: G. Camodeca, Sulle proprietà imperiali in Campania, in D. Pupillo
(ed.), *Le proprietà imperiali nell'Italia romana. Economia, produzione, amministrazione*, Le
Lettere: Firenze 2007, pp. 143–167.
Camodeca 2008: G. Camodeca, *I ceti dirigenti di rango senatorio equestre e decurionale della*
Campania romana 1, Satura: Napoli 2008.
Campesi 2009: G. Campesi, *Genealogia della pubblica sicurezza. Teoria e storia del moderno*
dispositivo poliziesco, Ombre Corte: Verona 2009.
Canfora 1993: L. Canfora, Il processo di Cremuzio Cordo (Annali IV, 34–35), in *Studi*
di storia della storiografia romana, Edipuglia: Bari 1993, pp. 221–246.
Capogrossi Colognesi 2009: L. Capogrossi Colognesi, *Storia di Roma tra diritto e potere*,
Il Mulino: Bologna 2009.
Capponi, Mengozzi 1993: S. Capponi, B. Mengozzi, *I vigiles dei Cesari. L'organizzazione*
antincendio nell'antica Roma, Pieraldo: Roma 1993.
Carnabuci 2006: E. Carnabuci, La nuova Forma del Foro di Augusto: considerazioni
sulle destinazioni d'uso degli emicicli, in R. Meneghini, R. Santangeli Valenzani
(eds), *Formae Urbis Roma: nuovi frammenti di piante marmoree dallo scavo dei Fori*
Imperiali, L'Erma di Bretschneider: Roma 2006, pp. 173–195.
Carriè 1993: J.M. Carrié, Eserciti e strategie, in A. Momigliano, A. Schiavone (eds),
Storia di Roma, III, 1: *l'età tardoantica, crisi e trasformazioni*, Torino: Einaudi 1993,
pp. 83–154.
Carriè, Rousselle 1999: J.M. Carriè, A. Rousselle, *L'Empire Romain en mutation des*
Sévères à Constantin, Paris: Seuil 1999.
Caruso 2000: G. Caruso, R. Volpe, Preesistenze e persistenze delle Terme di Traiano,
in E. Fentress (ed.), *Romanization and the City: Creation, Dynamics and Failures*, (JRA
Suppl. 38), 2000 pp. 42–56.
Cascione 1999: C. Cascione, *Tresviri capitales. Storia di una magistratura minore*, Editoriale
Scientifica s.r.l.: Napoli 1999.
Cascione 2016: C. Cascione, Polizia, giurisdizione, corruzione: prospettive (e un caso)
della Roma repubblicana, in A. Palma (ed.), *Civitas e civilitas. Studi in onore di*
F. Guizzi, I, Giappichelli: Torino 2016, pp. 187–214.
Cassieri 2008: N. Cassieri, La villa 'Spelunca' di Tiberio a Sperlonga, in Valenti 2008,
pp. 11–26.
Cassieri, Ghini 1990: N. Cassieri, G. Ghini, La cosidetta villa degli Antonini al XVIII
miglio della via Appia, in *Arch. Laz.* 10, 1, 1990, pp. 168–178.
Cattaneo 2011: C. Cattaneo, *Salus publica populi Romani*, Victrix: Forlì 2011.
Cavallaro 1975–1976: M.A. Cavallaro, Un liberto 'prega' per Augusto: CIL, VI 30975,
in *Helikon* 15–16, 1975–1976, pp. 179–180.
Cavillier 2007: G. Cavillier, Ricerche sui castra praetoria. riflessioni su di un modello
di architettura militare di età imperiale attraverso la rilettura di alcuni elementi, in
M. Mayer Olivé, G. Baratta, A. Guzmán Almagro (eds), *Acta XII Congressus inter-*
nationalis epigraphiae Graecae et latinae. Acta 1, Institut d'Estudis Catalans: Barcelona
2007, pp. 259–268.
Chastagnol 1960: A. Chastagnol, *La préfecture urbaine à Rome sous le Bas Empire*, Presses
Universitaires de France: Paris 1960.
Chastagnol 1992–1993: A. Chastagnol, Trente ans après. Les préfects de la ville de
Rome, in *Scienze dell'Antichità: Storia, Archeologia, Antropologia* 6–7, 1992–1993,
pp. 487–497.

Chastagnol 1997: A. Chastagnol, Le fonctionnement de la préfecture urbaine, in *La Rome impériale. Demographie et logistique* (CEFR 230), École française de Rome: Rome 1997, pp. 111–119.

Chausson 2012: F. Chausson, La fausse immobilité du Prince. Remarques préliminaires sur la présence du Prince à Rome et dans ses environs, in A. Hostein, S. Lalanne (eds), *Les voyages des empereurs dans l'orient romain: époques antonine et sévérienne*, Editions Errance: Arles 2012, pp. 17–35.

Chillet 2012: C.L. Chillet, Vers de nouvelles formes de décisions: Auguste, Mécène et la préfecture de la Ville, in Rivière 2012 (ed.), pp. 185–222.

Christol 2012: M. Christol, La collégialité après Plautien. La préfecture du prétoire jusqu'à la fin du règne de Caracalla, in E. Chevreau, D. Kremer, A. Laquerrière-Lacroix (eds), *Carmina iuris. Mélanges en l'honneur de Michel Humbert*, De Boccard: Paris 2012, pp. 87–100.

Cifani, Fusco, Munzi 1999–2000: G. Cifani, U. Fusco, M. Munzi, Indagini topografiche nel suburbio di Grumentum. Le dinamiche insediative, in *Archeologia Classica* 51, 1999–2000, pp. 439–445.

Classen 1991: C.J. Classen, Virtutes imperatoriae, in *Arctos* 25, 1991, pp. 17–39.

Clauss 1973: M. Clauss, Untersuchungen zur den Principales des römischen Heeres von Augustus bis Diokletian. Cornicularii, speculatores, frumentarii, diss. Bochum 1973, pp. 87–90.

Cloud 1963: D. Cloud, The Text of Digest XLVIII, 4: ad legem Iuliam maiestatis, in *Zeitschrift der Savigny-Stiftung für Rechtsgeschichte* 80, 1963, pp. 206–232.

Cloud 1969: J.D. Cloud, The Primary Purpose of the lex Cornelia de sicariis et veneficiis, in *Zeitschrift der Savigny-Stiftung für Rechtsgeschichte* 86, 1969, pp. 258–286.

Cloud 1988: J.D. Cloud, Lex Iulia de vi: Part I, in *Athenaeum*, 66, 1988, pp. 579–595.

Cloud 1989: J.D. Cloud, Lex Iulia de vi: Part II, in *Athenaeum*, 67, 1989, pp. 427–465.

Cloud 1994: D. Cloud, The Constitution and Public Criminal Law, in J.A. Crook, A. Lintott, E. Rawson (eds), *Cambridge Ancient History IX: The Last Age of the Roman Republic, 146–43 BC*, Cambridge University Press: Cambridge 1994, pp. 491–530.

Coarelli 1986: F. Coarelli, L'Urbs e il suburbio, in *Società romana e impero tardoantico*, II, *Roma: politica, economia, paesaggio urbano*, Laterza: Bari, Roma 1986, pp. 1–56.

Coarelli 1987: F. Coarelli, La situazione edilizia di Roma sotto Severo Alessandro, in *L' Urbs. Espace urbain et histoire* (CEFR, 98), École française de Rome: Roma 1987, pp. 429–456.

Coarelli 1993: F. Coarelli, Castra urbana, in *LTUR* I, 254–255.

Coarelli 1999: F. Coarelli, Tellus, aedes, in *LTUR* V, pp. 24–25.

Coarelli 2001: F. Coarelli, *Roma. Guida archeologica*, Laterza: Roma-Bari 2001.

Coarelli 2008: F. Coarelli, *Roma*, Laterza: Bari 2008 (edizione riveduta e aggiornata).

Coarelli 2009: F. Coarelli, Il pomerio di Vespasiano e Tito, in *La lex de imperio Vespasiani e la Roma dei Flavi* (Atti del Convegno, Roma 2008), L'Erma di Bretschneider: Roma 2009, pp. 299–309.

Coarelli 2010: F. Coarelli, La Basilica di Massenzio e la Praefectura Urbis, in G. Bonamente, R. Lizzi Testa (eds), *Istituzioni, carismi ed esercizio del potere* (IV–VI secolo d.C.), Edipuglia: Bari 2010, pp. 133–146.

Corbier 1983: M. Corbier, Fiscus and Patrimonium: The Saepinum Inscription and Transhumance in the Abruzzi, in *Journal of Roman Studies* 73, 1983, pp. 126–131.

Corbier 1991: M. Corbier, *La romanisation du Samnium aux IIe et Ier siècles av. J.-C.* (Actes du colloque. Naples – Centre Bérard, 4–5 November 1988), Bibliothèque de l'Institute français de Naples: Naples 1991, pp. 169–176.

Corbier 1999: M. Corbier, Maiestas domus Augustae, in *Bulletin de la Société nationale des antiquaires de France*, 1999, pp. 261–274.

Corbier 2001: Maiestas domus Augustae, in M.G. Angeli Bertinelli, A. Donati (eds), *Varia Epigraphica* (Atti del Convegno Internazionale Borghesi, Bertinoro 8–10 June 2000), Fratelli Lega: Faenza 2001, pp. 155–199.

Corbier 2006: M. Corbier, *Donner à voir, donner à lire. Mémoire et communication dans la Rome ancienne*, CNRS Éditions: Paris 2006.

Corbier 2007: M. Corbier, Proprietà imperiale e allevamento transumante in Italia, in D. Pupillo (ed.), *Le proprietà imperiali nell'Italia romana. Economia, produzione, amministrazione*, Le Lettere: Firenze 2007, pp. 1–48.

Coriat 1997: J.-P. Coriat, *Le prince législateur: la technique législative des Sévères et les méthodes de création du droit impérial à la fin du Principat* (BEFAR 294), École française de Rome: Roma 1997.

Coriat 2007: J.-P. Coriat, Les préfets du prétoire de l'époque sévérienne: un essai de synthèse, in *Cahiers du Centre Glotz* 18, 2007, pp. 179–198.

Coroi 1915: J. Coroi, *La violence en droit criminel romain*, Pion Nourrit: Paris 1915.

Cosme 2011: P. Cosme, Les Bataves au centre et à la périphérie de l'Empire: quelques hypothèses sur les origines de la révolte de 69–70, in O. Hekster, T. Kaizer (eds), *Frontiers in the Roman World* (Proceedings of the IX Workshop of the International Network Impact of Empire), Brill: Leiden, Boston 2011, pp. 305–320.

Cosme 2012: P. Cosme, Les réformes militaires augustéennes, in Rivière 2012 (ed.), pp. 171–184.

Correnti 1990: F. Correnti, Centumcellae: la villa, il porto, la città, in A. Maffei, E. Nastasi, *Caere e il suo territorio. Da Agylla a Centumcellae*, Libreria dello Stato: Roma, 1990, pp. 209–214.

Coulston 2000: J.C.N. Coulston, Armed and Belted Men: The Soldiery of Imperial Rome, in J.C.N. Coulston, H. Dodge (eds), *Ancient Rome: The Archaeology of the Eternal City*, Oxford University School of Archaeology: Oxford, 2000, pp. 76 118.

Cramer 1954. F.H. Cramer, *Astrology in Roman Law and Politics*, American Philosophical Society: Philadelphia 1954.

Crawford 1976: D.J. Crawford, Imperial Estates, in M.I. Finley (ed.), *Studies in Roman Property*, Cambridge University Press: Cambridge, 1976, pp. 35–70.

Crawford 1996 (ed.): D.J. Crawford (ed.), with contributions by J.D. Cloud, R.G. Coleman, M.H. Crawford et al., *Roman Statutes*, I–II, Institute of Classical Studies: London 1996.

Crea 2006: S. Crea, Aree sepolcrali all'interno di fondi imperiali. Il caso del sepolcreto nella villa dell'Albanum, in *Rendiconti dell'Accademia dei Lincei*, 9.17, 2006, pp. 145–170.

Crimi 2008: G. Crimi, Iscrizioni inedite (o quasi) di pretoriani da Roma, in *Aquila legionis* 10, 2008, pp. 23–38.

Crimi 2009a: G. Crimi, P. Vennonius L.f. Ste.: uno speculator originario di Augusta Taurinorum? in *Epigraphica* 71, 2009, pp. 360–365.

Crimi 2009b: G. Crimi, Un 'nuovo' pretoriano di Fanum Fortunae, in *Picus* 29, 2009, pp. 173–178.

Crimi 2009–2010: G. Crimi, Tituli militum praetorianorum. *Ricerche sulle coorti pretorie, 70 anni dopo le opere di Marcel Durry e Alfredo Passerini*. Tesi di dottorato i, discussa a Sapienza Università di Roma a.a. 2009–2010.

Crimi 2010: G. Crimi, Tribù e origo nelle iscrizioni dei pretoriani e urbaniciani arruolati in Italia: tre nuove attestazioni epigrafiche, in M. Silvestrini (ed.), *Le tribù romane (Atti della XVI Rencontre sur l'épigraphie du monde romaine)*, Edipuglia: Bari 2010, pp. 329–336.

Crimi 2012: G. Crimi, Il mestiere degli speculatores: nuovi dati e ricerche dopo gli studi di Manfed Clauss, in C. Wolff (ed.), *Le métier de soldat dans le monde romaine (Actes du Ve congrès de Lyon*, 23–25 September 2010), De Boccard: Paris, 2012, pp. 491–499.

Cristilli 2011: A. Cristilli, Surrentum ductus amoenum. Sculture in marmo dalla 'c.d. Villa di Agrippa Postumo' a Sorrento, in *Oebalus*, 6, 2011, pp. 179–213.

Crogiez 2002: S. Crogiez, Le cursus publicus et la circulation des informations officielles par voie de mer, in *L'information et la mer dans le monde antique* (CEFR, 297), a cura di J. Andreau e C. Virlouvet, École française de Rome: Roma, 2002, pp. 55–67.

Daguet–Gagey 2000: A. Daguet–Gagey, I grandi servizi pubblici, in E. Lo Cascio (ed.), *Roma imperiale: Una metropoli antica*, Carocci: Roma 2000, pp. 71–102.

Dahmen, Ilisch 2006: K. Dahmen, P. Ilisch, Securitas: A New Revival of a Probus Reverse-Type in the Gold Coinage of Constantine I, in *The Numismatic Chronicle* 166, 2006, pp. 229–231.

Dana, Ricci 2014: D. Dana, C. Ricci, Divinità provinciali nel cuore dell'impero. Le dediche dei militari traci nella Roma imperiale, in *Mélanges de l'École française de Rome: Antiquité* 126.2, 2014. http://mefra.revues.org/2569 (accessed 3 September 2017).

D'Arms 1970: J.H. D'Arms, *Romans on the Bay of Naples: A Social and Cultural Study of the Villas and their Owners from 150 BC to AD 400*, Harvard University Press: Harvard-Cambridge, 1970.

D'Arms 1975: J.H. D'Arms, Tacitus, Annals 13, 48 and a New Inscription from Puteoli, in B. Levick (ed.), *The Ancient Historian and his Materials. Essays C.E. Stevens*, Westmead: Farnborough 1975, pp. 155–165.

D'Arms 1977: J.H. D'Arms, *Proprietari e ville nel golfo di Napoli*, in *I Campi Flegrei nell'archeologia e nella storia* (Convegno internazionale, Roma, 4–7 May 1976), Accademia Nazionale dei Lincei: Roma 1977, pp. 347–363.

D'Arms 2003: J. H. D'Arms, *Romans on the Bay of Naples and Other Essays on Roman Campania*, Laterza: Roma-Bari 2003.

Darwall-Smith 1994: R.H. Darwall-Smith, Albanum and the Villas of Domitian, in *Pallas*, 40, 1994, pp. 145–165.

Darwall-Smith 1996: R.H. Darwall-Smith, *Emperors and Architecture: A Study on Flavian Rome* (Collection Latomus 231), Latomus: Bruxelles 1996.

De Blois (ed.) 2007: L. De Blois (ed.), *The Impact of the Roman Army (200 B.C.–A.D. 476). Economic, Social, Political, Religious and Cultural Aspects* (Proceedings of the VI Workshop of the International Network Impact of Empire), Brill: Leiden, Boston 2007.

De Blois, Erdkamp, Hekster, de Kleijn, Mols (eds) 2003: L. De Blois, P. Erdkamp, O. Hekster, G. de Kleijn, S. Mols (eds), *The Representation and Perception of the Roman Imperial Power*, Gieben: Amsterdam 2003.

De Caro 1975: S. De Caro, Scavi nell'area fuori porta Nola a Pompei, in *Cron.Pomp.*, 5, 1979, pp. 61–101.

De Caro, Vecchio 1994: S. De Caro, G. Vecchio, Pausilypon: la villa imperiale, in F. Zevi (ed.), *Neapolis*, Banco di Napoli: Napoli 1994, pp. 83–94.

De Kleijn 2003: G. De Kleijn, The Emperor and Public Works in the City of Rome, in De Blois, Erdkamp, Hekster, de Kleijn, Mols (eds) 2003, pp. 207–2014.

Della Corte 1980: F. Della Corte, La breve praefectura urbis di Messalla Corvino, in *Philìas charin. Miscellanea Manni*, Giorgio Bretschneider: Roma 1980, pp. 669–677.

Del Re 1972: N. Del Re, *Monsignor Governatore di Roma*, Libreria Editrice Vaticana: Città del Vaticano 1972.

Del Re 1993: N. Del Re, *La Curia Capitolina e tre altri antichi organi giudiziari romani*, Fondazione Marco Besso: Roma 1993.

De Robertis 1937: F.M. De Robertis, La repressione penale nella circoscrizione dell'Urbe. Il praefectus urbi e le autorità concorrenti, Bari 1937 ora in *Scritti vari di diritto romano* 3, Diritto penale, Cacucci: Bari 1987, pp. 35–104.

De Robertis 1974: F.M. De Robertis, *Storia delle corporazioni e del regime associativo nel mondo romano*, Adriatica editrice: Bari 1974.

De Robertis 1987: F.M. de Robertis, *Scritti varii di diritto romano*, II, Cacucci: Bari, 1987.

Devillers 2009: O. Devillers, Sed aliorum exitus, simul cetera illius aetatis, memorabo (An., III, 24, 2). *Le régne d'Auguste et le projet historiographique de Tacite*, in F. Hurlet, B. Mineo (eds), *Le principat d'Auguste: réalités et représentations du pouvoir: autour de la 'Res publica restituta'* (Actes du colloque de l'Université de Nantes, 1–2 June 2007), Presse Unversitaire: Rennes 2009, pp. 309–324.

Dietz 1980: K. Dietz, *Senatus contra principem. Untersuchungen zur senatorischen Opposition gegen Kaiser Maximinus Thrax*, C.H. Beck: München 1980.

Dietz 1983: K. Dietz, Caracalla, Fabius Cilo und die Urbaniciani. Unerkannt gebliebene Suffektconsuln des Jahres 212 d.C., in *Chiron* 13, 1983, pp. 381–404.

Di Giuseppe 2007: H. Di Giuseppe, I Bruttii Praesentes, proprietari e produttori in Val d'Agri, in H. Di Giuseppe, M.P. Gargano, A. Russo, A. (eds), *Dalla villa dei Bruttii Praesentes alla proprietà imperiale: il complesso archeologico di Marsicovetere–Barricelle* (PZ), in *Siris* 8, 2007, pp. 81–119.

Di Giuseppe 2010: H. Di Giuseppe, I Bruttii Praesentes: interessi politici ed economici di un'importante famiglia lucana, in F. Tarlano (ed.), *Il territorio grumentino e la valle dell'Agri nell'antichità* (Atti della Giornata di Studi Grumento Nova–Potenza, 25 April 2009), BraDypUS: Bologna 2010, pp. 39–47.

Dobson 1978: B. Dobson, Die Primipilares. *Entwicklung umd Bedeutung, Laufbahnen nmd Persönlichkeiten eines römischen Offiziersranges* (Beihbl. Bonn.Jahrbuch. 37), R. Habelt Verlag: Köln, 1978.

Dorandi (ed.) 1982: T. Dorandi (ed.), *Filodemo. Il buon re secondo Omero*, Bibliopolis: Napoli 1982.

Dorandi 1997: T. Dorandi, Lucrèce et les Épicuriens de Campanie, in K.A. Algra, M.H. Koenen, P.H. Schrijvers (eds), *Lucretius and his Intellectual Background* (Proceedings of the colloquium Amsterdam, 26–28 June 1996), Royal Netherland Academy: Amsterdam 1997, pp. 35–48.

Dupré Raventós, Remolà (eds) 2000: X. Dupré Raventós, J.A. Remolà (eds), *Sordes Urbis: la eliminación de residuos en la ciudad romana* (Actas de la Reunión de Roma, 15–16 November 1996), L'Erma di Bretschneider: Roma 2000.

Durry 1938: M. Durry, *Les cohortes prétoriennes*, Bibliothèque des Écoles françaises d'Athènes et de Rome: Paris 1938.

Duval 1993: N. Duval, À propos de la garnison de Lyon: le problème de la composition de la garnison de Carthage, in Y. Le Bohec (ed.), *Militaires romains en Gaule civile* (Actes de la Table Ronde, Lyon 1991), Université de Lyon III; Lyon 1993, pp. 23–27.

Echols 1958: E. Echols, The Roman City Police: Origin and Development, in *Classical Journal* 53.8 (May 1958), pp. 377–385.

Echols 1961: E. Echols, The Provincial Urban Cohorts, in *Classical Journal* 57, 1961, pp. 25–28.

Eck 1979: W. Eck, Die staatliche Organisation Italiens in der hohen Kaiserzeit, Beck: München 1979 (now also W. Eck, *L'Italia nell'Impero Romano. Stato e amministrazione in epoca imperiale*, 2. edizione rivista, Edipuglia: Bari 1999.

Eck 1986: W. Eck, Augustus' administrative Reformen: Pragmatismus oder systematisches Planen? in *Acta Classica* 29, 1986, pp. 105–120 (now also Eck 1995, pp. 83–102; and Eck 2009).

Eck 1995: W. Eck, *Die Verwaltung des Römischen Reiches in der Hohen Kaiserzeit. Ausgewählte Beiträge*. 2 *Bände*, Reinhardt: Basel 1995.

Eck 1996a: W. Eck, I sistemi di trasmissione delle comunicazioni d'ufficio in età altoimperiale, in M. Pani (ed.), *Epigrafia e territorio, politica e società: Temi di antichità romane*, IV, Edipuglia: Bari, 1996, pp. 331–352.

Eck 1996b: W. Eck, *Horti: P. Cn. Dolabella*, in LTUR III, p. 58.

Eck 2001: W. Eck, Die Sonderregelungen für Soldatenkinder seit Antoninus Pius. Ein niederpannonisches Militärdiplom vom 11.Aug.146, in *Zeitschrift für Papyrologie und Epigraphik* 135, 2001, pp. 195–208.

Eck 2007: W. Eck, Die Veränderungen in Konstitutionen und Diplomen unter Antoninus Pius, in M.A. Speidel, H. Lieb (eds), *Militärdiplome. Die Forschungbeiträge der Berner Gespräche von 2004* (Mavors, 15), F. Steiner Verlag: Stuttgart 2007, pp. 87–104.

Eck 2009: W. Eck, The Administrative Reforms of Augustus: Pragmatism or Systematic Planning? in J. Edmondson (ed.) *Augustus, Readings on the Ancient World Series*, Edinburgh University Press: Edinburgh 2009, pp. 229–249.

Eck 2016: W. Eck, Herrschaftssicherung und römische Heer unter Augustus, in G. Negri, A. Valvo (eds), *Studi su Augusto in occasione del XX centenario della sua morte*, G. Giappichelli: Torino 2016, pp. 78–93.

Eich 2013: P. Eich, Politik und Administration unter den Severern, in N. Sojc, A. Winterling, U. Wulf-Rheidt (eds), *Palast und Stadt im Severischen Rom*, F. Steiner Verlag: Stuttgart 2013, pp. 85–104.

Ennabli, Ben Abdallah 1998: L. Ennabli, Z. Benzina ben Abdallah, Listes militaires découvertes dans la Basilique de Carthagenne, in *Epigraphica* 60, 1998, pp. 135–164.

Ensslin 1954: W. Ensslin, Praefectus vigilum, in RE XXII.2, 1954, pp. 1340–1347.

Erdkamp 1998: P. Erdkamp, *Hunger and the Sword: Warfare and Food Supply in Roman Republican Wars (264–30 B.C.)*, J.C. Gieben: Amsterdam 1998.

Fabia 1918: Ph. Fabia, *La garnison romaine de Lyon*, Cumin et Masson: Lyon 1918.

Färber 2012: R. Färber, Die Amtssitze der Stadtpräfekten im spätantiken Rom und Konstantinopel, in F. Arnold, A. Busch, R. Haensch, U. Wulf-Rheidt (eds), *Orte der Herrschaft. Charakteristika von antiken Machtzentren*, VML Verlag Marie Leidorf: Rahden/Westfalen 2012, pp. 49–71.

Fagan 2011: G.J. Fagan, Violence in Roman Social Relations, in *The Oxford Handbook of Social Relations in the Roman World*, Oxford University Press: Oxford 2011 pp. 468–495.

Faure 2001: P. Faure, Iulius Proculus, centurion frumentaire a peregrinis, in *Cahiers Centre Glotz* 12, 2001, pp. 307–308.

Faure 2003: P. Faure, Les centurions frumentaires et le commandement des castra peregrina, in *Mélanges de l'École française de Rome: Antiquité* 115, 2003/1, pp. 377–427.

Favro 1996: D. Favro, *The Urban Image of Augustan Rome*, Cambridge University Press: Cambridge 1996.

Favro 2006: D. Favro, Making Rome a World City, in K. Galinsky (ed.), *The Cambridge Companion of the Age of Augustus*, Cambridge University Press: Cambridge 2006, pp. 234–263.

Fears 1981: J. R. Fears, The Cult of Virtues and Roman Imperial Ideology, in ANRW II.17.2, 1981, pp. 827–948.

Federico, Miranda 1998: E. Federica, E. Miranda, *Capri antica. Dalla preistoria alla fine dell'età romana*, La Conchiglia: Roma 1998.

Ferrary 1983: J.-L. Ferrary, Les Origines de la loi de majesté à Rome, *in Comptes rendus de l'Académie des inscriptions et belles-lettres* 127, 1983, pp. 556–572.

Ferrary 1991: J.-L. Ferrary, Lex Cornelia de sicariis et veneficiis, in *Athenaeum* 79, 1991, pp. 417–434.

Ferrary 2001: J.-L. Ferrary, À propos des pouvoirs d'Auguste, in *Cahiers du Centre Gustave Glotz*, 12, 2001, pp. 101–154.

Fink 2012: M. Fink, Eine Villa des Domitian? Kampagne zur Aufnahme und Analyse architektonischer Strukturen im Circeo (Latium), in *Kölner und Bonner Archaeologica*, 22, 2012, pp. 141–155.

Flaig 1992: E. Flaig, *Den Kaiser herausfordern. Die Usurpation im Römischen Reich*, Campus Verlag: Stuttgart, New York 1992.

Flower 2001: H.I. Flower, A Tale of Two Monuments: Domitian, Trajan and Some Praetorians at Puteoli, in *American Journal of Archaeology*, 105, 2001, pp. 625–648.

Fora 1996: M. Fora, *Epigrafia anfiteatrale dell'occidente romano, IV Regio Italiae I: Latium (Vetera, 11)*, Quasar: Roma 1996.

Fosi 2007: L. Fosi, *La giustizia del papa. Sudditi e tribunali nello Stato Pontificio in età moderna*, Laterza: Bari, Roma 2007.

Foucault 1994: M. Foucault, Le jeu de Michel Foucault, in *Dits et écrits* I–IV, Gallimard: Paris 1994.

Foucault 2004: M. Senellart (ed.), M. Foucault, *Sicurezza, territorio, popolazione. Corso al Collège de France (1977–1978)*, Feltrinelli: Milano 2004.

Foucault 2015. P.A. Rovatti (ed.), M. Foucault, *Lezioni sulla volontà di sapere. Corso al Collège de France (1970–1971)*, Feltrinelli: Milano 2015.

Fraschetti 1990: A. Fraschetti, *Roma e il principe*, Laterza: Bari-Rome 1990.

Fraschetti 2000: A. Fraschetti, La città di Roma in epoca augustea. Amministrare sorvegliando, in C.l. Nicolet, R. Ilbert, J.-Ch. Depaule (eds), *Mégalopoles méditerranéennes. Geographie urbaine rétrospective* (Actes Colloque Roma, 8–11 May 1996), École française de Rome: Roma 2000, pp. 724–731.

Fraschetti 2008: A. Fraschetti, *Marco Aurelio o la miseria della filosofia*, Laterza: Roma-Bari 2008.

Fredrick 2003: D. Fredrick, Architecture and Surveillance in Flavian Rome, in A.J. Boyle, W.J. Dominik (eds), *Flavian Rome: Culture, Image, Text*, Brill: Leiden, Boston 2003, pp. 199–227.

Freis 1967: H. Freis, *Die cohortes urbanae: Epigraphische Studien* 2, Böhlau: Köln, Graz 1967.

Fuhrmann 2012: Ch. J. Fuhrmann, *Policing the Roman Empire: Soldiers, Administration, and Public Order*, Oxford University Press: Oxford 2012.

Fuhrmann 2016: C.H. Fuhrmann, Police Functions and Public Order, in C. Ando, P.J. du Plessis, K. Tuori (eds), *Oxford Handbook of Roman Law and Society*, Oxford University Press: Oxford 2016, pp. 297–309.

Gabba 1991: E. Gabba, L'impero di Augusto, in G. Clemente, F. Coarelli, E. Gabba (eds), *Storia di Roma II. L'impero mediterraneo, 2. I principi e il mondo*, Einaudi: Torino 1991, pp. 9–28.

Gabba 1994: E. Gabba, *Italia romana*, Biblioteca di Athenaeum: Como 1994.

Gabba, Pasquinucci 1979: E. Gabba, M. Pasquinucci, *Strutture agrarie e allevamento transumante nell'Italia romana* (III–I sec. a. C.), Giardini: Pisa 1979.

Galieti 1911: A. Galieti, Memorie dell'heracleion lanuvino a Civita Lavinia, in *Bollettino dell'Associazione Archeologica Romana* 1, 1911, pp. 25–43.

Galieti 1935: A. Galieti, Le ville suburbane dei colli lanuvini: II. Ville ricordate dagli autori antichi, in *Bullettino della Commissione Archeologica Comunale di Roma*, 69, 1935, pp. 13–144.

Galimberti 2014: A. Galimberti, *Erodiano e Commodo. Traduzione e commento storico al primo libro della Storia dell'Impero dopo Marco*, Vandenhoeck & Ruprecht, Göttingen 2014.

Galinsky 1996: K. Galinsky, *Augustan Culture: An Interpretive Introduction*, Princeton University Press: Princeton 1996.

Galsterer 1980: H. Galsterer, Politik in römischen Städten: die 'seditio' des Jahres 59 n. Chr, in Pompei, in W. Eck, H. Galsterer, H. Wolff (eds), *Studien zur antiken Sozialgeschichtes. Festschrift F. Vittinghoff*, Böhlau: Köln-Wien 1980, pp. 323–338.

Gatti 2005: S. Gatti, La villa imperiale di Palestrina, in L. Quilici, S. Quilici Gigli (eds), *La forma della città e del territorio*, 2 (ATTA 14), L'Erma di Bretschneider: Roma 2005, pp. 67–90.

Gex 2013: N. Gex, Les inscriptions des castra praetoria: les laterculi, in F. Bertholet, C. Schmidt Heidenreich (eds), *Entre archéologie et épigraphie. Nouvelles perspectives sur l'armée romaine*, Peter Lang: Bern 2013, pp. 113–132.

Ghini 2001: G. Ghini, La villa degli Ottavi a Velletri, in *Augusto a Velletri* (Atti Convegno, Velletri 16 December 2000), Velletri, 2001, pp. 35–53.

Giardina 2000 (ed.): A. Giardina (ed.), *Roma antica*, Laterza: Roma-Bari 2000.

Gigante 1990: M. Gigante, *Filodemo in Italia*, Mondadori: Milano 1990.

Gigante 2004: M. Gigante, Vergil in the Shadow of Vesuvius, in D. Armstrong, J. Fish, P.A. Johnston, M.B. Skinner (eds), *Vergil, Philodemus, and the Augustans*, University of Texas Press: Austin 2004, pp. 85–99.

Gigante, Dorandi 1980: M. Gigante, T. Dorandi, Anassarco e Epicuro 'Sul regno', in F. Romano (ed.), *Democrito e l'atomismo antico* (Atti del Convegno Internazionale, Catania 18–21 April1979), Università degli Studi di Catania: Catania 1980, pp. 479–497.

Gottschall 1997: U. W. Gottschall, Securitas, in *Lexicon Iconographicum Mythologiae Classicae (LIMC)*, band VIII, Artemis Verlag: Zürich/München 1997, pp. 1090–1093.

Granino Cecere 1989: M.G. Granino Cecere, Una dedica a Giove nel Museo Nazionale di Palestrina, in *Miscellanea Greca e Romana*, 14, 1989, pp. 145–156.

Granino Cecere 1995: M. G. Granino Cecere, Villa Mamurrana, in *Rendiconti dell'Accademia Nazionale dei Lincei* ser. 9, 6, 1995, pp. 361–386.

Granino Cecere, Ricci 2006: M.G. Granino Cecere, C. Ricci, Dalle sponde del Reno a quelle dell'Aniene. Marinai e navigazione fluviale, in *Zeitschrift für Papyrologie und Epigraphik* 157, 2006, pp. 237–246.

Granino Cecere, Ricci 2015: M.G. Granino, C. Ricci, Il porto di Centumcellae (Civitavecchia) e la sua epigrafia, in C. Zaccaria (ed.) *L'Epigrafia dei porti (Actes de la XVIIᵉ Rencontre franco-italienne sur l'épigraphie du monde romaine*, Aquileia 14–16 October 2010), Editreg: Trieste 2015, pp. 123–136.

Graverini 1997: L. Graverini, Un secolo di studi su Mecenate, in *Rivista di Storia dell'Antichità* 27, 1997, pp. 231–289.

Graverini 2006: L. Graverini, Mecenate, mecenatismo e studi augustei, in *Annali aretini* 12, 2006, pp. 49–71.

Griffin 1995: M. Griffin, Tacitus, Tiberius *and the Principate*, in I. Malkin, Z.W. Robinson (eds), *Leaders and Masses in the Roman World*, Brill: Leiden, New York, Köln 1995, pp. 33–57.

Grimal 1986: P. Grimal, Les éléments philosophiques dans l'idée de monarchie à Rome à la fin de la République, in H. Flashar, O. Gigon (eds.), *Aspects de la philosophie hellénistique*, Fondation Hardt: Vandœuvres-Genève 1986, pp. 233–273.

Gros 1976: P. Gros, Aurea Templa. *Recherches sur l'Architecture Religieuse de Rome à l'époque d'Auguste* (BEFAR 231), École Française de Rome: Roma 1976.

Grosso 1964: F. Grosso, *La lotta politica al tempo di Commodo*, Accademia delle Scienze di Torino, 1964.

Grosso 1965: F. Grosso, Il diritto latino ai militari di età flavia, in *Ricerche di Cultura Classica e Medievale*, 7, 1965, pp. 541–560.

Grosso 1968: F. Grosso, Il papiro Oxy. 2565 e gli avvenimenti del 222–224, in *Rendiconti dell'Accademia Nazionale dei Lincei* 23, 1968, pp. 205–211.

Gradel 2002: I. Gradel, *Emperor Worship and Roman Religion*, Oxford University Press: Oxford 2002.

Grünewald 2004: T.H Grünewald, *Räuber, Rebellen, Rivalen, Rächer. Studien zu* latrones *im römischen Reich*, Stuttgart 1999 (see now *Bandits in the Roman Empire: Myth and Reality*, translated by J.F. Drinkwater, Routledge: London, New York 2004).

Gundel 1963: H.G. Gundel, Der Begriff Maiestas im politischen Denken der römischen Republick, in *Historia* 12.3, 1963, pp. 283–320.

Halfmann 1986: H. Halfmann, *Itinera Principum. Geschichte und Typologie der Kaiserreisen im römischen Reich*, F. Steiner Verlag: Stuttgart 1986.

Hamilton 2013: J.T. Hamilton, Securitas. *Politics, Humanity and the Philology of Care*, Princeton University Press: Princeton 2013.

Hano 1986: M. Hano, À l'origine du culte impérial· les autels des Lares Augusti. Recherches sur les thèmes iconographiques et leur signification, in ANRW, II, 16, 3, 1986, pp. 2333–2380.

Harries 2008: J. Harries, *Law and Crime in the Ancient World*, Cambridge University Press: Cambridge 2008.

Harrison 2011: J.R. Harrison, *Paul and the Imperial Authorities at Thessalonica and Rome*, Mohr Siebeck: Tübingen 2011.

Hartmann 1921: R. Hartmann, Securitas, in RE II.A.1, 1921, cols 1000–1003.

Hekster 2007: O.J. Hekster, Fighting for Rome: The Emperor as a Military Leader, in De Blois (ed.) 2007, pp. 91–105.

von Hesberg 2001: H. von Hesberg, E cornu taurum. Zu fragment von Staatsrelief aus dem Albanum Domitians, in *Rome et ses provinces: Hommages Balty*, Latomus: Bruxelles, 2001, pp. 245–265.

von Hesberg 2009: H. von Hesberg, Le ville imperiali dei Flavi: Albanum Domitiani, in F. Coarelli (ed.), *Divus Vespasianus: Il bimillenario dei Flavi*, Electa: Milano, 2009, pp. 326–333.

Hickson 1991: F.V. Hickson, Augustus triumphator: manipulation of the triumphal themes in the political program of Augustus, in *Latomus* 50, 1991, pp. 125–138.

Hinard 2003: F. Hinard, Entre République et Principat. Pouvoir et urbanité, in T.H. Hantos (ed.), *Laurea internationalis. Festschrift für Jochen Bleicken*, F. Steiner Verlag: Stuttgart 2003, pp. 331–358.

Hirschfeld 1891: O. Hirschfeld, Die Sicherheitspolizei im römischen Kaiserreich, in *Sitzungsberichte der Königlich Preussischen Akademie der Wissenschaften zu Berlin* 39, 1891, pp. 845–877 (now *Kleine Schriften*, Berlin 1913, pp. 576–613).

Hirschfeld 1892: O. Hirschfeld, Die ägyptische Polizei der römischen Kaiserzeit nach Papyrus-urkunden, in *Sitzungsberichte der Königlich Preussischen Akademie der Wissenschaften zu Berlin* 40, 1892, pp. 815–824 (now *Kleine Schriften*, Berlin 1913, pp. 613–623).

Hirschfeld 1902: O. Hirschfeld, Der Grundbesitz der römischen Kaiser in den ersten drei Jahrhunderten, in *Klio*, 2, 1902, pp. 45–72 and 284–315.

Hölscher 1984: T. Hölscher, *Staatsdenkmal und Publikum. Vom Untergang der Republik bis zur Festigung des Kaisertums in Rom*, Konstanz Universitätsverlag: Konstanz 1984.

Hölscher 1988: T. Hölscher, *Historische Reliefs*, in *Kaiser Augustus und die verlorene Republik. Eine Ausstellung im Martin Gropius Bau* (Berlin, 7 June–14 August 1984), Philipp von Zabern: Berlin 1988, pp. 351–400.

Hoffmann, Wulf (eds) 2004: A. Hoffmann, U. Wulf, *Die Kaiserpaläste auf dem Palatin in Rom: das Zentrum der römischen Welt und seine Bauten*, Philipp von Zabern: Berlin 2004.

Hunter, V. 1994: V. Hunter, *Policing Athens: Social Control in the Attic Lawsuits, 420–320 BC*, Princeton University Press: Princeton 1994.

Instinsky 1952: H.U. Instinsky, *Sicherheit als politisches problem des römischen Kaisertums*, Verlag für Kunst und Wissenschaft: Baden-Baden 1952.

Isaac 1992: B. Isaac, *The Limits of Empire: The Roman Army in the East*, Clarendon Press: Oxford 1992 (revised edn).

Jacopi 1973: I. Jacopi, *Il santuario della Fortuna Primigenia e il Museo Archeologico Prenestino*, Istituto Poligrafico dello Stato: Roma 1973.

Jaja 2008: A.M. Jaja, *Anzio, La villa imperiale*, in Valenti 2008, pp. 73–80.

Jallet-Huant 2004: M. Jallet-Huant, *La garde prétorienne dans la Rome antique*, Presse de Valmy: Charenton le Pont 2004.

Jolivet 2007: V. Jolivet, La localisation des toponymes de la Rome antique à partir des Régionnaires: une étude de cas, in A. Leone, D. Palombi (eds), *Res bene gestae. Ricerche di storia urbana su Roma antica in onore di Eva Margareta Steinby*, Quasar: Roma 2007, pp. 103–125.

Jones 1992: B. W. Jones, *The Emperor Domitian*, Routledge: London and New York, 1992.

Jory 1984: J. Jory, The Early Pantomime Riots, in A.M. Offatt (ed.), *Maistor: Classical, Byzantine and Renaissance Studies for Robert Browning*, University of Canberra: Canberra 1984, pp. 57–66.

Kayser 1990: F. Kayser, Les 'statores' en Egypte, in *Bulletin de l'Institut français d'archéologie orientale*, 90, 1990, pp. 241–246.

Keaveney 2007: A. Keaveney, *The Army in the Roman Revolution*, Routledge: London and New York 2007.

Kelly 1957: J.M. Kelly, *Princeps Iudex. Eine Untersuchung zur Entwicklung und zu den Grundlagen der kaiserlichen Gerichtbarkeit*, Hermann Böhlaus Nachfolger: Weimar 1957.

Kelly 2007: B. Kelly, Riot Control and Imperial Ideology in the Roman Empire, in *Phoenix* 61.1–2 (2007), pp. 150–176.

Kelly 2013: B. Kelly, Policing and Security, in P. Erdkamp (ed.), *The Cambridge Companion to Ancient Rome*, Cambridge University Press: Cambridge 2013, pp. 410–424.

Kennedy 1978: D.L. Kennedy, Some Observations on the Praetorian Guard, in *Ancient Society* 9, 1978, pp. 275–301.

Kennedy 1983: D. L. Kennedy, C. Velius Rufus, in *Britannia* 14, 1983, pp. 194–196.

Keppie 1984a: L. Keppie, Colonisation and Veteran Settlement in Italy in the First Century AD, in *Papers of the British School in Rome* 52, 1984, pp. 77–114.

Keppie 1984b: L. Keppie, *The Making of the Roman Army: From Republic to Empire*, Batsford: London 1984 (revised edn University of Oklahoma Press 2001).

Keppie 1996: L. Keppie, The Praetorian Guard before Sejanus, in *Athenaeum* 84, 1996, 101–124 = Keppie 2000, pp. 99–122.

Keppie 2000: L. Keppie, *Legions and Veterans. Roman Army Papers 1971–2000*, F. Steiner Verlag: Stuttgart 2000.

Keppie 2003: L. Keppie, 'Having Been a Soldier': The Commemoration of Military Service on Funerary Monuments of the Early Empire, in J.J. Wilkes (ed.), *Documenting the Roman Army: Essays in Honour of M. Roxan*, Institute of Classical Studies, School of Advanced Study: London 2003, pp. 21–35.

Kleijwegt 1994: M. Kleijwegt, Caligula's Triumph at Baiae, in *Mnemosyne*, 47, 1994, pp. 652–671.

Kneppe 1957: A. Kneppe, *Metus temporum. Zur Bedeutung von Angst in Politik und Gesellschaft der römischen Kaiserzeit des 1 und 2 Jhdts. n. Chr*, F. Steiner Verlag: Stuttgart 1994.

Köhler 1957: W. Köhler, Securitas, in *Enciclopedia dell'arte antica* VII 1957, p. 151. www.treccani.it/enciclopedia/securitas_(Enciclopedia-dell'-Arte-Antica)/ (accessed 3 September 2017).

Kolb 2000: A. Kolb, *Transport und und Nachrichtentransfer im römischen Reich*, Akademie Verlag: Berlin 2000.

Koloski-Ostrow 2015: A.O. Koloski-Ostrow, *The Archaeology of Sanitation in Roman Italy. Toilets, Sewers, and Water Systems*, The University of North Caroline Press: Chapel Hill 2015.

Konen 2001: H.C. Konen, Classis Germanica. *Die römische Rheinflotte im 1–3 Jahrlhundert n. Chr* (Pharos, 15), Scripta Mercaturae Verlag: St. Katharinen 2001.

Krause 2003: C. Krause, Villa Iovis. *Die Residenz des Tiberius auf Capri*, Philipp von Zabern: Mainz 2003.

Krause 2004: J.-U. Krause, *Kriminalgeschichte der Antike*, C.H. Beck: München 2004.

Krause, Mylonopoulos, Cengia (eds) 1998: J.U. Krause, J. Mylonopoulos, R. Cengia (eds), *Schichten, Konflikte, religiöse Gruppen, materielle Kultur*, F. Steiner Verlag: Stuttgart 1998.

Lafer 2001: R. Lafer, Securitas hominibus. Literarische Fiktion oder Realität? Die Bekämpfung von Räubern und Dieben im Imperium Romanum, in F. W. Leitner (ed.), *Carinthia romana und die römische Welt: Festschrift für Gernot Piccottini*, Verlag des Geschichtsvereines für Kärnten: Klagenfurt 2001, pp. 125–134.

Laffi 1965: U. Laffi, L'iscrizione di Sepino relativa ai contrasti fra le autorità municipali e i 'conductores' delle greggi imperiali con l'intervento dei prefetti del pretorio, in *Studi Classici e Orientali*, 14, 1965, pp. 177–200.

Lafon 2001: X. Lafon, *Villa Maritima. Recherches sur les uillas littorales de l'Italie romaine (IIIe siècle av.J.-C. / III siècle ap.J.-C.)* (BEFAR 307), École française de Rome: Roma 2001.

Lana 2001: I. Lana, Qualche riflessione sulla securitas secondo Seneca, in P. Fedeli (ed.), *Scienza, cultura, morale in Seneca. Atti del Convegno di Monte Sant'Angelo* (27–30 September 1999), Edipuglia: Bari 2001, pp. 35–53.

Lancel 2000. S. Lancel, *Carthage: da la colonie tyrienne à la mégalopole hellenistique*, in Nicolet, Ilbert, Depaule (eds) 2000, pp. 534–544.

La Rocca 2000: E. La Rocca, L'affresco con veduta di città dal Colle Oppio, in *Romanization and the City: Creation, Dynamics and Failures*, (JRA Suppl. 38), 2000, pp. 57–71.

Latteri 2002: N. Latteri, La statio dei pretoriani al III miglio dell'Appia antica e il loro sepolcreto 'ad catacumbas', in *Mélanges de l'École française de Rome: Antiquité* 114.2, 2002, pp. 739–757.

Le Bohec (ed.) 1994: Y. Le Bohec (ed.), *L'Afrique, la Gaule, la religion à l'époque romaine: mélanges à la mémoire de Marcel Le Glay*, Latomus: Bruxelles 1994.

Le Bohec, Ben Abdallah 1997: Y. Le Bohec, Z. Benzina Ben Abdallah, Nouvelles inscriptions d'Häidra concernant l'armée romaine, in *Mélanges de l'École française de Rome: Antiquité* 109, 1997.1, 41–82 (= *AE* 1997, 1621a).

Le Bohec, Duval, Lancel 1984: Y. Le Bohec, N. Duval, S. Lancel, Études sur la garnison de Carthage, in *Bulletin archéologique du Comité des travaux historiques et scientifiques. Afrique du Nord.*, n.s. 15–16B, Paris 1984, pp. 33–89.

Lelli 1999. P. Lelli, Considerazioni sulla guardia pretoria nel primo secolo, in *Atene e Roma* 44, 1999, pp. 9–13.

Lendon 2006: J.E. Lendon, Contubernalis, Commanipularis and Commilito in Roman Soldiers' Epigraphy: Drawing the Distinction, in *Zeitschrift für Papyrologie und Epigraphik* 157, 2006, pp. 270–276.

Lenski 2009: N. Lenski, Schiavi armati e formazione di eserciti privati nel mondo tardo antico, in G. Urso (ed.), *Ordine e sovversione nel mondo greco e romano* (Atti Convegno Internazionele, Cividale del Friuli, 25–27 September 2008), ETS: Pisa 2009, pp. 145–175.

Leppert 1974: M. Leppert, 23 Kaiservillen. *Vararbeiten zu Archäologie and Kulturgeschichte der Villegiatur der Hohen Kaiserzeit*. Doktoral-Dissertation, Freiburg 1974.

Leppin 1992: H. Leppin, *Histrionen. Untersuchungen zur sozialen Stellung von Bühnenkünstlern im Westen des Römischen Reiches zur Zeit der Republik und des Principats*, Rudolf Habelt: Bonn 1992.

Le Roux 1972: P. Le Roux, Recherches sur les centurions de la legio VII Gemina, in *Mélanges de la Casa de Velázquez* 8, 1972, pp. 89–147 (re-edited in P. Le Roux, *La toge et les armes*, Rennes 2011, pp. 287–343, with additions).

Le Roux 1990: P. Le Roux, L'amphithéâtre et le soldat sous l'Empire romain, in *Spectacula I, Gladiateurs et amphithéâtres* (Actes du colloque, Toulouse et Lattes, 26–29 May 1987), Editions Imago: Lattes 1990, pp. 203–215.

Le Roux 2002: P. Le Roux, Armées et ordre public dans le monde romain à l'époque impériale, in *Armée et maintien de l'ordre*, Centre d'études d'histoire de la Défense: Paris 2002, pp. 17–51 (re-edited in P. Le Roux, *La toge et les armes*, Rennes 2011, pp. 217–237).

Levi 1969: M.A. Levi, Maiestas e crimen maiestatis, in *Parola del passato* 24, 1969, pp. 81–96.

Levick 1999: B. Levick, *Vespasian*, Routledge: London, New York 1999.

Lewin 1993: A. Lewin, Ius armorum, polizie cittadine e grandi proprietari nell'Oriente tardoantico, in G. Crifò (ed.), *Atti del X convegno internazionale dell'Accademia romanistica internazionale*, Edizioni scientifiche italiane: Perugia 1993, pp. 375–386.

Liberati, Silverio 2010: A.M. Liberati, E. Silverio, *Servizi segreti in Roma antica*, L'Erma di Bretschneider: Roma 2010.

Liberati, Silverio 2013: A.M. Liberati, E. Silverio, Il sistema romano d'informazione e sicurezza nell'età del principato di Caligola. Aspetti militari e civili, in F. Coarelli,

G. Ghini (eds), *Caligola: la trasgressione al potere* (Catalogo della mostra, Nemi, Museo delle navi romane, 5 July–5 November 2013), Gangemi: Roma 2013, pp. 87–100.

Linderski 1991: J. Linderski, Si vis pacem para bellum: Concepts of Difensive Imperialism, in *Roman Questions: Selected Papers*, F. Steiner Verlag: Stuttgart 1995, pp. 217–223.

Linderski 1995: J. Linderski, *Roman Questions: Selected Papers*, I, F. Steiner Verlag: Stuttgart 1995.

Lintott 1968: A. Lintott, *Violence in Republican Rome*, Clarendon Press: Oxford 1968 (new edn 1998).

Lintott 1998: A. Lintott, *Violence in Republican Rome*, Clarendon Press: Oxford 1998.

Lintott 2008: A. W. Lintott, How High a Priority Did Public Order and Public Security Have Under the Republic? in Brélaz, Ducrey 2008, pp. 205–220.

Liou 2009: J. Liou, *Collegia Centonariorum: The Guilds of Textile Dealers in the Roman West* (Columbia Studies in the Classical Tradition 24), Brill: Leiden, Boston 2009.

Lissi Caronna 1993: E. Lissi Caronna, Castra praetoria in LTUR I, pp. 251–254.

Liverani 1988: P. Liverani, Le proprietà private nell'area lateranense fino all'età di Costantino, in *Mélanges de l'École française de Rome. Antiquité* 100, 1988, pp. 891–915.

Liverani 1988 (ed.): P. Liverani (ed.), *Laterano I. Scavi sotto la basilica di S. Giovanni in Laterano. I materiali*, Edizioni Musei Vaticani: Città del Vaticano 1988.

Liverani 1989: P. Liverani, *L'Antiquarium di Villa Barberini a Castel Gandolfo*, Edizioni Musei Vaticani: Città del Vaticano 1989.

Liverani 2008: P. Liverani, La villa di Domiziano a Castel Gandolfo, in Valenti 2008 (ed.), pp. 53–60.

Lo Cascio 1985–1990: E. Lo Cascio, I 'greges oviarici' dell'iscrizione di Sepino (CIL, IX 2438) e la transumanza in età imperiale, in Abruzzo 23/28, 1985/90, pp. 557–569 = Id., *Il princeps e il suo impero: studi di storia amministrativa e finanziaria romana*, Edipuglia: Bari 2000, pp. 151–161.

Lo Cascio 1991: E. Lo Cascio, Le tecniche dell'amministrazione, in F. Gabba, A. Schiavone (eds), *Storia di Roma II. L'impero mediterraneo. 2. I principi e il mondo*, Einaudi: Torino 1991, pp. 119–192.

Lo Cascio 1997: E. Lo Cascio, Le procedure di recensus dalla tarda repubblica al tardo antico e il calcolo della popolazione di Roma, in *La Rome impériale. Démographie et logistique* (Actes de la table ronde de Rome, 25 March 1994) (CEFR 230), École française de Rome: Roma 1997, pp. 3–76.

Lo Cascio 1998: E. Lo Cascio, Registri dei beneficiari e modalità delle distribuzioni nella Roma tardoantica, in *La mémoire perdue. Recherches sur l'administration romaine* (CEFR 243), École française de Rome: Roma 1998, pp. 365–385.

Lo Cascio 2000a: E. Lo Cascio, Patrimonium, ratio privata, res privata, in *Il princeps e il suo impero. Studi di storia amministrativa e finanziaria romana*, a cura di E. Lo Cascio, Bari, 2000 (Doc. St. Univ. St. Bari, 26), pp. 97–149.

Lo Cascio 2000b: E. Lo Cascio, Registrazioni di tipo censuale e stime della popolazione di Roma nell'antichità, in Nicolet, Ilbert, Depaule (eds) 2000, pp. 628–659.

Lo Cascio 2000c (ed.): E. Lo Cascio (ed.), *Roma imperiale. Una metropoli antica*, Carocci: Roma 2000.

Lo Cascio 2007: E. Lo Cascio, Il ruolo dei vici e delle regiones nel controllo della popolazione e nell'amministrazione di Roma, in R. Haensch, J. Heinrichs (eds), *Herrschen und Verwalten: der Alltag der römischen Administration in der Hohen Kaiserzeit*, Böhlau Verlag: Köln 2007, pp. 145–159.

Lo Cascio 2008: E. Lo Cascio, Vici, regiones e forme di interazione sociale nella Roma imperiale, in Royo, Hubert, Bérenger (eds) 2008, pp. 65–76.

Łoś 1995: A. Łoś, Quand et pourquoi a-t-on envoyé des prétoriens à Pompei, in *Nvnc de Svebis dicendvm est: studia archaeologica et historica Georgio Kolendo ab amicis et discipulis dicata*, Uniwersytet Warszawa Instytut Archeologii: Warczawa, 1995, pp. 165–170.

Loschiavo: L. Loschiavo, Autodifesa, vendetta, repressione poliziesca. La lotta al brigantaggio nel passaggio dalle province tardo-imperiali ai regni romano barbarici, in F. Botta (ed.), *Il diritto giustinianeo fra tradizione classica e innovazione*, Giappichelli: Torino 2003, pp. 105–133.

Lott 2004: J.B. Lott, *The Neighborhoods of Augustan Rome*, Cambridge University Press: Cambridge, New York 2004.

Lovato, Puliatti, Solidoro Maruotti 2014: A. Lovato, S. Puliatti, L. Solidoro Maruotti, *Diritto privato romano*, Giappichelli: Torino 2014.

Lovotti 2000: F. Lovotti, L'arco di Cirta: considerazioni sulle epigrafi onorarie, in M. Khanoussi, P. Ruggeri, C. Vismara (eds), *L'Africa romana. Geografi, viaggiatori, militari nel Maghreb: alle origini dell'archeologia nel Nord Africa* (Atti del XIII convegno di studio, Djerba, 10–13 December 1998), vol. II, Carocci: Roma 2000, pp. 1603–1612.

Luc 2004: I. Luc, *Oddziały pretorianów w starożytnym Rzymie: rekrutacja, struktura, organizacja*, Wydawn. Uniw. Marii Curie: Skłodowskiej Lublin 2004.

Lugli 1917: G. Lugli, La villa di Domiziano sui Colli Albani, in *Bullettino della Commissione Archeologica Comunale di Roma* 47, 1917 pp. 5–54.

Lugli 1918: G. Lugli, La villa di Domiziano sui Colli Albani (seconda parte), in *Bullettino della Commissione Archeologica Comunale di Roma* 48, 1918, pp. 3–68.

Lugli 1940: G. Lugli, Saggio sulla topografia dell'antica Antium, *in Rivista dell'Istituto di Archeologia e Storia dell'Arte di Roma*, 7, 1940, pp. 153–188.

McGinn 1998: A.J. McGinn *Prostitution, Sexuality, and the Law in Ancient Rome*. Oxford University Press: Oxford New York 1998.

MacMullen 1966: R. MacMullen, *Enemies of the Roman Order: Treason, Unrest and Alienation in the Empire*, Harvard University Press: Cambridge, MA 1966 (Routledge: London, New York 1992, new edn).

MacMullen 1974: R. MacMullen, *Roman Social Relations 50 BC to AD 284*, Yale University Press: New Haven 1974.

MacMullen 1988: R. MacMullen, *Corruption and the Decline of Rome*, Yale University Press: New Haven 1988.

MacMullen 2006: R. MacMullen, Soldiers in the Cities of Roman Empire, in REMA 3, 2006, pp. 123–130.

Magalhaes 2003: M.M. Magalhaes, *Storia, istituzioni e prosopografia di Surrentum romana. La collezione epigrafica del Museo Correale di Terranova*, N. Longobardi: Castellamare 2003.

Magi 1945: F. Magi, *I rilievi flavi del palazzo della Cancelleria*, G. Bardi editore: Roma 1945.

Maiuro 2012: M. Maiuro, *Res Caesaris. Ricerche sulla proprietà imperiale nel principato*, Edipuglia: Bari 2012.

Maiuro 2015: M. Maiuro, Ulpian and Public Uses of Imperial Properties: A Note On Dig. 30.39.7–10, in *Riv.Ital.Fil.Class.* 143, 2015, pp. 362–379.

Makhlayuk 1996: A.V. Makhlayuk, Military Comradeship and Corporativeness in the Imperial Roman Army, in *Vestn.Drev.Ist.* 216, 1996, pp. 18–37.

Manacorda 1999: D. Manacorda, Per l'edizione del secondo colombario Codini: il problema epigrafico nel contesto archeologico, in *Atti del XI Congresso Internazionale di Epigrafia Greca e Latina* (Roma, 18–24 September 1997), vol. 2, Edizioni Quasar: Roma 1999, pp. 249–261.

Manders 2012: E. Manders, *Coining Images of Power: Patterns in the Representation of Roman Emperors on Imperial Coinage, A.D. 193–284* (Impact of Empire, 15), Brill: Leiden, Boston 2012.

Manderscheid 2000: H. Manderscheid, Uberlegungen am Wasserarchitektur und ihrer Funktion in der Villa Adriana, in *Römische Mitteilungen*, 107, 2000, pp. 109–140.

Manderscheid 2002: H. Manderscheid, L'architettura dell'acqua a Villa Adriana con particolare riguardo al Canopo, in A.M. Reggiani, *Villa Adriana. Paesaggio antico e ambiente moderno: elementi di novità e ricerche in corso* (Atti Convegno, Roma, 23–24 giugno 2000), Electa: Milano 2002, pp. 84–89.

Maniscalco 1997: G. Maniscalco, *Ninfei ed edifici marittimi severiani del Palatium imperiale di Baia*, Massa: Napoli 1997.

Mantovani 1988: D. Mantovani, Sulla competenza penale del praefectus urbi attraverso il liber singularis di Ulpiano, in A. Burdese (ed.), *Idee vecchie e nuove sul diritto criminale romano*, Pubblicazioni della Facoltà di giurisprudenza dell'Università di Padova: Padova 1988, pp. 171–223.

Mantovani 2008: D. Mantovani, Leges et iura populi Romani restituit. Principe e diritto in un aureo di Ottaviano, in *Athenaeum* 96, 2008, pp. 5–54 (ora anche in F. Milazzo (ed.), *I tribunali dell'impero* (Convegno internazionale di diritto romano, Copanello giugno 2006), Giuffré editore: Milano 2015, pp. 41–105).

Mantovani 2011: D. Mantovani, Rivelazioni sulla salus principis e tattiche per sottrarsi al processo. Il secondo editto di Augusto ai Cirenei alla luce del de officio proconsulis di Ulpiano, in E. Stolfi (ed.), *Giuristi e officium. L'elaborazione giurisprudenziale di regole per l'esercizio del potere fra II e III secolo d.C.*, Edizioni Scientifiche Italiane: Napoli 2011, pp. 195–214.

Mar 2009: R. Mar, La Domus Flavia, utilizzo e funzioni del Palazzo di Domiziano, in F. Coarelli (ed.), *Divus Vespasiano. Il bimillenario dei Flavi*, Electa: Milano 2009, pp. 250–263.

Marcattili 2005: F. Marcattili, Compitum, in *Thesaurus Cultus et Rituum Antiquorum* (ThesCRA), vol. IV 2005, The J. Paul Getty Museum: Los Angeles 2005, pp. 222–224.

Marchese 2007: M.E. Marchese, *La Prefettura Urbana a Roma. Un tentativo di localizzazione attraverso le epigrafi*, in *MEFRA* 119, 2007, pp. 613–634.

Mari 1991: Z. Mari, *Tibur, pars quarta*, Olschki: Firenze, 1991.

Mari 2008: Z. Mari, Il Sublaqueum: la villa di Nerone a Subiaco, in Valenti 2008, pp. 43–52.

Mastino 2006: A. Mastino, A. Ibba, L'imperator pacator orbis, in *Diritto e Storia* 5, 2006. www.dirittoestoria.it/5/Tradizione-Romana/Mastino-Ibba-Imperatore-pacator-orbis.htm (accessed 6 September 2017).

Mastrocinque 2007: A. Mastrocinque, Giulio Cesare e la fondazione della colonia di Grumentum, in *Klio* 89, 2007, pp. 118–124.

Mastrocinque 2009 (ed.): A. Mastrocinque (ed.), *Grumentum romana* (Convegno di Studi Grumento Nova–Potenza, 28–29 June 2008), Valentina Porfidio: Meliterno 2009.

Mastrodonato 1999–2000: V. Mastrodonato, Una residenza imperiale nel suburbio di Roma. La villa di Lucio Vero in località Aquatraversa, in *Archeologia Classica*, 51, 1999–2000, pp. 157–235.

Mattingly 1937: H. Mattingly, The Roman Virtues, in *The Harvard Theological Review* 30, 1937, pp. 103–117.

Maxfield 1981: V. Maxfield, *The Military Decorations of the Roman Army*, Batsford: London 1981.

Meloni 1958: P. Meloni, *L'amministrazione della Sardegna da Augusto all'invasione vandalica*, L'Erma di Bretschneider: Roma 1958.

Meloni 1988: P. Meloni, Storia dei Sardi, in M. Guidetti (ed.), *Storia dei Sardi e della Sardegna* I, Jaca Book: Milano 1988, pp. 235–261.

Ménard 2004: H. Ménard, *Maintenir l'ordre à Rome (IIe–IVe siècles ap. J.-C.)*, Champvallon: Paris 2004.

Mench 1968: F. C. Mench, The Cohortes Urbanae of Imperial Rome: An Epigraphic Study, Diss. Yale University 1968.

Menéndez Argüín 2006: A.R. Menéndez Argüín *Pretorianos. La guardia imperiale de la antigua Roma*, Almena: Madrid 2006.

Meylan Krause 2002: M.-F. Meylan Krause, *Domus tiberiana: analyses stratigraphiques et céramologiques*, BAR: Oxford, 2002.

Millar 1964: F. Millar, *A Study on Cassius Dio*, Oxford University Press: Oxford, 1964.

Millar 1977: F. Millar, *The Emperor in the Roman World (31 BC–337 AD)*, Duckworth: London, 1977.

Millar 1986: F. Millar, Italy and the Roman Empire: Augustus to Constantine, in *Phoenix* 40, 1986, pp. 295–318.

Millar 1994: F. Millar, *The Crowd in Rome in the Late Republic*, Ann Arbor: Michigan University Press 1998 (slightly expanded version of the five Jerome Lectures given at the University of Michigan at Ann Arbor in 1993 and at the American Academy in Rome in 1994).

Millar 2000: F. Millar, Trajan: Government by Correspondence, in J. Gonzalez (ed.), *Trajano Emperador de Roma*, L'Erma di Bretschneider: Roma, 2000, pp. 363–388 (reprinted in F. Millar, H.M. Cotton, G. MacLean Rogers (eds), *Government, Society, and Culture in the Roman Empire*, Eurospan, University of North Carolina Press: Chapel Hill, London 2004, pp. 23–46).

Mócsy 1968: A. Mócsy, Pannonici nelle flotte di Ravenna e di Miseno, in *Atti del Convegno Internazionale di Studi sulle antichità di Classe*, Fratelli Lega: Faenza, 1968, pp. 305–312.

Modugno Tofini 1989: S. Modugno Tofini, Osservazioni su alcune iscrizioni edite di Albano, in *Documenta Albana*, 11, 1989, pp. 55–61.

Molle 2012: C. Molle, Hic hospitati sunt homi(nes). Graffiti parietali antichi a Villa Adriana (Tibur), in *Sylloge Epigraphica Barcinonensis* 10, 2012, pp. 389–404.

Momigliano 1938: A. Momigliano, I problemi delle istituzioni militari di Augusto, in *Augustus. Studi in occasione del bimillenario augusteo*, Accademian Nazionale dei Lincei: Roma 1938, pp. 195–216.

Montanari 2006: E. Montanari, Il concetto originario di 'Pax' e la 'Pax deorum', in P. Catalano, P. Siniscalco (eds), *Concezioni della Pace (VII Seminario di Studi Storici. Da Roma alla Terza Roma)*, Herder editrice: Roma 2006, pp. 39–50.

Moreau 2005: Ph. Moreau, La domus Augusta et les formations de parenté à Rome, in *Cahiers du Centre Glotz* 16, 2005, pp. 7–23.

Morizio 1995: V. Morizio, I Lusii a Lorium, in G. Paci (ed.), *Epigrafia romana in area adriatica* (Actes de la IVe Rencontre franco-italienne sur l'épigraphie du monde romaine, Macerata 10–11 November 1995), Istituti editoriali e Poligrafici Internazionali: Pisa-Roma, 1998, pp. 419–433.

Morretta 2007: S. Morretta, Roma. indagini archeologiche nell'area dei Castra Praetoria (angolo sud-ovest), in Fasti online 2007. www.fastionline.org/docs/foldeR-it-2007-101.pdf (accessed 3 September 2017).

Morricone Matini, Scrinari 1975: M.L. Morricone Matini, V.S.M. Scrinari, *Mosaici antichi in Italia, Regione prima*. *Antium*, Istituto Poligrafico dello Stato: Roma 1975.

Müller 1996: F.L. Müller, Herodian, *Geschichte des Kaisertums nach Marc Aurel*, F. Steiner Verlag: Stuttgart 1996.

Muzzioli 1970: M.P. Muzzioli, *Praeneste, pars altera* (Forma Italiae, 1, 8), Arbor Sapientiae: Roma 1970.

Napoli 2003: P. Napoli, *Naissance de la police moderne. Pouvoirs, normes, société*, Éditions la Découverte: Paris 2003.

Nasti 1999: F. Nasti, Curatores regionum urbis e il cursus honorum di C. Caelius Censorinus, in *Atti dell'XI Congresso Internazionale di Epigrafia Greca e Latina* (Roma, 18–24 settembre 1997), Edizioni Quasar: Roma 1999, pp. 533–544.

Nasti 2006: F. Nasti, *L'attività normativa di Severo Alessandro 1. Politica di governo, riforme amministrative e giudiziarie*, Satura Editrice: Napoli 2006.

Nelis-Clément 2000: J. Nelis-Clément, *Les Beneficiarii: militaires et administrateurs au service de l'empire*, Ausonius Editions: Bordeaux 2000.

Neri 1990: V. Neri, Verso Ravenna capitale: Roma, Ravenna e le residenze tardo-antiche, in G. Susini (ed.), *Storia di Ravenna I. L'evo antico*, Marsilio: Venezia 1990, pp. 535–584.

Nicolet 1988: Cl. Nicolet, De Vérone au Champ de Mars: Chorographia et Carte d'Agrippa, in *Mélanges de l' École française de Rome: Antiquité* 100.1, 1988, pp. 127–138.

Nicolet 1991: C. Nicolet, Space, *Geography and Politics in the Early Roman Empire*, The University of Michigan Press: Ann Arbor 1991.

Nicolet 2000: Cl. Nicolet, Fragments pour une géographie urbaine comparée: à propos d'Alexandrie, in Nicolet, Ilbert, Depaule (eds) 2000, pp. 245–252.

Nicolet, Ilbert, Depaule (eds) 2000: Cl. Nicolet, R. Ilbert, J.-Ch. Depaule (eds), *Mégalopoles méditerranéennes. Geographie urbaine rétrospective. Actes Coll.* 8–11 May 1996 (CEFR 261), Maisonneuve & Larose: Paris 2000.

Nippel 1984: W. Nippel, Policing Rome, in *Journal of Roman Studies* 74, 1984, pp. 20–29.

Nippel 1988: W. Nippel, *Ausfuhr und 'Polizei' in der römischen Republik*, Kröner: Stuttgart 1988.

Nippel 1995: W. Nippel, *Public Order in Ancient Rome*, Cambridge University Press: Cambridge 1995.

Noreña 2011: C.F. Noreña, *Imperial Ideals in the Roman West: Representation, Circulation, Power*, Cambridge University Press: Cambridge 2011.

Nugent 1990: S.G. Nugent, Tristia 2. Ovid and Augustus, in K. A. Raaflaub, M. Toher (eds), *Between Republic and Empire: Interpretations of Augustus and His Principate*, University of California Press: Berkeley, Los Angeles, Oxford 1990, pp. 239–257.

Okamura 1988: L. Okamura, Social Disturbances in Late Roman Gaul: Deserters, Rebels and Bagaudae, in T. Yuge, M. Doi (eds), *Forms of Control and Subordination in Antiquity*, Brill: Leiden, Boston 1988, pp. 288–294.

Oliver 1949: G.H. Oliver, On Edict II and the Senatus Consultum at Cyrene, in *Memoirs of American Academy in Rome* 19, 1949, pp. 107–108.

Oliverio 1927: G. Oliverio, La stele di Augusto rinvenuta nell'agorà di Cirene, in *Notiziario Archeologico* 4, 1927, pp. 13–67.

Orlandi 2011: S. Orlandi, Appendice: l'iscrizione del praefectus urbi F. Felix Passifilus Paulinus, in R. Egidi, F. Filippi, S. Martone (eds), *Archeologia e infrastrutture. Il tracciato fondamentale della linea C della metropolitana di Roma*, Olschki: Firenze 2010, pp. 124–127.

Orlandi 2013: S. Orlandi, Le testimonianze epigrafiche, in *Bollettino di Archeologia on line*. Direzione Generale per le Antichità IV, 2013. https://romethe imperialfora19952010.files.wordpress.com/2015/01/4-orlandi_def.pdf (accessed 3 September 2017).

Packer 1997: J.A. Packer, *The Forum of Trajan in Rome, I*, University of California Press: Berkeley 1997.

Palazzolo 2015: N. Palazzolo, Dalle cognitiones alla cognitio: principe e giuristi verso la costruzione del nuovo sistema processuale, in F. Milazzo (ed.), *I tribunali dell'impero* (Convegno internazionale di diritto romano, Copanello giugno 2006), Giuffré editore: Milano 2015, pp. 217–240.

Palombi 1997: D. Palombi, *Tra Palatino ed Esquilino. Velia Carinae Fagutal Storia urbana di tre quartieri di Roma antica*, Istituto Nazionale di Archeologia e Storia dell'arte: Roma 1997.

Palombi 2013: D. Palombi, Roma. La città imperiale prima dei Severi, in Sojc, Winterling, Wulf-Rheidt (eds) 2013, pp. 23–60.

Panciera 1968: S. Panciera, Gli schiavi nelle flotte augustee, in *Atti del Convegno Internazionale di Studi sulle Antichità di Classe* (Ravenna 1967), A. Longo: Ravenna 1968, pp. 313–330 (now updated in Panciera 2006, pp. 1283–1294).

Panciera 1970: S. Panciera, Regiones, Vici e Iuventus. Tra epigrafia e topografia I.3, in *Archeologia Classica* 22, 1970, pp. 151–163 (now updated in Panciera 2006, pp. 173–182).

Panciera 1974: S. Panciera, Equites singulares Augusti. Nuove testimonianze epigrafiche, in *Rivista di Archeologia cristiana* 50, 1974, pp. 221–247 (now updated in Panciera 2006, pp. 1309–1326).

Panciera 1974–1975: S. Panciera, Altri pretoriani di origine veneta, in *Aquileia Nostra* 45–46, 1974–1975, cols 163–182 (now updated in Panciera 2006, pp. 1295–1306).

Panciera 1978: S. Panciera, Invigulantes pro vicinia, in *Scritti storico-epigrafici in memoria di Marcello Zambelli*, Centro editoriale internazionale: Roma 1978, pp. 315–320 (now updated in Panciera 2006, pp. 265–268).

Panciera 1984: S. Panciera, Le stele dei militari dal sepolcreto presso Ponte Milvio, with A. Ambrogi and M. E. Micheli, in A. Giuliano (ed.), *Museo Nazionale Romano. Le sculture* I, 7.1, De Luca: Roma 1984, 158–178 (now updated in Panciera 2006, pp. 1391–1409).

Panciera 1987: S. Panciera, La cura regionum Urbis e Alessandro Severo, in *L' Urbs. Espace urbain et histoire* (CEFR, 98), École Française de Rome: Roma 1987, pp. 78–80 (now updated in Panciera 2006, pp. 357–358).

Panciera 1993: S. Panciera, Soldati e civili nei primi tre secoli dell'impero, in W. Eck (ed.), *Prosopographie und Sozialgeschichte. Studien zur Methodik und Erkenntnismöglichkeit der kaiserzeitlichen Prosopographie-Kolloquium Köln 1991*, Böhlau Verlag: Köln 1993, pp. 261–276 (now updated in Panciera 2006, pp. 1441–1452).

Panciera 1994: S. Panciera, Signis legionum. Insegne, immagini imperiali e centurioni frumentari a peregrinis, in Le Bohec (ed.) 1994, pp. 610–623 (now updated in Panciera 2006, pp. 1453–1463).

Panciera 1995: S. Panciera, Una diciannovesima coorte pretoria? in *Römische Inschriften, Neufunde, Neulesungen und Neuinterpretationen. Festschrift für Hans Lieb (AREA 2)*, Reinhardt Verlag: Basel 1995, pp. 113–121 (now updated in Panciera 2006, pp. 1465–1470).

Panciera 1989: S. Panciera, Genio castrorum peregrinorum, in *Acta Arch. Budapest* 41, 1989, pp. 365–383 (now updated in Panciera 2006, pp. 1421–1439).

Panciera 2000: S. Panciera, Nettezza urbana a Roma. Organizzazione e responsabili, in Dupré Raventós, Remolà (eds) 2000, pp. 95–105 (now updated in Panciera 2006, pp. 479–490).

Panciera 2003: S. Panciera, Sulle vicende di un sacrarium di Liber Pater nel suburbio di Roma in età tardoantica, in J.-M. Carrié, R. Lizzi Testa (eds), *Humana sapit. Études d'antiquité tardive offertes a Lellia Cracco Ruggini*, Brépols: Turnhout, 2003, pp. 63–74 (now updated in Panciera 2006, pp. 505–520).

Panciera 2004: S. Panciera, Altri pretoriani a Roma. Nuove iscrizioni e vecchie domande, in *Cahiers du Centre Glotz* 15, 2004 [2006], pp. 281–316 (now updated in Panciera 2006, pp. 1492–1523).

Panciera 2006: S. Panciera, *Epigrafi, epigrafia, epigrafisti. Scritti vari editi e inediti (1956–2005) con note complementari e indici* (Vetera, 16), Edizioni Quasar: Roma 2006.

Panciera 2012: S. Panciera, Religio militum. Due inediti da Roma, in B. Cabouret, A. Groslambert, C. Wolff (eds), *Visions de l'Occident Romain. Hommages à Y. Le Bohec*, De Boccard: Paris 2012, pp. 554–567.

Pani 2003: M. Pani, *La corte dei Cesari*, Laterza: Roma-Bari 2003.

Pani 2013: M. Pani, *Augusto e il principato*, Il Mulino: Bologna 2013.

Parma 1994: A. Parma, Classiarii, veterani e società cittadina a Misenum, in *Ostraka* 3, 1994, pp. 43–59.

Parma 1995: A. Parma, Una nuova iscrizione di Misenum con un veterano duovir della città, in *Ostraka* 4, 1995, pp. 301–306.

Parma 2002: A. Parma, *Stabiae e la classis Misenensis, in Stabiae: storia e architettura* (Convegno Internazionale Castellammare di Stabia, 25–27 March 2000), L'Erma di Bretschneider: Roma 2002, pp. 185–188.

Passerini 1939: A. Passerini, *Le coorti pretorie*, A. Signorelli editore: Roma 1939.

Patsch 1893: K. Patsch, Die Garnison von Praeneste. Zu Tacitus, Ann., XV, 46, in *Römische Mitteilungen* 8, 1893, pp. 219–221.

Patterson 1992: J. Patterson, The City of Rome: From Republic to Empire, in *Journal of Roman Studies* 82, 1992, pp. 186–215.

Pavan 1962: M. Pavan, Iscrizioni latine ad Albano laziale, in *Athenaeum* 40, 1962, pp. 85–93.

Pavis d'Escurac 1976: H. Pavis d'Escurac, *La préfecture de l'annone: service administratif impérial d'Auguste à Constantin* (BEFAR 226), École Française de Rome: Roma 1976.

Peachin 1996: M. Peachin, Iudex vice Caesaris. *Deputy Emperors and the Administration of Justice during the Principate*, F. Steiner Verlag: Stuttgart 1996.

Peachin 2015: M. Peachin, Augustus' Emergent Judicial Powers, the Crimen Maiestatis and the Second Cyrene Edict, in J.-L. Ferrary, D. Mantovani (eds), *Il princeps romano. Autocrate o magistrato? Fattori giuridici e fattori sociali del potere imperiale da Augusto a Commodo*, Iuss Press: Pavia, 2015, pp. 497–553.

Pensabene, Milella 1989: P. Pensabene, M. Milella, Foro Traiano. Contributi per una ricostruzione storica e architettonica. Introduzione storica e quadro architettonico, in *Archeologia classica* 41, 1989, pp. 33–54.

Pera 2012: R. Pera, In trono, a destra: nota iconografica su Securitas nelle emissioni neroniane, in R. Pera (ed.), *Il significato delle immagini: numismatica, arte, filologia, storia* (Atti del secondo incontro internazionale di studio del Lexicon Iconographicum Numismaticae, Genova, 10–12 November 2005), Giorgio Bretschneider: Roma 2012, pp. 345–364.

Perrin 2003a: Y. Perrin, Aux marches du palais: Les accès au Palatium de 54 au 70, in De Blois, Erdkamp, Hekster, de Kleijn, Mols (eds) 2003, pp. 358–375.

Perrin 2003b: Imperii arx: métaphore ou réalité? Les fonctions de la domus tiberiana néronienne, in P. Defosse (ed.), *Hommages à Carl Deroux. III. Histoire et épigraphie, Droit*, Latomus: Bruxelles 2003, pp. 340–355.

Perrin 2012: Y. Perrin, Les soldats de Néron: quelques remarques sur les Augustiani et la société impériale, in B. Cabouret, A. Groslambert, C. Wolff (eds), *Visions de l'Occident romain, Hommages à Yann Le Bohec*, De Boccard: Paris 2012, pp. 569–579.

Petraccia Lucernoni 2001: M.F. Petraccia Lucernoni, *Gli stationarii in età imperiale (Serta antiqua et mediaevalia III)*, L'Erma di Bretschneider: Roma 2001.

Petz 2000: G. Petz, Serviteurs d'Arès, serviteurs de Mars. Sur la coexistence de deux mondes séparés, in S. Follet (ed.), *L'Hellenisme dans l'époque romaine. Nouveaux documents, nouvelles approches* (Ier s. av. J.C.–IIIe s. ap. J.C.) (Actes Coll. L. Robert, Paris 7–8/7/2000), De Boccard: Paris 2000, pp. 287–295.

Pferdehirt 2002: B. Pferdehirt, *Römische Militärdiplome und Entlassungsurkunden in der Sammlung des Römisch-Germanischen Zentralmuseum*, Verlag des Römisch-Germanischen Zentralmuseum: Mainz 2004.

Pflaum 1960–1961: H.G. Pflaum, *Les carrières procuratoriennes équestres sous le Haut-Empire ramain* (Bibl. Arch. Hist., 57), P. Geuthner: Paris, 1960–1961.

Pflaum 1966: H.G. Pflaum, *Les sodales Antoniniani de l'époque de Marc Aurèle. Mémoires présentés par divers savants à l'Académie des inscriptions et belles-lettres de l'Institut de France*. Première série, Sujets divers d'érudition. Tome 15, 2e partie, Imprimerie Nationale: Paris 1966, pp. 141–235.

Pisani Sartorio 1988: G. Pisani Sartorio, Compita Larum. Edicole sacre nei crocicchi di Roma antica, in *Bollettino dell'Unione Storia e Arte* 1–4, 1988, pp. 23–34.

Pisani Sartorio 1993: G. Pisani Sartorio, Compitum Acilium, in LTUR I, pp. 314–315.

Pollard 1996: N. Pollard, The Roman Army as a 'Total Institution' in the Near East? Dura Europos as a Case Study, in D.L. Kennedy (ed.), *The Roman Army in the East* (*Journal of Roman Archeology*, suppl. 18), Portsmouth 1996, pp. 221–227.

Pollard 2000: N. Pollard, *Soldiers, Cities and Civilians in Roman Syria*, University of Michigan Press: Ann Arbour 2000.

Pugliese 1939: G. Pugliese, *Appunti sui limiti dell' 'imperium' nella repressione penale; a proposito della 'Lex Iulia de vi publica'*, Giappichelli: Torino 1939.

Pugliese 1982: G Pugliese, Linee generali dell'evoluzione del diritto pubblico durante il principato, in ANRW II.14, 1982, pp. 723–789.

Quilici 2009: L. Quilici, Praetorium Speluncae. *Ricerca sui confini della proprietà imperiale* (*Atlante tematico di topografia antica*, 19), L'Erma di Bretschneider: Roma 2009.

Raaflaub 1987: K.A. Raaflaub, Die militärreformen, in G. Binder (ed.), *Saeculum Augustum*, vol. I, Wissenschaftliche Buchgesellschaft: Darmstadt 1987, pp. 246–307.

Rainbird 1986: J.S. Rainbird, The Fire Stations of Imperial Rome, in *Papers of the British School in Rome* 54, 1986, pp. 147–169.

Rankov 1994: B. Rankov, *The Praetorian Guard*, Osprey: Oxford 1994.

Rankov 2006: B. Rankov, Les frumentarii et la circulation de l'information entre les empereurs romains et les provinces, in L. Capdetrey, J. Nelis-Clément (eds), *La circulation de l'information dans les états antiques*, Ausonius: Bordeaux 2006, pp. 129–140.

Reddé 1986: M. Reddé, *Mare nostrum. Les infrastructures, le dispositif et l'histoire de la marine militaire sous l'Empire Romain* (BEFAR, 260), École Française de Rome: Roma 1986.

Reuter 1999: M. Reuter, Die frumentarii – neugeschaffene 'Geheimpolizei' Traians, in E. Schallmeyer (ed.), *Traian in Germanien, Traian im Reich. Bericht des Dritten Saalburgkolloquiums*, Römerkastell Saalburg: Bad Homburg 1999, pp. 77–81.

Rich 2003: J. Rich, Augustus, War and Peace, in De Blois, Erdkamp, Hekster, de Kleijn, Mols (eds) 2003, pp. 329–357.

Ricci 1994: C. Ricci, *Soldati delle milizie urbane fuori di Roma* (Opuscula Epigraphica, 5), Edizioni Quasar: Roma 1994.

Ricci 2000: C. Ricci, Legio II Parthica: una messa a punto, in Y. Le Bohec, C. Wolff (eds). *Les légions de Rome sous le Haut-Empire* (*Actes du Congrés de Lyon, sept. 1998*), De Boccard: Lyon 2000, pp. 397–410.

Ricci 2003: C. Ricci, In ordinem redigere. Polizia e ordine pubblico nella Roma imperiale, in *Zapruder* 1, 2003, pp. 13–28.

Ricci 2004: C. Ricci, Praetoria e pretoriani: viaggi imperiali e servizi di sicurezza, in *Cahiers du Centre Glotz* 14, 2004 [2006], pp. 317–334.

Ricci 2008: C. Ricci, Una dedica militare del 168 d.C. e gli statores, in M.L. Caldelli, G.L. Gregori, S. Orlandi (eds), *Epigrafia 2006. Atti della XIVᵉ Rencontre sur l'Épigraphie in onore di Silvio Panciera con altri contributi di colleghi, allievi e collaboratori (Tituli, 9)*, Edizioni Quasar: Roma 2008, pp. 1227–1240.

Ricci 2009: C. Ricci, Veterani Augusti. Studio sulla nascita e sul significato di una formula, in *Aquila legionis* 12, 2009, pp. 7–39.

Ricci 2010: C. Ricci, *Soldati, ex soldati e vita cittadina: l'Italia romana* (Urbana Species, 1), Edizioni Quasar: Roma 2010.

Ricci 2011a: C. Ricci, In custodiam urbis. Notes on the cohortes urbanae (1968–2010), in *Historia* 60, 2011, pp. 484–508.

Ricci 2011b: C. Ricci, Note sull'iconografia dei soldati delle coorti urbane, in *Sylloge epigraphica Barcinonensis* 11, 2011, pp. 131–148.

Ricci 2014: C. Ricci, The Urban Troops Between the Antonines and Severus, in *Öffentlichkeit – Monument – Text: Akten den XIV Congressus Internationalis Epigraphiae Latinae, 27 31 August 2012*, De Gruyter: Berlin 2014, pp. 471–473.

Ricci 2015: C. Ricci, Pro bona valetudine. Considerazioni sul personale addetto all'infermeria e sui valetudinaria di Roma, in *Humanitas* 70, 2015, pp. 361–374.

Ricci (2017 in print): C. Ricci, L'imperatore viaggia. La sicurezza nelle città della Campania e del Latium da Augusto a Domiziano, in J. Barbier, F. Chausson, S. Destephen (eds), *Gouvernement et déplacement*, I. *Antiquité*, Paris in print 2017 (French version).

Riccobono 1945: S. Riccobono, *Acta Divi Augusti I*, Regia Accademia d'Italia: Roma 1945.

Riess 2001: W. Riess, *Apuleius und die Räuber. ein Beitrag zur historischen Kriminalitätsforschung*, F. Steiner Verlag: Stuttgart 2001.

Riess 2005: W. Riess, Randgruppen: Banditen, Zöllner und andere, in K. Noethlichs, J. Zangenberg (eds), *Neues Testament und Antike Kultur*, vol. 2: *Familie–Gesellschaft–Wirtschaft*, Neukirchener Verlagsgeschaft: Neukirchen–Vluyn 2005, pp. 100–104.

Riess 2007: W. Riess, Hunting Down Robbers in 3rd-Century Central Italy, in C. Wolff (ed.), *Les exclus dans l'Antiquité* (Actes du colloque organisé à Lyon les 23–24 September 2004), De Boccard: Paris 2007, pp. 195–213.

Riess 2010: W. Riess, Banditry and Brigandage, Roman, in M. Gagarin (ed.), *The Oxford Encyclopedia of Ancient Greece and Rome*, Oxford University Press: Oxford 2010, pp. 359–361.

Rivière 1994: Y. Rivière, Carcer et vincula. La detention publique à Rome sous la République et le Haut Empire, in *Mélanges de l' École française de Rome: Antiquité* 106.2 1994, pp. 579–652.

Rivière 2004: Y. Rivière, Les batailles de Rome. Présence militaire et guérilla urbaine à l'époque impériale, in *Histoire urbaine* 10, 2004, pp. 63–87.

Rivière 2008a : Y. Rivière, Décrire l'enfermement et le système de pénalité antiques, in *Hypothèses 2007. Travaux de l'École doctorale d'histoire de l'Université Paris I Pantheon-Sorbonne*. Publications de la Sorbonne: Paris 2008, pp. 203–2011.

Rivière 2008b: Y. Rivière, L'Italie, les îles et le continent. Recherches sur l'exil et l'administration du territoire impérial (Ier–IIIe siècles) in Brélaz, Ducrey (eds) 2008, pp. 261–310.

Rivière 2009: Y. Rivière, Compétence territoriale, exercice de la coercition, et pouvoirs juridictionnels du préfet de la Ville (I–IV siècle ap. J.-C.), in *Mediterraneo antico* 12, 2009, pp. 227–256.

Rivière (ed.) 2012: Y. Rivière, *Des réformes augustéennes. Etudes réunies par Y. Rivière* (CEFR 458), École Française de Rome: Roma 2012.

Robinson 1984: O.F. Robinson, Baths: An Aspect of Roman Government Law, in *Sodalitas. Scritti in onore di Antonio Guarino*, vol. III, Jovene: Napoli 1984, pp. 1065–1082.

Robinson 1992: O.F. Robinson, *Ancient Rome: City Planning and Administration*, Routledge: London, New York 1992.

Robinson 2000: O.F. Robinson, The *Criminal Law of Ancient Rome*, John Hopkins University Press: Baltimore 2000.

Rocco 2012: M. Rocco, *L'esercito romano tardoantico: persistenze e cesure dai Severi a Teodosio*, Libreria Universitaria.it 2012.

Rocco 2016: M. Rocco, Ottaviano Augusto praesens deus: echi letterari di un sincretismo epicureo? in I. Baglioni (ed.), *Saeculum Aureum. Tradizione e innovazione nella religione romana di epoca augustea. Vol I: Augusto, da uomo a dio*, Edizioni Quasar: Roma 2016, pp. 179–192.

Roddaz 1984: J.M. Roddaz, *Marcus Agrippa*, De Boccard: Paris 1984.

Romanelli 1961: P. Romanelli, La funzione del porto di Centumcellae, in *Civitavecchia. Pagine di storia e di archeologia, Tipografia 'L'Etruria'*: Civitavecchia 1961.

Rossignol 2007: B. Rossignol, Les préfets du prétoire de Marc Aurèle, in *Cahiers du Centre Glotz* 18, 2007, pp. 141–177.

Rostovzev 1905: M. Rostovzev, Die Domänenpolizei in dem römischen Kaiserreiche, in *Philologus* 64, 1905, pp. 297–307.

Roymans 2009: N. Roymans, Hercules and the Construction of a Batavian Identity in the Context of the Roman Empire, in T. Derks, N. Roymans (eds), *Ethnic Constructs in Antiquity: The Role of Power and Tradition*, Amsterdam University Press: Amsterdam 2009, pp. 219–238.

Royo 1999: M. Royo, *Domus imperatoriae: topographie, formation et imaginaire des palais impériaux du Palatin* (IIe siècle av. J.-C.–Ier siècle ap. J.-C.) (BEFAR 303), École française de Rome: Roma 1999.

Royo, Hubert, Bérenger (eds) 2008: M. Royo, É. Hubert, A. Bérenger (eds), *Rome des quartiers: des Vici aux Rioni. Cadres institutionnels, pratiques sociales et requalifications entre antiquité et époque moderne* (Actes Coll. Intern. Sorbonne, 20–21 May 2005), De Boccard: Paris 2008.

Rucinski 2003a: S. Rucinski, Le rôle du préfect des vigiles dans le maintien de l'ordre public dans la Rome impériale, in *Eos* 90, 2003, pp. 261–274.

Rucinski 2003b: S. Rucinski, Position des curatores regionum dans la Hierarchie administrativfe de la ville de Rome, in *Eos* 91, 2004, pp. 108–119.

Rucinski 2009: S. Rucinski, *Praefectus urbi. Le Gardien de l'ordre public à Rome sous le Haut-Empire romain*, Contact: Poznan 2009.

Rüpke 1990: J. Rüpke, *Domi militiae: Die religiöse Konstruktion des Krieges in Rom*, F. Steiner Verlag: Stuttgart 1990.

Rüpke 1998: J. Rüpke, Les archives des petites colleges. Le cas des vicomagistri, in *La memoire perdue: recherches sur l'administration romaine*, École française de Rome: Rome 1998, pp. 33–35.

Russo Ruggeri 2006: C. Russo Ruggeri, *Quaestiones ex libero homine. La tortura degli uomini liberi nella repressione criminale romana dell'età repubblicana e del I secolo dell'impero*, Giuffré: Milano 2006.

Sabbatini 1988: R. Sabbatini Tumolesi, *Epigrafia Anfiteatrale dell'Occidente Romano*, I. *Roma*, Quasar: Roma 1988.

Sablayrolles 1996: R. Sablayrolles, *Libertinus miles. Les cohortes des vigils* (CEFR 224), École française de Rome: Roma 1996.

Sablayrolles 2001: E. Sablayrolles, La rue, le soldat et le pouvoir: la garnison de Rome de César à Pertinax, in *Pallas* 55, 2001, pp. 127–158.

Saddington 1987: D.B. Saddington, Military Praefecti with Administrative Functions, in *Actes du IX Congrès International d'Epigraphie grecque et latine, vol. 1*, Centrum Historiae Terra Antiqua Balcanica: Sofia 1987, pp. 268–274.

Salza Prina Ricotti 1978–1980: E. Salza Prina Ricotti, Cucine e quartieri servili in epoca romana, in *Rendiconti della Pontificia Accademia Romana d'Archeologia* 51–52, 1978–1980, pp. 237–294.

Salza Prina Ricotti 1982: E. Salza Prina Ricotti, Villa Adriana nei suoi limiti e nella sua funzionalità, in *Monumenti della Pontificia Accademia Romana d'Archeologia*, ser. 3, 14, 1982, pp. 22–55.

Salza Prina Ricotti 1992–1993: E. Salza Prina Ricotti, Nascita e sviluppo di Villa Adriana, in *Rendiconti della Pontificia Accademia Romana d'Archeologia* 65, 1992–1993, pp. 41–73.

Salza Prina Ricotti 2001: E. Salza Prina Ricotti, *Villa Adriana. Il sogno di un imperatore*, L'Erma di Bretschneider: Roma 2001.

Sanders 1997: G. Sanders, Echos épigraphiques d'une assurance dite 'd'après-mort', in Y. Le Bohec (ed.) 1994, pp. 841–862.

Santalucia 1990: B. Santalucia, La repressione penale e le garanzie del cittadino, in A. Momigliano, A. Schiavone (eds), *Storia di Roma II. L'impero mediterraneo. 1. La repubblica imperiale*, Einaudi: Torino 1990, pp. 535–556 (now also in *Altri studi di diritto penale romano*, Cedam: Padova 2009, pp. 35–61).

Santalucia 1992: B. Santalucia, La giustizia penale nel Principato, in E. Gabba, A. Schiavone (eds), *Storia di Roma II. L'impero mediterraneo. 3. La cultura e l'impero*, Einaudi: Torino 1992, pp. 211–236 (now also in *Altri studi di diritto penale romano*, Cedam: Padova 2009, pp. 63–91).

Santalucia 1994: B. Santalucia, *Processo penale*, in *Studi di diritto penale romano*, L'Erma di Bretschneider: Roma 1994, pp. 145–231.

Santalucia 1998: B. Santalucia, *Diritto e processo penale nell'antica Roma*, Giuffrè: Milano: 1998.

Santalucia 1999: B. Santalucia, *Augusto e i 'iudicia publica'. Gli ordinamenti giudiziari di Roma imperiale. Princeps e procedura dalle leggi Giulie ad Adriano* (Atti del Convegno Intern. di diritto romano, Copanello 1996), Edizioni Scientifiche Italiane: Napoli 1999, pp. 261–277.

Sartre 2000: M. Sartre, Antioche: capitale royale, ville impériale, in Nicolet, Ilbert, Depaule (eds) 2000, pp. 492–505.

Schiller 1949: A.A. Schiller, The Jurists and the Praefects of Rome, in L. Caes, R. Dekkers, R. Henrion (eds), *Mélanges F. De Visscher*, vol. 2, Office international de librairie: Bruxelles 1949, pp. 319–359.

Schillinger-Häfele 1977: U. Schillinger-Häfele, Vierter Nachtrag zu CIL XIII und zweiter Nachtrag zu Fr. Vollmer, Inscriptiones Bavariae Romanae. Inschriften aus dem deutschen Anteil der germanischen Provinzen und des Trevererebietes sowie Rätiens und Noricums, in *Bericht der Römisch-Germanischen Kommission* 58.2, 1977, pp. 447–604.

Schmidt-Dick: F. Schmidt-Dick, *Typenatlas der römischen Reichsprägung von Augustus bis Aemilianus*, Verlag der Österreichische Akademie der Wissenschaften: Wien 1, 2002; 2, 2011.

Schrimm-Heins 1991: A. Schrimm-Heins, Gewißheit und Sicherheit. Geschichte und Bedeutungswandel der Begriffe certitudo und securitas, in *Archiv für Begriffsgeschichte* 34, 1991, pp. 123–224; e 35, 1992, pp. 115–213.

Sedley 2008: D. Sedley, Epicureanism in the Roman Republic, in J. Warren (ed.), *The Cambridge Companion to Epicureanism*, Cambridge University Press: Cambridge 2009, pp. 29–45.

Segenni 2004: S. Segenni, *La proprietà imperiale nell'Abruzzo antico (sec. I.–II. d.C.)*, in M. Pani (ed.), *Epigrafia e territorio, politica e società. Temi di antichità romane*, Edipuglia: Bari 2004, pp. 123–148.

Shaw 1984: B. Shaw, Bandits in the Roman Empire, in *Past and Present* 105.1, 1984, pp. 3–52.

Sherk 1970: R.K. Sherk, *Roman Documents from the Greek East. Senatus Consulta and Epistulae to the Age of Augustus*, John Hopkins Press: Baltimore 1970.

Simelon 1993: P. Simelon, *La propriété en Lucanie depuis les Gracques jusqu'à l'avènement des Sévères. Étude épigraphique*, Latomus: Bruxelles 1993.

Sinnigen 1962: W.G. Sinnigen, The Origins of the Frumentarii, in *Memoirs of American Academy in Rome* 27, 1962, pp. 213–224.

Sirago 1983–1984: V.A. Sirago, Funzione politica della flotta misenate, in *Puteoli* 7–8, 1983–1984, pp. 93–112.

Smith 1972: R.E. Smith, The Army Reforms of Septimius Severus, in *Historia* 21, 1972, pp. 481–500.

Sojc, Winterling, Wulf-Rheidt (eds) 2013: N. Sojc, A. Winterling, U. Wulf-Rheidt (eds), *Palast und Stadt im severischen Rom*, F. Steiner Verlag: Stuttgart 2013.

Solidoro Maruotti 1993: L. Solidoro Maruotti, Aspetti della giurisdizione civile del prefetto urbano nell'età severiana, in *Labeo* 39, 1993, pp. 174–233.

Solin 2000: H. Solin, Antium et les légions. Nouveaux témoignages, in Y. Le Bohec, C. Wolff (eds), *Les légions de Rome sous le Haut-Empire* (Actes Congr. Lyon, 17–19 September 1998), De Boccard: Lyon, 2000, pp. 639–644.

Spagnuolo Vigorita 2010: Casta domus. *Un seminario sulla legislazione matrimoniale augustea*, Jovene: Napoli 2010.

Spanu 1998: P.G. Spanu, *La Sardegna bizantina tra VI e VII secolo*, Editrice S'Alvure: Oristano 1998.

Speidel 1965: M.P. Speidel, Die Equites singulares Augusti: *Begleittruppe der römischen Kaiser des zweiten und dritten Jahrhunderts*, R. Habelt Verlag: Bonn 1965.

Speidel 1979: M.P. Speidel, An Urban Cohort for the Mauretanian Kings, in *Antiquités Africaines* 14, 1979, pp. 121–122 (= *AE* 1979, 683).

Speidel 1994a: M.P. Speidel, *Die Denkmäler der Kaiserreiter: Equites singulares Augusti*, Rheinland Verlag: Köln 1994.

Speidel 1994b: M.P. Speidel, *Riding for Caesar: The Roman Emperors' Horse Guards*, Batsford: London 1994.

Speidel 2009: M.A. Speidel, Augustus'militärische Neuordnung und ihr Beitrag zum Erfolg des Imperium Romanum, in M.A. Speidel (ed.), *Herr und*

Herrschaft im Römischen Reich der Hohen Kaiserzeit, F. Steiner Verlag: Stuttgart 2009, pp. 19–51.

Spera 2002: L. Spera, Il territorio della via Appia. Forme trasformative del paesaggio nei secoli della tarda antichità, in Ph. Pergola, R. Santangeli Valenzani, R. Volpe (eds), *Suburbium. Il suburbio di Roma dalla crisi del sistema delle ville a Gregorio Magno* (CEFR 311), École française de Rome: Roma 2002, pp. 267–330.

Spurza 2002: J. Spurza, The Emperors at Ostia and Portus: Imperial Villas and Accomodation, in Ch. Bruun, A. Gallina Zevi (eds), *Ostia e Portus nelle loro relazioni con Roma* (Atti del Convegno, Roma 3–4 December 1999), Edizioni Quasar: Roma, 2002, pp. 123–134.

Starr 1941: C.G. Starr, *The Roman Imperial Navy 31 BC–AD 324*, Cornell University Press: New York 1941.

Starr 1960: C.G. Starr, *The Roman Imperial Navy, 31 BC–AD 324*, W. Heffer: Cambridge 1960.

Stefan 2005: A. Stefan, *Les guerres daciques de Domitien et de Trajan: Architecture militaire, topographie, images et histoire* (CEFR 353), École française de Rome: Roma 2005.

Stein 1928: E. Stein, *Geschichte des spätrömischen Reiches*, vol. 1, Seidel & Son: Wien 1928.

Stek 2008: T.D. Stek, *Sanctuary and Society in Central-Southern Italy (3rd–1st centuries BC): A Study into Cult Places and Cultural Change after the Roman Conquest of Italy*. PhD thesis, University of Amsterdam 2008.

Strobel 1984: K. Strobel, *Untersuchungen zu den Dakerkriegen Trajans. Studien zur Geschichte des mittleren und unteren Donauraumes in der Hohen Kaiserzeit*, R. Habelt Verlag: Bonn 1984.

Strobel 1986: K. Strobel, Zur Rekonstruktion der Laufbahn des C. Velius Rufus, in *Zeitschrift für Papyrologie und Epigraphik* 64, 1986, pp. 265–280.

Swan 2004: P.M. Swan, *The Augustan Succession: An Historical Commentary on Cassius Dio's Roman History, Books 55–56 (9 BC–AD 14)*, Oxford: Oxford University Press 2004.

Suspène 2009: A. Suspène, Aspects numismatiques de la Res publica restituta augustéenne, in F. Hurlet, B. Mineo (eds), *Le principat d'Auguste: réalités et représentations du pouvoir: autour de la 'Res publica restituta'* (Actes du colloque de l'Université de Nantes, 1–2 June 2007), Presse Universitaire: Rennes 2009, pp. 145–167.

Tacoma 2016: L.E. Tacoma, *Moving Romans: Migration to Rome in the Principate*, Oxford University Press: Oxford 2016.

Tarpin 2002: M. Tarpin, *Vici et pagi dans l'Occident romain* (CEFR 299), École française de Rome: Roma 2002.

Tarpin 2008: M. Tarpin, Les vici de Rome, entre sociabilité de voisinage et organisation administrative, in Royo, Hubert, Bérenger (eds) 2008 pp. 35–64.

Timpe 2000: D. Timpe, Der Epikureismus in der römischen Gesellschaft der Kaiserzeit, in M. Erler, R. Bees (eds), *Epikureismus in der späten Republik und der Kaiserzeit* (Akten der 2. Tagung der Karl-und-Gertrud-Abel-Stiftung vom 30 September–3. October 1998 in Würzburg), F. Steiner Verlag: Stuttgart 2000, pp. 42–63.

Todisco 1999: E. Todisco, *I veterani nell'Italia imperiale*, Edipuglia: Bari 1999.

Tomei, Filetici (eds) 2011: M.A. Tomei, M.G. Filetici (eds), *Domus Tiberiana: scavi e restauri 1990–2011*, Electa: Milano 2011.

Tomei, Rea (eds) 2011: M.A. Tomei, R. Rea, *Nerone* (Catalogo della Mostra, Roma, April 2011), Electa: Milano 2011.

Tortorici 1975: E. Tortorici, *Castra Albana* (Forma Italiae, Regio 1, 11), Arbor Sapientiae: Roma 1975.

Valenti 2008: M. Valenti, La proprietà imperiale nel Tuscolano nel I sec. d.C., in Valenti (ed.) 2008, pp. 61–72.

Valenti (ed.) 2008: M. Valenti (ed.), *Residenze imperiali nel Lazio* (Atti della Giornata di Studio, Monte Porzio Catone, 3 April 2004), Tuscolana – Quaderni del Museo di Monte Porzio Catone: Monte Porzio Catone 2008.

Vermeule 1987: C. Vermeule, *The Cult Images of Imperial Rome*, L'Erma di Btretschneider: Roma 1987.

Veyne 1976: P. Veyne, *Le pain et le cirque. Sociologie historique d'un pluralisme politique*, Le Seuil: Paris 1976.

Veyne 2007: P. Veyne, Che cos'era un imperatore romano?, in *L'impero greco-romano*, Rizzoli: Milano 2007, pp. 10–67.

Vigneaux 1896: P.E. Vigneaux, *Essai sur l'histoire de la praefectura urbis à Rome et à Constantinople*, A. Fontemoing: Paris 1896.

Villedieu (ed.) 2001: F. Villedieu (ed.), *Il giardino dei Cesari: dai palazzi antichi alla Vigna Barberini, sul Monte Palatino: scavi dell'École française de Rome 1985–1999 (Guida alla mostra)*, Edizioni Quasar: Roma 2001.

Villedieu 2006: F. Villedieu, La Vigna Barberini sul Palatino, in A.M. Tomei (ed.), *Roma. Memorie dal sottosuolo. Ritrovamenti archeologici 1980/2006* (Catalogo mostra Roma 2006), Electa: Milano 2006 pp. 58–61.

Villedieu 2013: F. Villedieu, La Vigna Barberini à l'époque séveriénne, in Sojc, Winterling, Wulf-Rheidt (eds) 2013, pp. 157–180.

Villedieu, André 2003: F. Villedieu, N. André, Proposition pour une reconstitution de l'édifice flavien et de l'ensemble monumental tardif de la Vigna Barberini (Rome, Palatin), in *Rom An 2000. Ville, maquette et modèle virtuel*, Presses universitaires de Caen: Caen 2003, pp. 361–376.

Villedieu, Antré, Del Trento 2007: F. Villedieu, N. Antré, M.L. Del Trento, *La Vigna Barberini II. Domus, palais impérial et temples. Stratigraphie du secteur nord-est du Palatin, Roma antica 6*, École française de Rome: Roma 2007.

Virlouvet 1985: C. Virlouvet, *Famines et émeutes à Rome des origines de la République à la mort de Néron* (CEFR 87), École française de Rome: Roma 1985.

Virlouvet 1991: C. Virlouvet, La plèbe frumentaire à l'époque d'Auguste. Une tentative de définition, in A. Giovannini (ed.), *Nourrir la plèbe* (Actes du colloque, Genève 28–29/9/1989 en hommage à Denis van Berchem), Rheinardt: Basel 1991, pp. 43–65.

Vitucci 1956: G. Vitucci, *Ricerche sulla Praefectura Urbi in età imperiale (sec. I–III)*, L'Erma di Bretschneider: Roma 1956.

Vitucci 1962–1963: G. Vitucci, La prefettura urbana nella cornice delle riforme augustee, in *Cultura e scuola* 26, 1962–1963, pp. 94–101 (now also in G. Vitucci, *Scritti minori*, edited by A. Pasqualini, Tored: Tivoli 2005, pp. 139–149).

Volkmann 1935: H. Volkmann, *Zur Rechtsprechung im Principat des Augustus*, Beck: München 1935.

Volpe 2007–2008: G. Volpe, Forme d'integrazione-scontro tra pastori-briganti e agricoltori in Italia centro-meridionale in età romana, in *Boletín Arkeolan* 15, 2007–2008, pp. 11–24.

Volpe 2000: R. Volpe, *Paesaggi urbani tra Oppio e Fagutal*, in MEFRA 112, 2000, pp. 511–556.

Wallace-Hadrill 1981: A. Wallace-Hadrill, The Emperor and his Virtues, in *Historia* 30, 1981, pp. 298–323.

Wallace-Hadrill 1993: A. Wallace-Hadrill, *Augustan Rome* (Classical World Series), Bristol Classical Press: London 1993.

Wallace-Hadrill 1997: A. Wallace-Hadrill, Mutatio morum: The Idea of a Cultural Revolution in T. Habinek, A. Schiesaro (eds), *The Roman Cultural Revolution*, Cambridge University Press: Cambridge 1997, pp. 3–22.

Wallace-Hadrill 2003: A. Wallace-Hadrill, The Streets of Rome as a Representation of Imperial Power, in De Blois, Erdkamp, Hekster, de Kleijn, Mols (eds) 2003, pp. 189–206.

Wallace-Hadrill 2007: A. Wallace-Hadrill, Mutatas Formas: The Augustan Transformation of Roman Knowledge, in *The Cambridge Companion to the Age of Augustus*, Cambridge University Press: Cambridge, 2007, pp. 55–84.

Wallace-Hadrill 2008: A. Wallace-Hadrill, *Rome's Cultural Revolution*, Cambridge University Press: Cambridge/New York 2008.

Wardle 2014: D. Wardle, *Suetonius: Life of Augustus*, Oxford University: Oxford 2014.

Watson 1994: A.J.M. Watson, Maecenas' Administration of Rome and Italy, in *Akroterion* 39, 1994, pp. 98–104.

Weaver 1972: P.R.C. Weaver, *Familia Caesaris: A Social Study of the Emperor's Freedmen and Slaves*, Cambridge University Press: Cambridge 1972.

Weber, Zimmermann (eds) 2003: G. Weber, M. Zimmermann (eds), *Propaganda, Selbstdarstellung, Repräsentation im römischen Kaiserreich des 1. Jhs. n. Chr.*, F. Steiner Verlag: Stuttgart 2003.

Weiler 1988: I Weiler, Die Gefährung des securitas: Angst von Angehörigen sozialer Randgruppen der römischen Kaiserzeit am Beispiel von Philosophen, Astrologen, Magiern, Schauspielern und Räubern, in A. Kneppe (ed.), *Soziale Randgruppen und Aussenseiter in Altertums*, Leykam: Graz 1988, pp. 165–176.

Welwei 1988: K.W. Welwei, *Umfreie im antiken Kriegsdienst, III: Rom*, F. Steiner Verlag: Stuttgart 1988.

Wheeler 1993: E. Wheeler, Methodological Limits and the Mirage of Roman Strategy, Parts I–II, in *Journal of Military History* 57, 1993, pp. 7–41 and 215–240.

Wheeler 2007: E. Wheeler, The Army and the Limes in the East, in P Erdkamp (ed.), *A Companion to the Roman Army*, Blackwell: Malden, MA 2007, pp. 235–266.

Will 2000: E. Will, Antioche, la métropole de l'Asie, in Nicolet, Ilbert, Depaule (eds) 2000, pp. 482–491.

Wistrand 1992: M. Wistrand, *Entertainment and Violence in Ancient Rome: The Attitudes of Roman Writers of the First Century A.D.*, Acta Universitatis Gothoburgensis: Göteborg 1992.

Wojciech 2010: K. Wojciech, *Die Stadtpraefektur im Prinzipat*, R. Habelt Verlag: Bonn 2010.

Wood 1988: N. Wood, *Cicero's Social and Political Thought*, University of California Press: Berkeley, LA and London 1988.

Wulf-Rheidt 2012: U. Wulf-Rheidt, Nutzungbereiche des flavischen Palastes auf dem Palatin in Rom, in F. Arnold, A. Busch, R. Haensch, U. Wulf-Rheidt (eds), *Orte der Herrschaft. Charakteristika von antiken Machtzentren*, VML Verlag Marie Leidorf: Rahden/Westfalien 2012, pp. 97–112.

Wulf-Rheidt 2013: U. Wulf-Rheidt, Die Bedeutung der severischen Paläste für spätere Residenzbauten, in Sojc, Winterling, Wulf-Rheidt (eds) 2013, pp. 287–306.

Yannakopoulos 2003: N. Yannakopoulos, Preserving the Pax Romana: The Peace Functionaries in Roman East, in *Mediterraneo Antico* 6, 2003, pp. 825–905.

Yavetz 1984: Z. Yavetz, *La Plébs et le prince. Foule et vie politique sous le Haut-Empire romaine*, Editions La Découverte: Paris 1983 (1st edn: *Plebs and Prince*, Oxford University Press: Oxford 1969).

Yavetz 1986: Z. Yavetz, The Urban Plebs in the Days of the Flavians, Nerva and Trajan, in A. Giovannini, D. van Berchem (eds), *Opposition et résistance à l'empire d'Auguste à Trajan: neuf exposés suivis de discussions* (Entretiens Hardt 33), Fondation Hardt: Vandoeuvres-Genève 1988, pp. 135–186.

Zamai 2001: A. Zamai, Gli irenarchi d'Asia minore, in *Patavium* 17, 2001, pp. 53–73.

Zanker 1987: P. Zanker, *Augustus und die Macht der Bilder*, C.H. Beck: München 1987 (English translation by A. Shapiro, University of Michigan Press: Ann Arbour 1988).

Zanker 1988: P. Zanker, *The Power of Images in the Age of Augustus* (translated by A. Shapiro), The University of Michigan Press: Ann Arbor 1988.

Zanker 2004: P. Zanker, Domitians Palast auf dem Palatin al Monument kaiserlicher Selbstdarstellung, in A. Hoffmann, U. Wulf (eds.), *Die Kaiserpaläste auf dem Palatin in Rom*, von Zabern: Mainz 2004, pp. 86–99.

Zarecki 2014: J. Zarecki, *Cicero's Ideal Statesman in Theory and Practice*. Bloomsbury Academic: London, New York 2014.

Zecchini 1997: G. Zecchini, *Il pensiero politico romano. Dall'età arcaica alla tarda antichità*, La Nuova Italia Scientifica: Roma, 1997.

Zevi 1976: F. Zevi, Proposta per un'interpretazione dei rilievi Grimani, in *Prospettiva* 7, 1976, pp. 38–41.

Zevi 2000: F. Zevi, Traiano e Ostia, in J. Gonzàlez (ed.), *Trajano. Emperador de Roma*, L'Erma di Bretschneider: Roma 2000, pp. 509–547.

Zevi (ed.) 2009: F. Zevi (ed.), *Museo archeologico dei Campi Flegrei. Catalogo generale, 2*, Electa Napoli: Napoli 2009.

Zoz 2001: M.G. Zoz, I soprusi degli organi di polizia municipali (magistrati e 'stationarii' in un caso di tangentopoli dell'antichità), in *Iuris Vincula. Studi in onore di M. Talamanca, VIII*, Jovene editore: Napoli 2001, pp. 515–526.

Zucca 1987: R. Zucca, *Le* Civitates Barbariae e l'occupazione militare della Sardegna: aspetti e confronti con l'Africa, in *L'Africa romana: atti del 5. Convegno di studio*, 11–13 dicembre 1987, Università degli Studi di Sassari: Sassari, pp. 349–373.

Index

Note: Page numbers in italic refer to illustrations; page numbers in bold refer to tables. Books referenced in the text are listed here by title and author.